# Polymeric Biomaterials

Biomaterials include a versatile group of molecules that have been designed to interact with biological systems for various applications and polymeric biomaterials are being designed based on their availability and compatibility. This book summarizes fabrication techniques, features, usage, and promising applications of polymeric biomaterials in diversified areas including advantageous industrial applications. Each chapter exclusively covers a distinct application associated with major classes of polymeric biomaterials.

Features:

- Provides platforms related to fabrication and advancement of all categories of polymeric biomaterials.
- Explores advancement of pertinent biomedical and drug-delivery systems.
- Includes a wide range of biomaterials and their applications in diversified fields.
- Gives environmental justification to green biopolymers and their applications in water remediation.
- Discusses advanced applications of biocomposite polymers viz. food packaging and anticorrosive coatings.

This book is aimed at researchers in Polymer Sciences, Biomaterials, Chemical/Bio Engineering, Materials Chemistry, and Biotechnology.

**Emerging Materials and Technologies**
*Series Editor: Boris I. Kharissov*

The *Emerging Materials and Technologies* series is devoted to highlighting publications centered on emerging advanced materials and novel technologies. Attention is paid to those newly discovered or applied materials with potential to solve pressing societal problems and improve quality of life, corresponding to environmental protection, medicine, communications, energy, transportation, advanced manufacturing, and related areas.

The series takes into account that, under present strong demands for energy, material, and cost savings, as well as heavy contamination problems and worldwide pandemic conditions, the area of emerging materials and related scalable technologies is a highly interdisciplinary field, with the need for researchers, professionals, and academics across the spectrum of engineering and technological disciplines. The main objective of this book series is to attract more attention to these materials and technologies and invite conversation among the international R&D community.

**Sustainable Nanomaterials for the Construction Industry**
*Ghasan Fahim Huseien and Kwok Wei Shah*

**4D Imaging to 4D Printing**
Biomedical Applications
*Edited by Rupinder Singh*

**Emerging Nanomaterials for Catalysis and Sensor Applications**
*Edited by Anitha Varghese and Gurumurthy Hegde*

**Advanced Materials for a Sustainable Environment**
Development Strategies and Applications
*Edited by Naveen Kumar and Peter Ramashadi Makgwane*

**Nanomaterials from Renewable Resources for Emerging Applications**
*Edited by Sandeep S. Ahankari, Amar K. Mohanty, and Manjusri Misra*

**Multifunctional Polymeric Foams**
Advancements and Innovative Approaches
*Edited by Soney C George and Resmi B. P.*

**Nanotechnology Platforms for Antiviral Challenges**
Fundamentals, Applications and Advances
*Edited by Soney C George and Ann Rose Abraham*

**Carbon-Based Conductive Polymer Composites**
Processing, Properties, and Applications in Flexible Strain Sensors
*Dong Xiang*

**Nanocarbons**
Preparation, Assessments, and Applications
*Ashwini P. Alegaonkar and Prashant S. Alegaonka*

For more information about this series, please visit:
www.routledge.com/Emerging-Materials-and-Technologies/book-series/CRCEMT

# Polymeric Biomaterials

## Fabrication, Properties and Applications

Edited by
Pooja Agarwal, Divya Bajpai Tripathy,
Anjali Gupta and Bijoy Kumar Kuanr

CRC Press is an imprint of the
Taylor & Francis Group, an **informa** business

First edition published 2023
by CRC Press
6000 Broken Sound Parkway NW, Suite 300, Boca Raton, FL 33487-2742

and by CRC Press
4 Park Square, Milton Park, Abingdon, Oxon, OX14 4RN

*CRC Press is an imprint of Taylor & Francis Group, LLC*



ISBN: 9781032146652 (hbk)
ISBN: 9781032147444 (pbk)
ISBN: 9781003240884 (ebk)

DOI: 10.1201/9781003240884

Typeset in Times
by Newgen Publishing UK

# Contents

# Editor Biographies

**Pooja Agarwal** is currently Professor in the Division of Chemistry, School of Basic and Applied Sciences, Galgotias University, Greater Noida, India. She has research and teaching experience of more than 12 years. She has been awarded DST-funded research projects and a Senior Research Fellowship from CSIR and many more grants. She has more than 10 international research publications in reputed journals/book chapters to her credit. She obtained her doctorate from Indian Institute of Technology (IIT), Varanasi, India.

**Divya Bajpai Tripathy** is currently a full-time Professor in the Department of Chemistry, School of Basic and Applied Sciences, Galgotias University, Greater Noida, India. She has research and teaching experience of more than 10 years. She has more than 30 research publications in reputed journals/book chapters/conference proceedings to her credit. She has mentored seven Masters research students. Currently four doctorate students have been registered under her supervision. She has been principal investigator in a DST-funded research project.

**Anjali Gupta** is currently working as Professor in the Department of Chemistry, School of Basic & Applied Sciences, Galgotias University, Greater Noida, India. She has research and teaching experience of around 11 years. She is recipient of the Young Scientist Award by Department of Science & Technology and Dr. D.S. Kothari postdoctoral fellowship by the University Grants Commission, and Senior Research Fellowship from CSIR, India. Her research area is *in silico* screening and synthesis of naturally occurring bio-active analogs. She has four published patents and around 16 research publications in journals/book chapters/conference proceedings to her credit. She obtained her graduation, post-graduation and doctorate from University of Delhi, Delhi, India.

**Bijoy Kumar Kuanr** is currently working as a full-time Professor at Special Centre for Nanoscience, Jawaharlal Nehru University, New Delhi, India. He has research and teaching experience of more than 28 years in various senior positions. He was a recipient of DAAD postdoctoral fellowship at University of KÖLN, Germany, in 1994, Rühr-University, Bochum, Germany, in 1999 and Research Center Jülich, Germany, in 1999. He has been Adjunct Research Faculty at University of Colorado, USA, from 2001–2013 and also worked as Co-researcher with Nobel Laureate Professor Peter Grünberg – Physics in 2007. He has been honored by being listed in the "*Who's Who in America – 66th Edition, 2011*" and "*Who's Who in Science and Engineering – 10th Edition, 2007.*" He has more than 150 publications in journals/book chapters/conference proceedings to his credit. He has been principal investigator in many sponsored research projects and contributed to technology development, namely "Giant-Magneto-Resistance (GMR) Sensor – READ/WRITE HEAD of computer Jülich, Germany" and "GMR bio-Sensor; Low cost, portable, point-of care device for early detection of emerging viral diseases" (Dengue, Chikungunya, H1N1 Swine flu), JNU, New Delhi, to name a few. Prof. Bijoy has been a member of many academic/research bodies of governmental organizations in India and abroad.

# Contributors

**Pooja Agarwal**
School of Basic and Applied Sciences
Galgotias University
Greater Noida, Uttar Pradesh, India

**Shahzad Ahmad**
Department of Chemistry
Zakir Husain Delhi College
University of Delhi
Delhi, India

**Vikas Bhardwaj**
Department of Applied Sciences
Jai Parkash Mukand Lal Institute of Technology
Radaur, Yamunanagar, Haryana, India

**Indu Bhushan**
Department of Electrical & Electronics Engineering
Lloyd Institute of Engineering and Technology
Greater Noida, India

**Anila Dhingra**
Poornima College of Engineering
Jaipur, Rajasthan, India

**Neha Dhingra**
Department of Chemistry
Arwachin Bharti Bhawan
Delhi, India

**Priyanka Dhingra**
Independent Researcher
United States

**Anjali Gupta**
School of Basic and Applied Sciences
Galgotias University
Greater Noida, Uttar Pradesh, India

**Garima Gupta**
Department of Biotechnology
IILM College of Engineering and Technology
Greater Noida, India

**Kalpitha M. Kumar**
School of Basic and Applied Sciences
Dayananda Sagar University
Bengaluru, Karnataka, India

**Kempahanumakkagari Sureshkumar**
Department of Chemistry
BMS Institute of Technology and Management
Bengaluru, Karnataka, India

**Sahana M.**
School of Basic and Applied Sciences
Dayananda Sagar University
Bengaluru, Karnataka, India

**Yogita Madan**
Amity School of Applied Sciences
Amity University
Jaipur, Rajasthan, India

**Shumaila Masood**
Department of Chemistry
Jamia Millia Islamia
New Delhi, India

**Shilpa Borehalli Mayegowda**
Department of Psychology
CHRIST-Deemed to be University
Bangalore Kengeri Campus, Kanmanike, Kumbalgodu
Bangalore, Karnataka, India

**Kavita Nagshetty**
Department of Microbiology
Gulbarga University
Gulbarga, Karnataka, India

**Nabya Nehal**
Department of Biotechnology
Faculty of Engineering and Technology
Rama University
Kanpur, Uttar Pradesh, India

**Manjula N.G.**
School of Basic and Applied Sciences
Dayananda Sagar University
Bengaluru, Karnataka, India

**Nahid Nishat**
Department of Chemistry
Jamia Millia Islamia
New Delhi, India

**Nidhi Puri**
Department of Applied Science & Humanities
Lloyd Institute of Engineering and Technology
Greater Noida, Uttar Pradesh, India

**Shristi Ram**
School of Basic and Applied Sciences
Dayananda Sagar University
Bengaluru, Karnataka, India

**Thippeswamy Ramakrishnappa**
BMS Institute of Technology and Management
Yelahanka, Bengaluru, India

**Bhoomika S.**
Department of Microbiology
School of Basic and Applied Sciences
Dayananda Sagar University
Bengaluru, Karnataka, India

**Shashank Sharma**
School of Basic Sciences & Research
Sharda University
Greater Noida, Uttar Pradesh, India

**Priyanka Singh**
Institute of Allied Medical Science and Technology
NIMS, University
Rajasthan, India

**Surabhi Singh**
Department of Chemistry
K. K. P. G. College
Etawah, Uttar Pradesh, India

**Divya Bajpai Tripathy**
School of Basic and Applied Sciences
Galgotias University
Greater Noida, Uttar Pradesh, India

**Jyoti Tyagi**
Department of Chemistry
Zakir Husain Delhi College
University of Delhi
New Delhi, India

**Fahmina Zafar**
Department of Chemistry
Jamia Millia Islamia
New Delhi, India

# Preface

Over the last few decades, novel polymeric biomaterials have emerged as a new tool in biomedical applications. Polymeric biomaterials have been fabricated to be used as in-body medical systems like absorbable closures or joints, transitory orthopaedic discs, as well as drug-delivery systems. This book provides exhaustive information on polymeric biomaterials, correlates their specific characteristics with their structural properties, and discusses their potential industrial applications.

Starting with the first chapter as an overview, the book covers the introduction, chronology of their development, types, common synthetic approach, and properties. The second chapter discusses the role of polymeric biomaterials as drug-delivery systems. This chapter emphasizes the biocompatibility of these materials and their use as target delivery of drugs. Both in vitro and in vivo case studies of commercially available products were also investigated.

As biomedical applications, biopolymers have been evaluated in Chapter 3 for tissue engineering, dentistry, prostheses, implantation devices, and in mini- and nanobots. The impacts of pH and temperature on the properties and application in delivery systems and tissue regeneration has been described in Chapter 4.

Sensitivity and conductive characteristics of biopolymers enable them to be exploited as sensors and plasma polymers in biomedical devices used in optometry and other medical fields, as discussed in Chapter 5.

Biocompatibility and non-toxicity of such materials make them one of the best options for the food packaging industry, ,s discussed in Chapter 6. Moreover, antimicrobial biocomposites have received more attention at a commercial level as they increase the shelf life of food items. Biopolymers are also used as corrosion inhibitors in coating materials. The syntheses and applications of biopolymers as corrosion protectors and in the textile industry are covered in Chapters 7 and 8, respectively. Furthermore, water and soil remediation and environmental acceptability of biopolymers are covered in the final three chapters.

**Divya Bajpai Tripathy**
**Pooja Agarwal**
**Anjali Gupta**
**Bijoy Kumar Kuanr**

# 1 Polymeric Biomaterials
## *An Overview*

*Pooja Agarwal, Divya Bajpai Tripathi, and Priyanka Singh*

**CONTENTS**



DOI: 10.1201/9781003240884-1

## 1.1 INTRODUCTION

The materials which can interact with biological systems are called biomaterials. These are biodegradable, biocompatible and easy to design. Biomaterials can be obtained from natural resources or can be synthesised by green methods. Biomaterials are mainly used for biosystems such as in medicines, manufacturing of body implants, tissue engineering, etc. [1]. Biomaterials is the term primarily used to indicate materials that possess biocompatibility. Biobased materials constitute the building blocks of substances which are acquired from living matter. These materials are available naturally or are required to be synthesised.

Polymeric biomaterials have been one of the most promising areas of interest for researchers in recent decades. Mostly, the development of novel polymeric biomaterials has been for various uses in biomedical applications. A great deal of research has been performed in the development of such new materials for therapies, biomedical applications, sensors, food packaging, textiles, etc. The special features of these materials include biocompatibility, explicit biological responses, identification of extrinsic signals, cell permeability along with wide fabrication opportunities. Advancements in synthesis and processing techniques have exposed their unexplored properties and structure functions. These have given a platform for the newly discovered biological functions related to tissue engineering, drug-delivery systems and greater expansion in commercially feasible and practically applicable therapies.

Biomaterials have been applied in medical application for a long time, with the use of biopolymers in resorbable sutures first reported in 1974. After this, various other biomedical applications such as resorbable bone screw and drug-delivery devices have been developed using many other biomaterials. The initially used biopolymers were poly(lactic acid), poly(glycolic acid), poly(caprolactone) (PCL) and then their copolymers [2]. This chapter provides an overview of their biopolymeric classification, characterisation and methods of fabrication along with the recent applications of biomaterials.

## 1.2 BIOPOLYMER CLASSIFICATION

Biopolymers can be categorised on the basis of their source, which can be naturally derived or artificially synthesised. Poly-lactic acid and poly-butylene succinate are among the examples of artificially synthesised biopolymers, while poly-3-hydroxybutyrate, chitosan, etc. are naturally derived [3]. Biopolymers contain long chains made up of molecules and can be linear, branched or cross-linked.

### 1.2.1 Polyesters

Polyesters are the more widely known group of aliphatic polymers [4]. They can be subcategorised into poly-(hydroxy acid)s and poly-(alkylene dicarboxylate)s. Poly-(hydroxy acid)s are made up of a sequence of hydroxyl acids, making hydroxyalkanoic acids which are connected through the hydroxyl group with respect to the carboxyl group. Poly-(3-hydroxybutyrate), poly-(4 hydroxybutyrate), poly-(3-hydroxybutyrate-co-4-hydroxybutyrate) and poly-(β-malic acid) are examples of commercially used PHAs. Poly-(alkylene dicarboxylate)s (PADC) are obtained from dihydroxy compounds and dicarboxylic acids. These can be categorised into (i) aliphatic polyesters; (ii) aliphatic-aromatic polyesters; and (iii) aromatic polyesters.

PAAs are poly-(alkylene alkanoate)s, which are a category of aliphatic polyesters. Aliphatic-aromatic copolymers are obtained from polycondensation of aliphatic-diols, aliphatic-dicarboxylic esters/acids and aromatic-dicarboxylic esters/acids. The most common example of aromatic polyesters is poly-(ethylene terephthalate) or PET. These are usually biobased polyesters.

### 1.2.2 Poly-(ether ester)s

Constituents of poly-(ether-ester) are an alkane diol, alkylene terephthalate and a poly-(alkylene glycol ether), and it can be obtained by a transesterification process [22]. These copolymers are widely used in tissue engineering and biomedical applications. This group of polymers has the required mechanical properties such as greater breaking stress, mechanical resistance, toughness, elasticity, etc. Therefore, it can work in a wide range of temperatures.

### 1.2.3 Aliphatic Polycarbonates

Aliphatic polycarbonates (APC) are obtained by ring-opening polymerisation of epoxy constituents or cyclic carbonate [5]. These have low functionality, along with low mechanical and thermal stability. For these reasons APCs have limited applications in several fields such as in biomedical applications, tissue engineering and drug delivery [6]. The most common examples of APCs are poly-(ethylene carbonate) and poly-(propylene carbonate)

### 1.2.4 Polyamides

Polyamides contain R-CO-NH-R moieties formed by polycondensation of either diamines and dicarboxylic acids or amino carboxylic acids. All polyamides absorb water, and this water acts as a plasticiser [7] and is responsible for regulating the plastic properties. The best known polyamide is Nylon, which has properties including good flexibility and strength. These properties make it suitable for several industries including use in plastics, textiles and other products. Polyamides are compatible and inert towards body fluids, and therefore are suitable for transfusion equipment [8].

### 1.2.5 Poly-(ester amide)s

The most promising method of synthesising poly-(ester amide)s is polycondensation of α-amino acids and α-hydroxy acids ,or ring opening polymerisation of depsipeptides [9]. These are very beneficial and useful categories because of their degradable behaviour, along with them having the required mechanical and thermal properties [10].

### 1.2.6 Poly-(ether amide)s

This is a group of polymers which are thermoplastic elastomeric (TPEs) in nature and can be produced by injection moulding and/or extrusion methods. These polymers are good in mechanical strength, lightweight and highly flexible. Polymers with different properties can be achieved by varying the sequence/ratio of monomers.

### 1.2.7 Polyurethanes

Polyurethanes (PU) are obtained as a result of the polyaddition process between a poly-isocyanate/s and a polyol/s with (-NH-CO-O-) in their structure. PU can have linear, branched or cross-linked structures. Poly-(ester urethane)s, poly-(carbonate urethane)s, poly-(siloxane urethane)s and poly-(ether urethane)s [11] are some of the categories of PU. Polyurethanes are becoming very common in biomedical applications and implants due to their toughness, flexibility and surface properties.

### 1.2.8 Vinyl Polymer

Vinyl polymers show geometrical isomerism due to carbon–carbon double bonds and are formed through the polymerisation of monomers mainly by free radical polymerisation. To control the polymerisation reaction, inhibitors are used for the monomer units. Bio-derived vinyl polymers are a good substitute for conventional petroleum-based polymers. Several bio-based substances such as waste cooking oil and citric acid have been explored for the plasticisation of poly-(vinyl chloride) [12].

### 1.2.9 Polyanhydrides

These polymers are obtained at high temperature and vacuum by condensation polymerisation reaction of monomeric units of dicarboxylic acid and acetic anhydride. Anhydride monomer units in the copolymers are responsible for the mechanical properties [13]. The rate of degradation can be controlled by altering the composition of hydrophilic and hydrophobic components. Since these polymers are non-toxic and degraded naturally into diacid monomers [14], polyanhydrides have their main applications in the pharmaceuticals and medical equipment industries.

## 1.3 CHEMICAL CHARACTERISATION OF BIOMATERIALS

The chemical characterisation of biomaterials provides information about their molecular structure, which is essential to establish its applications. Various chemical characterisations include FT-IR, NMR, MS spectra, XPS, UV-vis, etc.

### 1.3.1 FTIR Spectroscopy

FTIR spectroscopy is a potent method used by researchers for the identification of the chemical structure and functional groups present in biomaterials, which are the key factors in designating their applications [15]. Many researchers have used FTIR to identify the polymerisation process and properties of dental composites [16]. Attenuated total reflection FTIR (ATR-FTIR) spectroscopy is also used for the characterisation of biomaterials as this method has the benefit over FTIR that the sample does not require any pre-treatment or destruction of the specimen [17].

### 1.3.2 Raman Spectroscopy

Raman spectra are also non-destructive spectroscopic methods. They can be applied to any form of biomaterials (solid, liquid or gas) and provide additional and complementary information to IR spectra.

Raman spectra provide the information for the molecular structure confirmation by providing knowledge about the chemical species. The groups C–S, C=C, N=N, and S–S are very prominently detected through Ramana spectra [18].

### 1.3.3 X-ray Photoelectron Spectroscopy (XPS)

X-ray photoelectron spectroscopy (XPS) is an efficient technique that is used for the surface (1–10 nm) characterisation of solid surfaces. XPS has potential in analysing the functionality and extent of binding of molecules such as protein, lipids and enzymes on different surfaces. Therefore, it helps in deciding the designated purpose, such as for biosensors, bioarrays and biofouling controls for medical applications [19].

### 1.3.4 Scattering Techniques

These techniques are efficient tools to investigate and confirm the structure of biomaterials. They are non-destructive techniques that can be used to investigate a wide range of scales. The structural properties of biomaterials can be investigated through neutron- and light-scattering methods [20].

#### 1.3.4.1 X-ray Diffraction (XRD)

Small-angle X-rays and neutron-scattering and light-scattering techniques are very efficient techniques to identify the structure and related properties of biomaterials. These techniques are also competent to study the physiological conditions under biological systems [21].

#### 1.3.4.2 Atomic Force Microscopy (AFM)

Atomic force microscopy (AFM) is one of the most significant tools used to identify the structure along with the mechanical behaviour of biomaterials. Controlled applied force provides the high-resolution nanostructured details and its responses, which provide the information about its mechanical characterisation. AFM is considered to be a complete scientific instrument since it provides a high-resolution view of surface topographies [15].

#### 1.3.4.3 Scanning Electron Microscopy (SEM)

Scanning electron microscopy (SEM) is used to analyse morphology in the range of micro- and nanometres. SEM provides an in-depth and higher resolution of biomaterials. Electrically conductive samples of polymeric biomaterials are prepared through sputtering of gold or graphite for SEM analysis [22].

### 1.3.5 Molecular Weight Measurements

Macromolecular materials are a major area of interest for researchers nowadays due to their mechanical strength, flexibility, resistance to deformation, etc. Applications of polymeric materials can be defined on the basis of the above-described properties. Good molecular weight control, i.e., molecular weight and molecular weight distribution of polymers, are important parameters that affect the structures of polymers and their physicochemical properties.

The molecular weight of polymers defines the length of the polymer chains. Average molecular weights describe polymeric macromolecular systems; there are two main types of average molecular weight system, i.e., number average molecular weight (Mn) and weight average molecular weight (Mw) [23].

### 1.3.6 Wettability Behaviour

Adequate surface properties are required features for biomaterials as they interact with living cells. Wettability is one of the mandatory properties for biomaterials for the interactions and responses in the biological system. Better wettability behaviour of biomaterials ensures adequate dispersal of liquids on the surfaces of materials. Wettability behaviour is estimated by contact angle measurement and is associated with intermolecular forces. The contact angle (θ) of a biomaterial surface is the angle formed by the intersection of the solid–liquid and liquid–vapor interface, and is obtained by drawing a tangent along the liquid–vapor interface [24].

## 1.4 FABRICATION OF BIOPOLYMERS WITH SPECIAL REFERENCE TO TISSUE ENGINEERING AND MEDICAL DEVICES

One of the most recent advancements in the field of biopolymer development of polymeric biomaterials is their 3D structure modifications by means of a number of advanced instrumental techniques and methods with special prominence in numerous tissue engineering sectors.

A number of approaches for 3D scaffold fabrication are available that encompass printing techniques, lithography and patterning through self-administration of polymers, self-association of peptides, and cellular compatibility of hydrogels and polymer-based biomaterials. The exploitation of polymeric biomaterials in tissue engineering, specifically to regenerate the cartilage, bone and neural tissue is well known.

### 1.4.1 Tissue Engineering (TE)

Tissue engineering is used to substitute, renovate or restore the injured tissues, or to construct artificial tissues to be used as implants. This process has become essential where normal physiological procedures fail with the only option of surgical operations [25].
Some of the commonly available approaches of TE involve:

1. The explanation of cells from individual, this can be done *in vitro* in order to distinguish and subsequently transform them genetically with expansion before their infusion preferentially, into the parent individual.
2. In another approach, the explanted cells were engineered and then exposed to all the mechanical and molecular signals of the human body.
3. Another tissue engineering approach involves the use of substances that can induce the tissue *in vitro* cultivations (C-1) before their reinfusion into the exposed body cells.

Nowadays, two altogether different standards are employed: autografts and allografts. Both methods have numerous limitations. The main limitation associated with autografts is the donor-site, whereas in allografts it is the coupled latent risk of the transmission of diseases. In current research, great efforts are being made all over the world to overcome the inherent limitations of current standards and to modify the biomedical technology by employing 3D biomaterial scaffold-based TE strategies.

### 1.4.2 3D Scaffold Fabrication for TE

This biocompatible 3D scaffold methods of fabrication of biopolymers can be of two types:

1. Conventional methods: These types of fabrication methods are generally unable to deliver adequate physical and mechanical characteristics, and subsequently the scaffolds result in deformation due to the motility of the cells. This method encompasses porogen leaching, phase separation, fibre meshing, gas foaming and supercritical fluid processing.
2. Rapid prototyping: Rapid prototyping is an advanced fabrication technique capable of delivering all the indispensable properties that specify the TE application. Commonly used examples of rapid prototyping techniques include 3D printing, selective laser sintering and lithography.

### 1.4.3 3D Scaffold Fabrication by Lithography and Printing Techniques

Advancements in the field of 3D fabrication techniques have given rise to progress in hydrogel patterning through dip-pen lithography, photolithography, contact printing, nanoimprinting, robotic deposition and solid-free form, and their exploitation in TE [26].

#### 1.4.3.1 Photolithography

The photolithography fabrication method is one of the best-known advanced fabrication methods used to deliver 3D structures and desired patterns by modifying numerous polymers of varied molecular weights.

These patterns can be achieved in both polymeric as well as monomeric films. This technique of 3D scaffold generation comprises a 'two-step' procedure. The first step involves exposure of a polymer-coated surface to UV radiation, which results in photo cross-linking, photopolymerisation and/or other chemical alterations such as decomposition reactions, functionalisation or induced phase separation onto the surface. Subsequently, the leftover surface area of the polymer surface that was inhibited from UV exposure remains unchanged and can be eliminated using a suitable solvent, thus forming a 3D pattern surface [27].

#### 1.4.3.2 Nanoimprinting Lithography (NIL)

NIL is used to create cost-effective, 3D nano-structure surfaces with better resolution. In order to create a solid 3D pattern structure, this approach includes transferring polymerics or monomeric substrate into the mold, followed by a subsequent treatment at the required temperature, or UV exposure [28].

#### 1.4.3.3 Contact Printing

##### *1.4.3.3.1 Microcontact Printing (μCP) with UV-induced 3D Patterning*

One of the outstanding surface patterning techniques, developed about a decade ago, is microcontact printing which involves the specific resolution in nanometres. This method is used to achieve a high-resolution 3D pattern within suitable conditions without deformation of stamps, contamination or lateral dispersion of the ink.

##### *1.4.3.3.2 Contact Printing Without UV*

In contact printing without UV, two layers of glass slides are coated with a polymer, in which the lower slide is coated with ~ 1 μm thickness of chitosan and the upper slide is coated with either 1.2 μm or 2.4 μm of a polystyrene surface. The thicknesses of the polymers are observed using SEM.

The benefits of this technique over conventional methods include: (i) easy processing; (ii) the microwell dimensions can be smoothly controlled through regulating physical and chemical considerations; (iii) bulk flow of solvents is not required; and (iv) a high density of microwell features can be created in a one-step process.

#### 1.4.3.4 Solid Free-Foam 3D Patterning of Polymeric Materials by Ink-Jet Printing

This is a type of ink-jet printing technique which can be used to formulate 3D patterns of polymers over substrates either through 'drop-on-demand' or 'continuous' mode, in which a solution-based writing process is carried out on the substrates.

The 'drop-on-demand' method reads signals to control the ejection of each droplet, and is divided into three types:

1. Electromechanical
2. Electrothermal
3. Electrostatic vacuum.

meanwhile, in continuous mode, ink appears in a continuous mode from the nozzle under vacuum, and a jet is used to break the line of continuous droplets. This technique has two main divisions:

1. An electrical field-controlled ink-jet system
2. Hertz continuous, a mutual charged droplet repulsion type ink-jet system.

#### 1.4.3.5 Robotic Deposition

In robotic deposition, adequate printing materials such as biocomposites, biopolymers and dispersed materials are uninterruptedly doped onto the substrate either in the molten stage or as a solution to generate a 3D complex surface. This process enables the generation of 3D complex architectural scaffolds with varying pore sizes and porosities through software-controlled layer-by-layer printing and a solidifying procedure.

### 1.4.4 3D-scaffold Fabrication and Patterning by Self-organisation

Motivated by the 3D pattern prevailing in biological constructions, researchers working in the field of polymers and biomaterials have explored and developed 3D graded and erudite architecture in the range from micron- to nano-structures by exploiting the functional polymers and biomaterials as an alternate to the existing techniques. The main advantages of this technique involve:

1. The generation of a 3D scaffold under normal physiological conditions
2. Avoidance of toxic chemicals or initiators
3. No need for tedious reaction parameters such as high temperature or UV radiation for curing.

Hence, this technique has been found to be preferable in various biomedical applications.

### 1.4.5 3D Scaffolds by Self-assembly Peptides

Over the last few decades, a noteworthy development has occurred on the self-assembly peptides (SAPs) for generating 3D scaffolds. Currently, SAPs have reached a broad application range in the areas of drug delivery, biology, nanotechnology and nanoelectronics.
Some challenges have also been associated with this technique, such as:

1. Requirement of precise positioning of the peptide-based nanostructures
2. Control over their assembly and positioning
3. Their integration into microsystems.

### 1.4.6 Polymeric Biomaterial-Mediated Cell Manipulation

Numerous investigations have revealed the use of human embryonic stem cell (hESC) culturing in well-defined 3D settings through a number of scaffolds for the functioning of cells, their viability and lineage regulation. Disadvantages associated with these biological materials involve their high cost, lower reproducibility and vagueness in component identification.

### 1.4.7 Cartilage

Numerous studies have claimed that the culture of mesenchymal stem cells (MSCs) on a wide range of TE scaffolds expedites the chondrogenesis and generation of cartilage. However, some limitations have also been reported in attaining the similar characteristics as native cartilage. Moreover, the generation of functional cartilage by MSCs has been found to be difficult, as it depends on a viable cell source for extracellular cartilage matrix production, leading to high-quality cartilage regeneration [29].

It is noted that the uniform distribution of such a matrix generated by MSCs is essential for the optimum mechanical strength of the tissue. Therefore, appropriate design of 3D-structured biomaterials to support uniform distribution of formed tissue is essential for effective cartilage formation by MSCs.

Polymer substances of natural origin, such as collagen, alginate, silk fibroin, agarose, *etc.* were also used to design and fabricate scaffolds in a wide variety of forms, including meshes, sponges, foams, hydrogels, glues, composite layers, biotextiles, nanofibres and microspheres [30–33].

Various synthetic polymeric materials have been used to fabricate scaffolds for cartilage repair. These have included PLA, PGA and PLGA copolymers, and PEG or PPO polymers. These were found to form gels, ceramic composites and hydrogels containing PEG polymer-based derivatives at different temperatures. A description of polymer scaffolds fabricated using a variety of techniques and used in pre-clinical animal and clinical human trials in cartilage tissue engineering as reported by many researchers [34,35].

### 1.4.8 Bone

Recently, several research groups have reported on a variety of biodegradable synthetic polymer scaffolds for bone TE, with such scaffolds including PCL, poly(lactic acid) and their copolymers [36].

It has been concluded that these polymer-based scaffolds have some advantages over ceramic and glass-based ones, primarily because the properties of the polymer-based scaffolds can easily be processed and tailored to obtain suitable geometry for implantation. The major drawbacks of polymer scaffolds are low mechanical strength, shape retention failure and insufficient cell adhesion and growth, and hence, they require surface modification with functional groups or incorporation of bioactive materials to form multicomponent biocompatible composite bone scaffolds to enhance osteogenesis for ultimate bone tissue engineering [37].

### 1.4.9 Neural Tissue Engineering (NTE)

Nowadays, the autologous nerve grafting method is employed for repairing nerve defects. However, the two major disadvantages associated with these techniques include:

1. They lead to functional loss in the donor nerve graft sensory dispersal
2. A mismatch in the geometry between the injured and grafted nerve

Therefore, there is a requirement for a neural TE approach in order for their advancement, and to focus on 3D scaffold production with constructive neural cell growth that has expedited renewal. Various scientists have exploited these scaffolds to enhance the renewal within the CNS, and have produced substantial results.

## 1.5 RECENT POLYMERIC BIOMATERIALS APPLICATIONS

### 1.5.1 Biophotonic Applications

Polymeric biomaterials have numerous applications in several fields, especially in the area of biomedical applications. Polymeric biomaterials are used due to their properties such as degradability, biological responses, structural inertness and processability [38].

### 1.5.2 Biomaterials for Implants

In the advancement of medical sciences, medical devices and implants have also been developed with greater efficiency. By improving their functionality and applications, quality of life can be improved. the basic purpose of these devices is to enhance and preserve human lives. Noteworthy applications of these objects include knee implants and synthetic blood vessels in which they can provide strength and support, pacemakers enhancing the functionality of body organs, etc. According to the position and requirement in the human body, these implants provide power to correct bodily functions. The working abilities of these implants should be sustained along with the biological system of the body. These properties of implants depend on the materials with which they are fabricated.

Polymeric biomaterials are used in implants and related devices, as they are flexible, biocompatible and easy to fabricate. Biomaterials should have considerable mechanical strength to sustain the lifetime of the implant in the environmental conditions of the human body. Composites of biomaterials with different materials show a wide range of chemically, mechanically and thermally suitable behaviours. Biomaterials should have suitable gas and water permeability so that they can protect the electronic casing of the device in the conditions of the human body [39].

Biopolymeric nanomaterials have promising scope in biological and medical sciences. They have potential applications in medicines, drug-delivery systems, medical devices, medical diagnosis and wound healing. Biopolymer-based nanomaterials are also used in several fields including the food industry, cosmetics and environmental remediation (water purification), etc. Satish et al. explored and reviewed Ag nanoparticle-incorporated biopolymer-based biomaterials with antimicrobial properties for wound-healing purposes [40].

### 1.5.3 Piezoelectric Biomaterials

There is a new class of functional materials known as piezoelectric biomaterials. These materials have the potential to translate mechanical deformation into electricity, and vice versa. They have applications in biosensors, bioactuators and tissue stimulators due to their specific material properties. Polyvinylidene fluoride (PVDF) and polar polymers are ferroelectric in nature and show piezoelectric behaviour after poling treatments [41]. Piezoelectric effects are also exhibited by optically active polymers after uniaxial elongation [42], like poly-(L-lactic acid) (PLLA) and poly-(D-lactic acid) (PDLA). Due to their easy fabrication behaviour and cost-effectiveness, polymer-based piezoelectric devices do not require advanced techniques and are less expensive. Polymeric piezoelectric biomaterials have excellent properties such as mechanical flexibility and biocompatibility and are an excellent choice for biosensing applications [43].

### 1.5.4 Sensors

Chetana Vaghela et al. explored the ammonia-sensing properties of composites of agarose–guar gum–polyaniline (A–G-PANI) composite films. Outstanding sensing properties were shown by synthesised composites of biopolymers over a wide range of 10–90,000 ppm (9%) in various sites on the interpenetrated network structure of polyaniline biocomposites that are responsible for the excellent performance of sensors [44].

### 1.5.5 Nanofibres for Tissue Engineering

Electrospinning is a versatile technique for the fabrication of biomaterial nano/microfibers. These fabricated matrices of biomaterials have a tremendously high surface-to-volume ratio, extremely connected pores and diversified fibrous morphologies. These advanced characteristics of electrospun nanofibres impart desirable properties and have applications in biomedical systems. Several applications such as tissue engineering and wound healing have been studied by researchers [45].

### 1.5.6 Hybrid Materials for Bone Tissue Engineering

Hybrid polymeric biomaterials are used in bone tissue engineering due to their osteoconductive, elastomeric and biomimetic properties. Researchers are aiming to improve these materials by enhancing their osteoconductive behaviour and mechanical strength with preservation of the elastomeric properties. Next-generation hybrid polymeric biomaterials will also possess antibacterial activity and bioimaging ability [46].

Biomaterials play a tremendous role in medical devices and healthcare systems. Polymeric biomaterials have a wide range of applications in medical and drug-delivery systems. They have very high qualities of properties which support specific conditions such as adhesion, drug release, etc., although the evaluation of biocompatibility in regulated environments is a challenging task which depends on various factors such as the chemical nature, physical properties of the biomaterial and duration of contact.

### 1.5.7 3D Printing

3D printing is a relatively new method which has the potential of fabricating scaffolds for tissue engineering. Researchers have developed new biocomposite materials which can be applied for 3D printing methods. Scaffolds prepared using the 3D printing method are more porous structures with complex geometries and more growth factors can be involved. This method allows the incorporation of more ceramics in polymers. 3D printed polymeric biomaterials have a wider range of applications in regenerative medicine.

### 1.5.8 Grafting Techniques

Functionalisation of polymeric biomaterial by grafting techniques is the latest advancement in chemical surface modification. The grafting-to method, grafting-from method and grafting-through method are the three main methods of grafting, among which the grafting-from method is most common. These methods are advantageous in the biomedical field and are applicable at the interfaces of biomacromolecules, tissues, cells and biomaterials. Recently introduced methods of functionalisation through surface modification include chemical grafting, plasma-induced graft polymerisation, radiation-induced graft polymerisation, ozone graft polymerisation and photo-induced graft polymerisation [47].

### 1.5.9 Nerve Tissue Engineering

Polymeric biomaterials are extensively used in biomedical applications, in which nerve tissue engineering is one of the most specialised fields. Spinal cord injury, brain trauma and ischemic stroke are extremely devastating and can seriously affect the quality of life of patients due to damaged nervous systems [48]. To encourage the efficiency of medical devices and biomaterials for the regeneration of injured nervous systems, research is on-going globally. The main challenges in nerve tissue engineering are the deficiency of bioactivity of the materials and the foreign body response. More smart polymeric biomaterials with efficient designs will be a future solution in this area.

### 1.5.10 Biomaterials in the Bio–Water Interface

Biomaterials have suitable properties to absorb water molecules on their surface. This bio–water interface can be controlled by altering the design of functional biomedical polymers. The intermediate water concept has been studied for different biocompatible polymers in hydrated conditions. Materials in hydrated conditions have been found to have better protein adsorption and cell adhesion due to the function of intermediate water contents [49].

### 1.5.11 Cross-linked Biomaterials

An emerging category of tuneable biomaterials has arisen through the cross-linking of polymers with supramolecules. The conductivity and dynamics of any polymeric system are governed by its thermal and kinetic behaviour which can be controlled by the moiety selected for cross-linking. The most prominent characteristics of biomaterials such as construction and morphology are accountable for their responsiveness, viscoelasticity, diffusion, etc., which are key parameters for the biomedical properties of materials. Therefore, supramolecular cross-linking chemistry and biomedically related macroscopic properties are found to be interconnected [50].

### 1.5.12 Optoelectronic Biomaterials

Optoelectronic biomaterials are created using a newly developed treatment method for use in neurodegenerative disorders. Bowie et al. developed a biocompatible fibrous film with photoactive polymers to enable neurogenesis of PC12 cells. Poly(3- hexylthiophene) (P3HT), polycaprolactone (PCL) and polypyrrole (PPY) are the key photoactive polymers used in this tricomponent polymer [51].

### 1.5.13 Cytocompatible Biomaterials

A cytocompatible biomaterial with intelligent interfaces of 2-methacryloyloxyethyl phosphorylcholine (MPC) has been developed by Yakugaku et al. [52]. MPC is a phospholipid polymer, composed of a methacrylate monomer unit that has applications in the pharmaceutical and biomedical industries. These phospholipid polymers have less cytotoxic behaviour, and enable drugs to penetrate into cells and nearby tissues by solubilising the drugs efficiently.

### 1.5.14 Electrically Conducting Polymeric Bionanocomposites (ECPBs)

Another emerging category of polymeric biomaterial is electrically conducting polymeric bionanocomposites (ECPBs). These have characteristics of plastic along with biodegradability. Due to mimicking the biological and electrical nature of tissues in the human system, ECPBs are applicable for use in tissue engineering and drug-delivery systems [53].

### 1.5.15 Light-Guiding Biomaterials

Light-guiding biomaterials that have suitable optical properties can consist of natural or synthetic polymers. The source of these materials includes light-guiding structures found in plants and animals which are the basis of motivation for biomaterial photonics engineering. These are transparent materials which are used in phytomedicines [54].

## 1.6 CONCLUSION

This chapter summarises the general overview of biopolymeric materials covering key areas including their classes, characterisation methods, fabrication techniques and recent applications. This classification involves the category of biopolymers obtained from bioresources (excluding plants). The main challenge in assigning the exact biomaterial for a specific application is exploring its chemical, physical and biological characteristics. The roles of characterisation techniques used for polymeric biomaterials, including UV-vis, FTIR, Raman, XPS, scattering methods, molecular weight assessment and wettability behaviour, have been summarised. Designing and fabrication methods of biopolymeric scaffolds are also important concerns and key matters to achieving the ultimate goal. Although polymeric biomaterials are a versatile category of compounds with extensive use in the healthcare system, these materials present many opportunities and future prospects. Therefore, a huge area of opportunities to enable the advancement and application of polymer biomaterials has been described in this chapter.

## REFERENCES

[1] T. Biswal, S. K. BadJena, and D. Pradhan, "Sustainable biomaterials and their applications: A short review," *Mater. Today Proc.*, vol. 30, pp. 274–282, 2020, doi: 10.1016/j.matpr.2020.01.437

[2] M. L. Becker and J. A. Burdick, "Introduction: polymeric biomaterials," *Chem. Rev.*, vol. 121, no. 18, pp. 10789–10791, 2021, doi: 10.1021/acs.chemrev.1c00354

[3] A. George, M. R. Sanjay, R. Srisuk, J. Parameswaranpillai, and S. Siengchin, "A comprehensive review on chemical properties and applications of biopolymers and their composites," *Int. J. Biol. Macromol.*, vol. 154, pp. 329–338, 2020, doi: 10.1016/j.ijbiomac.2020.03.120

[4] I. Vroman and L. Tighzert, "Biodegradable polymers," *Materials (Basel).*, vol. 2, no. 2, pp. 307–344, 2009, doi: 10.3390/ma2020307

[5] S. M. Guillaume and L. Mespouille, "Polycarbonates and green chemistry," *J. Appl. Polym. Sci.*, vol. 131, no. 5, 2014, doi: 10.1002/app.40081

[6] W. Yu, E. Maynard, V. Chiaradia, M. C. Arno, and A. P. Dove, "Aliphatic polycarbonates from cyclic carbonate monomers and their application as biomaterials," *Chem. Rev.*, vol. 121, no. 18, pp. 10865–10907, 2021.

[7] T. Hidaka and H. Sugiyama, "Chemical approaches to the development of artificial transcription factors based on pyrrole-imidazole polyamides," *Chem. Rec.*, vol. 21, no. 6, pp. 1374–1384, 2021.

[8] M. Feldmann, "Bio-based Polyamides," *Ind. Appl. Biopolym. their Environ. Impact*, vol. 1000, p. 94, 2020.

[9] A. C. Fonseca, M. H. Gil, and P. N. Simões, "Biodegradable poly(ester amide)s – a remarkable opportunity for the biomedical area: review on the synthesis, characterization and applications," *Prog. Polym. Sci.*, vol. 39, no. 7, pp. 1291–1311, 2014, doi: 10.1016/j.progpolymsci.2013.11.007

[10] S. K. Murase and J. Puiggalí, *Poly(Ester Amide)s: Recent Developments on Synthesis and Applications*. Elsevier, 2014.

[11] P. Parcheta and J. Datta, "Structure-rheology relationship of fully bio-based linear polyester polyols for polyurethanes – synthesis and investigation," *Polym. Test.*, vol. 67, pp. 110–121, 2018.

[12] G. Feng *et al.*, "An efficient bio-based plasticizer for poly (vinyl chloride) from waste cooking oil and citric acid: synthesis and evaluation in PVC films," *J. Clean. Prod.*, vol. 189, pp. 334–343, 2018, doi: 10.1016/j.jclepro.2018.04.085

[13] S. Karandikar, A. Mirani, V. Waybhase, V. B. Patravale, and S. Patankar, *Nanovaccines for Oral Delivery-Formulation Strategies and Challenges*. Elsevier, 2017.
[14] S. Samavedi, L. K. Poindexter, M. Van Dyke, and A. S. Goldstein, *Synthetic Biomaterials for Regenerative Medicine Applications*. Elsevier, 2014.
[15] M. Omidi *et al.*, *Characterization of Biomaterials*. Elsevier, 2017.
[16] A. H. S. Delgado and A. M. Young, "Modelling ATR-FTIR spectra of dental bonding systems to investigate composition and polymerisation kinetics," *Materials (Basel).*, vol. 14, no. 4, p. 760, 2021.
[17] J. Durner, J. Obermaier, M. Draenert, and N. Ilie, "Correlation of the degree of conversion with the amount of elutable substances in nano-hybrid dental composites," *Dent. Mater.*, vol. 28, no. 11, pp. 1146–1153, 2012, doi: 10.1016/j.dental.2012.08.006
[18] D. Bazin *et al.*, "Diffraction techniques and vibrational spectroscopy opportunities to characterise bones," *Osteoporos. Int.*, vol. 20, no. 6, pp. 1065–1075, 2009.
[19] S. L. McArthur, G. Mishra, and C. D. Easton, "Applications of XPS in biology and biointerface analysis," in Vincent S. Smentkowski (ed.) *Surface Analysis and Techniques in Biology*, Springer, 2014, pp. 9–36.
[20] A. Bandyopadhyay and S. Bose, *Characterization of Biomaterials*. Newnes, 2013.
[21] D. Lombardo, P. Calandra, and M. A. Kiselev, "Structural characterization of biomaterials by means of small angle x-rays and neutron scattering (saxs and sans), and light scattering experiments," *Molecules*, vol. 25, no. 23, p. 5624, 2020.
[22] N. Vyas, R. L. Sammons, O. Addison, H. Dehghani, and A. D. Walmsley, "A quantitative method to measure biofilm removal efficiency from complex biomaterial surfaces using SEM and image analysis," *Sci. Rep.*, vol. 6, no. 1, pp. 1–10, 2016.
[23] P. Snetkov, K. Zakharova, S. Morozkina, and R. Olekhnovich, "Hyaluronic acid: the influence of molecular weight and degradable properties of biopolymer." Int. J. Biol. Macromol., vol. 154, pp. 329–338, 2020.
[24] M. Zielecka, "Methods of contact angle measurements as a tool for characterization of wettability of polymers," *Polimery*, vol. 49, no. 5, pp. 327–332, 2004.
[25] T. Biswal, "Biopolymers for tissue engineering applications: a review," *Mater. Today Proc.*, vol. 41, pp. 397–402, 2021.
[26] R. Liang, Y. Gu, Y. Wu, V. Bunpetch, and S. Zhang, "Lithography-based 3D bioprinting and bioinks for bone repair and regeneration," *ACS Biomater. Sci. Eng.*, vol. 7, no. 3, pp. 806–816, 2020.
[27] P. Prosposito *et al.*, "Photolithography of 3D scaffolds for artificial tissue," *Materials Science Forum*, vol. 879, pp. 1519–1523, 2017.
[28] V. A. Seleznev and V. Y. Prinz, "Hybrid 3D–2D printing for bone scaffolds fabrication," *Nanotechnology*, vol. 28, no. 6, p. 64004, 2017.
[29] S. O. Ebhodaghe, "Natural polymeric scaffolds for tissue engineering applications," *J. Biomater. Sci. Polym. Ed.*, vol. 32, no. 16, pp. 2144–2194, 2021.
[30] M. Ahearne, J. Fernández-Pérez, S. Masterton, P. W. Madden, and P. Bhattacharjee, "Designing scaffolds for corneal regeneration," *Adv. Funct. Mater.*, vol. 30, no. 44, p. 1908996, 2020.
[31] Y. P. Singh, N. Bhardwaj, and B. B. Mandal, "Potential of agarose/silk fibroin blended hydrogel for in vitro cartilage tissue engineering," *ACS Appl. Mater. Interfaces*, vol. 8, no. 33, pp. 21236–21249, 2016.
[32] H. Khalid *et al.*, "Silk fibroin/collagen 3D scaffolds loaded with $TiO_2$ nanoparticles for skin tissue regeneration," *Polym. Bull.*, vol. 78, no. 12, pp. 7199–7218, 2021.
[33] F. Zhou *et al.*, "Silk fibroin-chondroitin sulfate scaffold with immuno-inhibition property for articular cartilage repair," *Acta Biomater.*, vol. 63, pp. 64–75, 2017.
[34] J. Gil, M. L. Tomov, A. S. Theus, A. Cetnar, M. Mahmoudi, and V. Serpooshan, "In vivo tracking of tissue engineered constructs," *Micromachines*, vol. 10, no. 7, p. 474, 2019.
[35] J. Y. Park, S. H. Park, M. G. Kim, S.-H. Park, T. H. Yoo, and M. S. Kim, "Biomimetic scaffolds for bone tissue engineering," *Biomim. Med. Mater.*, vol. 1064, pp. 109–121, 2018.
[36] P. N. Christy *et al.*, "Biopolymeric nanocomposite scaffolds for bone tissue engineering applications – a review," *J. Drug Deliv. Sci. Technol.*, vol. 55, p. 101452, 2020.
[37] R. S. Ambekar and B. Kandasubramanian, "Progress in the advancement of porous biopolymer scaffold: tissue engineering application," *Ind. Eng. Chem. Res.*, vol. 58, no. 16, pp. 6163–6194, 2019.
[38] Shan *et al.*, "Polymeric biomaterials for biophotonic applications," *Bioact. Mater.*, vol. 3, no. 4, pp. 434–445, 2018, doi: 10.1016/j.bioactmat.2018.07.001

[39] A. J. T. Teo, A. Mishra, I. Park, Y. J. Kim, W. T. Park, and Y. J. Yoon, "Polymeric biomaterials for medical implants and devices," *ACS Biomater. Sci. Eng.*, vol. 2, no. 4, pp. 454–472, 2016, doi: 10.1021/acsbiomaterials.5b00429

[40] S. S. D. Kumar, N. K. Rajendran, N. N. Houreld, and H. Abrahamse, "Recent advances on silver nanoparticle and biopolymer-based biomaterials for wound healing applications," *Int. J. Biol. Macromol.*, vol. 115, no. 2017, pp. 165–175, 2018, doi: 10.1016/j.ijbiomac.2018.04.003

[41] R. Kumar, P. Senthamaraikannan, S. S. Saravanakumar, A. Khan, K. Ganesh, and S. V. Ananth, "Electroactive polymer composites and applications," in *Polymer Nanocomposite-Based Smart Materials*, R. Kumar (ed.). Elsevier, 2020, pp. 149–156.

[42] S. Zhukov *et al.*, "Biodegradable cellular polylactic acid ferroelectrets with strong longitudinal and transverse piezoelectricity," *Appl. Phys. Lett.*, vol. 117, no. 11, p. 112901, 2020.

[43] M. T. Chorsi *et al.*, "Piezoelectric biomaterials for sensors and actuators," *Adv. Mater.*, vol. 31, no. 1, pp. 1–15, 2019, doi: 10.1002/adma.201802084

[44] C. Vaghela, M. Kulkarni, S. Haram, M. Karve, and R. Aiyer, "Biopolymer-polyaniline composite for a wide range ammonia gas sensor," *IEEE Sens. J.*, vol. 16, no. 11, pp. 4318–4325, 2016, doi: 10.1109/JSEN.2016.2541178

[45] J. Ding *et al.*, "Electrospun polymer biomaterials," *Prog. Polym. Sci.*, vol. 90, pp. 1–34, 2019, doi: 10.1016/j.progpolymsci.2019.01.002

[46] B. Lei, B. Guo, K. J. Rambhia, and P. X. Ma, "Hybrid polymer biomaterials for bone tissue regeneration," *Frontiers of Medicine*, vol. 13, pp. 189–201, 2018.

[47] W. Sun, W. Liu, Z. Wu, and H. Chen, "Chemical surface modification of polymeric biomaterials for biomedical applications," *Macromol. Rapid Commun.*, vol. 41, no. 8, pp. 1–26, 2020, doi: 10.1002/marc.201900430

[48] H. Amani, H. Kazerooni, H. Hassanpoor, A. Akbarzadeh, and H. Pazoki-Toroudi, "Tailoring synthetic polymeric biomaterials towards nerve tissue engineering: a review," *Artif. Cells, Nanomedicine Biotechnol.*, vol. 47, no. 1, pp. 3524–3539, 2019, doi: 10.1080/21691401.2019.1639723

[49] M. Tanaka *et al.*, "Design of polymeric biomaterials: the 'intermediate water concept,'" *Bull. Chem. Soc. Jpn.*, vol. 92, no. 12, pp. 2043–2057, 2019, doi: 10.1246/bcsj.20190274

[50] J. L. Mann, A. C. Yu, G. Agmon, and E. A. Appel, "Supramolecular polymeric biomaterials," *Biomater. Sci.*, vol. 6, no. 1, pp. 10–37, 2018, doi: 10.1039/c7bm00780a

[51] B. Yuan, M. R. F. Aziz, S. Li, J. Wu, D. Li, and R. K. Li, "An electro-spun tri-component polymer biomaterial with optoelectronic properties for neuronal differentiation," *Acta Biomater.*, vol. 139, pp. 82–90, 2022, doi: 10.1016/j.actbio.2021.05.036

[52] T. Konno, "Design of intelligent interface based on cytocompatible polymers for control on cell function," *Yakugaku Zasshi*, vol. 141, no. 5, pp. 641–646, 2021, doi: 10.1248/yakushi.20-00219-3

[53] C. I. Idumah, "Recent advancements in conducting polymer bionanocomposites and hydrogels for biomedical applications," *Int. J. Polym. Mater. Polym. Biomater.*, vol. 71, no. 7, pp. 513–530, 2022, doi: 10.1080/00914037.2020.1857384

[54] S. Shabahang, S. Kim, and S. H. Yun, "Light-guiding biomaterials for biomedical applications," *Adv. Funct. Mater.*, vol. 28, no. 24, 2018, doi: 10.1002/adfm.201706635

# 2 Polymeric Biomaterials in Drug-delivery Systems

*Surabhi Singh*

## CONTENTS



## 2.1 INTRODUCTION

Increasingly, polymer carriers are applied to current drug-delivery technologies in order to deliver pulsatile dosages and to deliver spatially and temporally regulated drug releases. A significant contribution to the development of intelligent drug-delivery systems has been the development of potent and specific biological therapeutics, but conventional formulations are still used extensively in the treatment of chronic diseases. In order for a delivery system to be truly intelligent, it must be able to participate in specific targeting, intracellular delivery, and biocompatibility.

Solvent-activated approaches have made significant improvements in the delivery of drugs. Using hydrogels and other polymer-based carriers, pharmaceuticals are transported safely through inhospitable body regions. Polymers can be engineered to respond to external conditions with a well-defined response by understanding the mechanisms of behavior transitions [1]. A variety of

DOI: 10.1201/9781003240884-2

biodegradable polymers can be included in therapeutics so as to improve the release kinetics and prevent carrier accumulation or biologically active polymers can be included so as to enhance their therapeutic features or provide their own therapeutic benefits [2]. There are many ways that pharmaceutical agents can be conjugated with polymers so that they can extend their half-lives, modify their transport properties, enhance their targeted properties, or enhance both their passive and active bioavailabilities. It is also noteworthy that the development of proprietary polymer carriers for the rapid delivery of therapeutics to the cytoplasm emerged out of the adoption of polymeric materials in the formulation of pharmaceuticals and their delivery [1,2].

It provides a unique analysis of polymers in drug delivery by discussing polymer therapeutics, responsive polymer technology, and advanced systems. On the basis of molecular recognition technology, this chapter discusses cellular recognition and intracellular drug delivery.

Drug-delivery materials that are able to transport drugs to target sites without difficulty or problems during and after delivery have become increasingly important as a result of side effects associated with conventional drugs. To aid in remuneration and speed up the healing process, patients are regularly administered drugs throughout their treatment. The result of this can be an increase in drug concentration in the body, which may cause problems from excess therapeutic concentrations if it exceeds the body's tolerance level [1]. Furthermore, the drug-delivery rate may be so fast that therapeutic effects are no longer achieved because the concentration of the drug at the delivery site is too low, which could occur because of drug metabolism, degradation, and transport out of the target area [1]. This phenomenon leads to the destruction of drugs and the loss of media, both of which can lead to adverse effects on cells, tissues, and organs. To effectively deliver drugs to targets, carriers must be capable of delivering slow release of the drug, and must allow for the completion of therapeutic recovery before allowing degradation and transportation of excess concentrations of the drug and carrier medium. Drugs, carriers, and formulations may be administered orally or by the parenteral route [3]. The onset of drug release can be controlled in the capsule by using a slow-dissolving cellulose coating, incorporating compounds that slow the quick dissolution of drugs, using compressed tablets, and by incorporating emulsions and suspensions. The release of drugs without changing or degrading has to be protected by materials with a long therapeutic window (days to years). It is possible to inject and/or implant these carriers directly into diseased tissues/cells for improved delivery [4]. Various affinity ligands deposited on the surfaces of biomaterials, allowing for a set retention and utilization by diseased tissues and cells, are commonly used to achieve targeted drug delivery. Apart from permitting surface modification with ligands, the design of biomaterials for drug carriers should also protect drugs from rapid breakdown and/or degradation within the target site.

A number of key parameters may be important: (i) maximizing drug encapsulation in the bioactive material for maximum drug release; (ii) maintaining drug stability and bioactivity throughout transport; and (iii) controlling the cost of biomaterial synthesis and fabrication.

## 2.2 DRUG DELIVERY BY NATURAL POLYMERS

### 2.2.1 Arginine Derivatives

It is important to know that arginine, or L-arginine, is an amino acid that is used to synthesize proteins. The molecule in Figure 2.1 contains a trio of amino groups, an amine group, and an aliphatic straight chain that contains three carbons in the form of guanidino groups [1]. At physiological pH, the carboxylic acid (COO-), the amino $(NH_3)^+$ group, and the guanidinium $(C\text{-}NH_2)^+$ group of a charged aliphatic amino acid are protonated, yielding arginine. In addition to side chains containing three carbons, their distal end consists of capping of the guanidinium group, which has a pKa of 12.48. Because the nitrogen lone pair and double bonds enable the delocalization of positive charges, the chemical structure is capable of forming a multitude of hydrogen bonds.

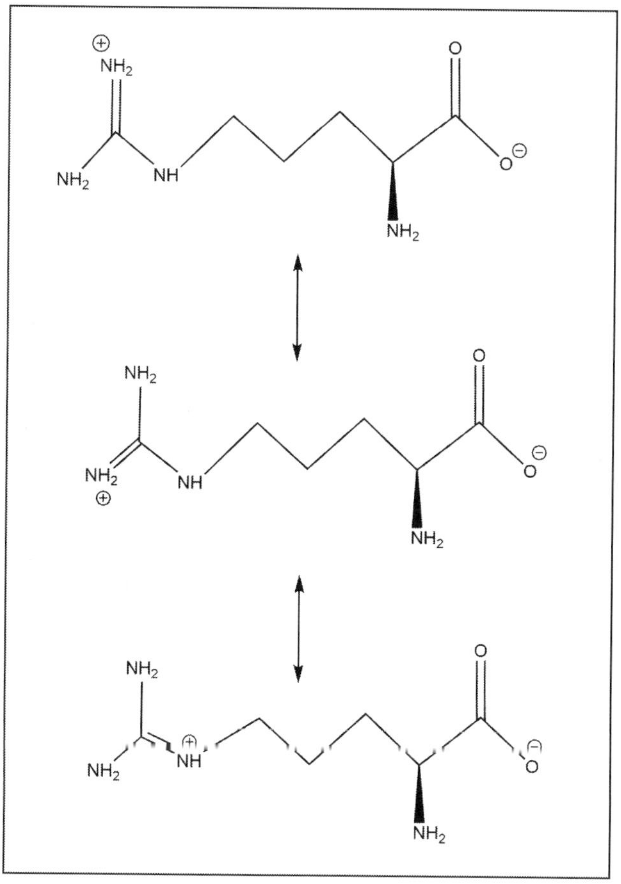

**FIGURE 2.1** Arginine group for polymeric drug delivery.

### 2.2.2 Products Derived from Chitosan

Among cationic polysaccharides, chitosan is derived from natural chitin. Polyelectrolytes complexed with polyanions and polycationic polymers have been widely used to form drug-delivery polyelectrolytes. Glucosamine and an N-acetylglucosamine unit are joined by a linkage of -(1,4) between 2-amino-2-deoxy-d-glucan. The deacetylated form of chitin is called chitosan (2-acetoxy-2-deoxy-d-glucan) (Figure 2.2a) [5,6]. This substance has been shown to possess a number of beneficial properties, including nontoxicity, mucoadhesiveness, biocompatibility, and biodegradability [15–17]. Due to the increasing interest in water-soluble chitosan derivatives such as salts (Figure 2.2b), zwitterionic chitosan and chitosan oligomers for biomedical applications, chitosan derivatives have undergone an increasing amount of research.

### 2.2.3 Delayed-release Cyclodextrins

A cyclic oligosaccharide composed of linked glucose-subunits, cyclodextrins are cyclic oligosaccharides. Cyclodextrin is useful as a molecular chelator [3,5]. Cyclodextrins are found in nature in three types (Figure 2.3). In addition to having a large cavity and an excellent ability to host complexing and loading agents, cyclodextrin is readily available and relatively inexpensive [22]. As a class of cyclodextrin, 2-hydroxylpropyl derivatives are effective in drug delivery because they are powerful solubilizers. They consist of hydrophilic outside chains and hydrophobic inside chains.

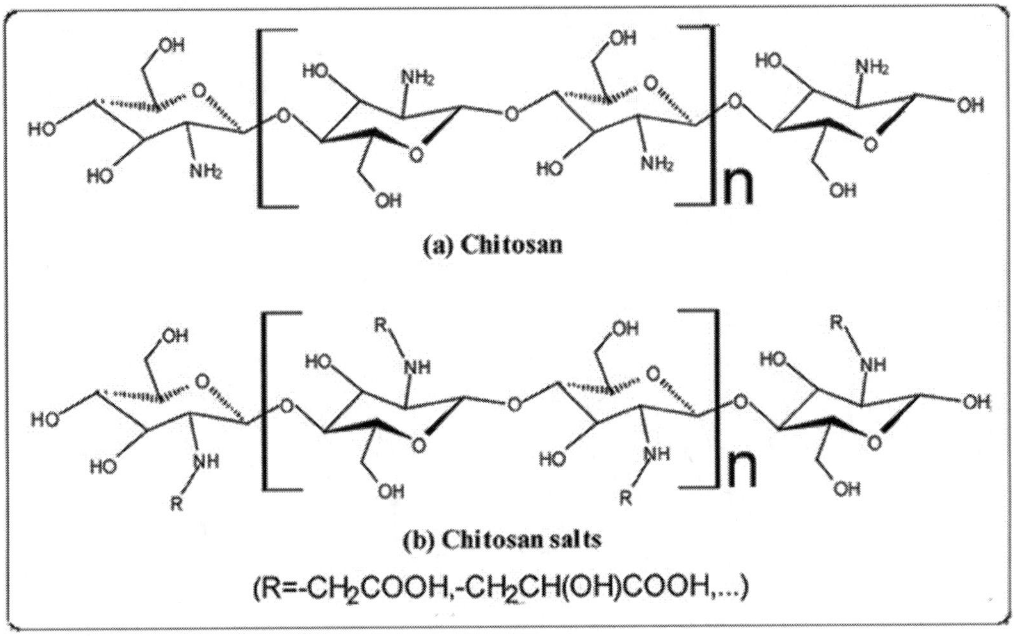

(a) Chitosan

(b) Chitosan salts

(R=-$CH_2$COOH,-$CH_2$CH(OH)COOH,...)

**FIGURE 2.2** Chitosan and chitosan salt chemical structure.

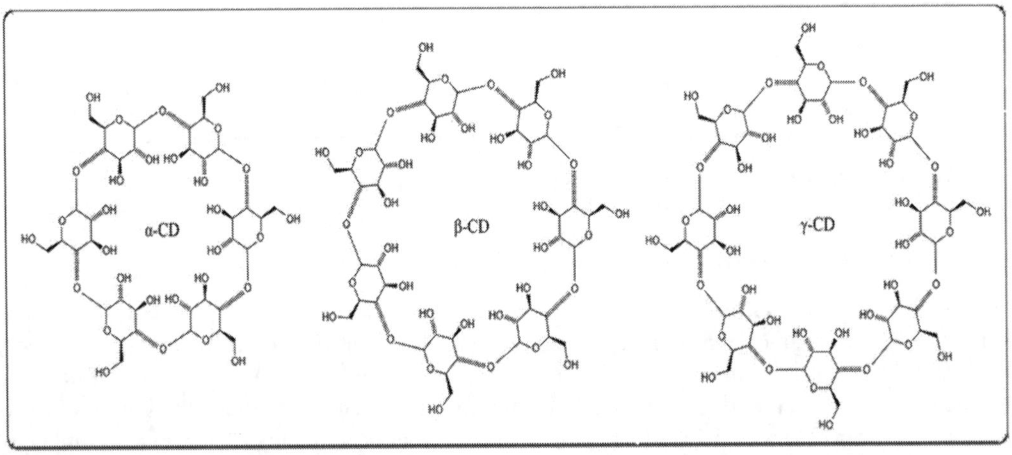

**FIGURE 2.3** Types of cyclodextrin.

Substances such as these help to protect drugs from degradation and improve their solubility and stability, resulting in greater bioavailability [24,25]. Such polymers are effective for practical drug-delivery methods.

Moreover, the FDA has approved polyglycolic acid (PGA) and polylactic acid (PLA) microparticle depots due to their versatility and the possibility to contain harmless degradation products (lactic acid and glycolic acid). They have been approved for treating Kaposi's sarcoma (approved in 1995) and recurrent ovarian cancer (approved 1997). Doxil [6] is a poly(ethylene glycol) (PEG)-coated (i.e., PEGylated) liposome that encapsulates doxorubicin. The half-life of the drug in circulation is extended and the tumor uptake of the drug is improved, as well as the toxicological consequences of using a free drug being reduced [7]. A liposome containing vincristine is used for treating rare types of leukemia, along with Abraxane, a nanoparticle of paclitaxel encapsulated in albumin, for treating rare forms of leukemia. Furthermore, there is OROS technology, which is an osmotically controlled delivery system that can deliver drugs via transdermal patches, which is used with various medications, such as Concerta [11]. Further, poly-carboxy phenoxy-propylene/sebacic acid wafers and carmustine are employed as implantable biomaterials in patients with recurrent glioblastoma multiforme [12–14].

In water or an aqueous solution, poly(2-hydroxyethyl methacrylate) can form hydrogels, which are pharmaceutical polymers. We have developed a poly(PHEMA) hydrogel for intraocular lenses using ethylene glycol dimethacrylate (EGDMA) or tri-ethylene glycol dimethacrylate (TEGMA) or ethylene glycol dimethacrylate (EGDMA) as a tert-butyl-amine catalyst, with sodium pyrosulfite (SMBS) as the cross-linker, and 2-hydroxyethylmethacrylate (HEMA) as the plasticizer. A polymer (HEMA) coating is commonly used on cell culture flasks to inhibit cell adhesion and induce spheroid formation. Agar and agarose gels are older alternatives to pHEMA [19]. Polymer (HEMA) gels cross-linked with tripropylene glycol diacrylate (TPGDA) have been tested at various TPGDA concentrations, including swelling equilibrium, structural characterization, and solute transport [15–24]. To discover how chemotherapeutic drugs interact with polymers, and how this interaction influences the release behavior of drug-eluting polymeric devices, pilocarpine was injected into poly(HEMA) hydrogels [16].

A polymer called poly(HEMA) hydrogel is commonly used in biomedical implants. A powerful coating for ventricular catheters with high hydrophilicity, poly(HEMA) resists protein fouling. Poly(N-isopropyl acrylamide) diluted in water exhibits a lower critical solution temperature (LCST). Researchers have been studying thermosensitive polymers since the 1960s. A thermosensitive poly(N-isopropyl arylamide) was characterized as having an LCST of 32°C. In order to estimate the thermodynamics of the system [39], phase diagrams and heat absorption during phase separation were used. In this case, N-isopropyl-acrylamide, a monomer, is polymerized by a free radical. In most cases, AIBN (azobisisobutyronitrile) is used to trigger radical polymerization. A growing number of research studies have focused on the potential for thermo-responsive polymers for applications in biology and medicine [10–15]. An increase in swelling was observed when poly(N,N′-alkyl substituted acrylamides) were exposed to temperature changes.

Several factors affect the size, configuration, and mobility of alkyl side-chain groups that provide swelling with water, including the degree of hydrophilic and hydrophobic bonds in polymers [13]. This technique enables the reversible conversion of a polymer with PNIPAAm grafts from hydrophilic to hydrophobic. Using NIPAAm+AAc copolymers, a temperature/pH-sensing hydrogel is produced [15]. In temperature/pH-sensitive hydrogels, polyelectrolytes have been studied for their effect on the LCST within the pH range of the swelling ratio. As a polyelectrolyte, poly(allyl amine) (PAA) was used for measuring swelling ratios in similar conditions to hydrogels.

Various in situ stimuli have been discussed including pH, redox, hypoxia, and nanoparticles that respond to the microenvironment surrounding the tumor. Water, ethanol, and chloroform are capable of dissolving linear poly(ethylenimine). Benzene, acetone, and ethyl ether are insoluble in cold water [18,19].

## 2.3 POLYMERS APPLIED TO DRUG DELIVERY CONVENTIONALLY

The pharmaceutical industry has been incorporating bioactive agents with polymers through compression, spray, and dip-coating techniques for more than 50 years. Most commonly, cellulose derivatives are used in these polymers, as well as poly(ethylene glycol) (PEG) and poly(N-vinyl pyrrolidone). Monolithic polymer devices are treated with different solvents to deliver drugs [5], solvent-activated devices at controllable swelling or osmotic pressure [5], biodegradable devices, and externally triggered devices at different pH, and temperature [4].

### 2.3.1 Systems Controlled by Diffusion

Diffusion-controlled carriers typically consist of simple and monolithic carriers. In this case, the polymer's solubility limit is exceeded, causing the matrix to be stable and nondegradable. A dissolved system's saturation concentration (C0 + CS) corresponds to the loading concentration (C0) and concentration (CS) [29]. If slab geometry is assumed, it is possible to solve for Fitzgerald's second law in order to find the expression for concentration Ci(x,t) [29,34]. Di is the diffusivity of the solute, and species i is the concentration of species i in the matrix [31]. This equation can be integrated if the boundary conditions at the interface, X, are appropriate, based on the cumulative mass or moles released (Figure 2.4) [6].

### 2.3.2 Systems that are Activated by Solvents

The two materials are simply packed together to form swellable systems. Swellable systems have traditionally been used to package small doses of drugs into hydrophilic polymers or hydrogels. Aqueous solvents do not provide a plasticizing effect, so these systems have very low diffusivities and glass transition temperatures. Hydrogels take up water and swell after coming into contact with an aqueous environment. Polymers that have not been cross-linked (or crystallized) dissolve and cause erosion fronts [8,10]. As a swelling-controlled system, drug-delivery devices change from glassy to rubbery as the polymer chains in the polymer relax and dissolve dispersed drug deposits. Two moving fronts are created simultaneously by this process, diffusion and swelling, as well as erosion, if present [36]. Using HPMC cylindrical sections containing floredil pyridoxal phosphate [9] as a model, researchers have demonstrated the importance of this. Where the localized volume fraction of solvent is higher at the dissolved–dispersed drug boundary than in the matrix core, the diffusion front is created. As water is absorbed into the matrix, the swelling front is formed, which increases chain motility. An entanglement gradient exists relative to the outside surfaces of

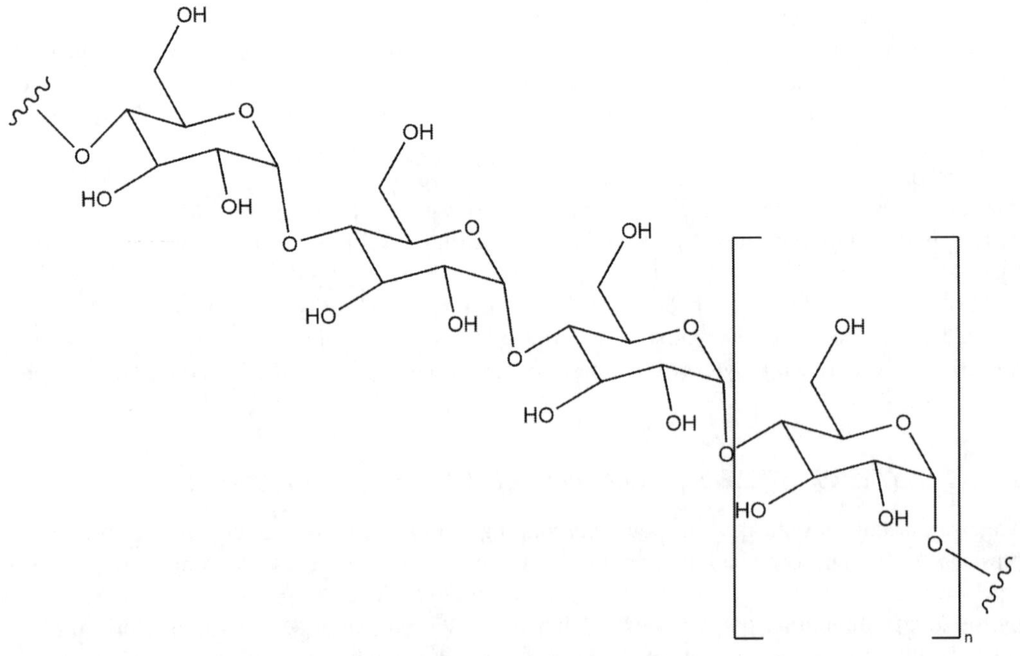

**FIGURE 2.4** System controlled by diffusion.

a polymer matrix starting at the center. Continually moving boundaries between dispersion and dissolution, and correspondingly changing diffusion lengths result from this process. According to the power law expression [10], we can use a popular empirical model to describe the behavior of swellable systems [10].

This expression consists of fractionizing a polymer matrix based on time. There are several factors that influence the value of N. These include polydispersity, geometry, and transport type. As the chain relaxes faster than the diffusion rate, diffusion in case I or Fickian diffusion occurs. This condition is correlated with n = 0.50 in thin-film geometries. These values are n = 0.45 for cylindrical geometries and 0.43 for spherical geometries [11,12]. The kinetic limit comes from chain relaxation, which is responsible for Case II diffusion. Thus, n equals 1. For systems with values of n (0.43 * n * 1), the diffusion and relaxation mechanisms follow similar rates. The use of lag times in the release of the signal and burst effects in this model allows us to model diffusion and Case II contributions separately. A review of several mathematical models of polymer drug release is provided by Arifin et al. [16] and Masaro and Zhu [17].

### 2.3.3 Poly(2-Hydroxyethyl Methacrylate)

Applied to water or an aqueous solution, poly(2-hydroxyethyl methacrylate) results in a hydrogel. The application of poly(PHEMA) hydrogels for intraocular lenses was reported using ethyleneglycoldimethacrylate (EGDMA) or triethyleneglycoldimethacrylate (TEGDMA) as cross-linking agents. Initial polymerization using azobisisobutyronitrile (AIBN), ammonium persulfate, or sodium pyrophosphate was followed by solution polymerization with azobisisobutyronitrile (AIBN) [35]. Cell culture flasks containing HEMA polymers are commonly used when studying cell adhesion and the formation of spheroid structures. Agar and agarose gels are older alternatives to pHEMA [33,34]. A broad range of tripropyleneglycol diacrylate (TPGDA) concentrations was not able to cause equilibrium swelling in poly(HEMA) gels cross-linked by TPGDA. Researchers focused on the drug–polymer interaction in order to discover how controlled release polymeric devices are affected by pilocarpine derived from poly(HEMA) hydrogels. Biomedical implants are commonly made with poly(HEMA) hydrogels. The hydrophilicity of poly(HEMA) coatings makes them an ideal coating for ventricular catheters, because they are highly resistant to protein fouling [37].

### 2.3.4 N-isopropylacrylamide Polymers

There is a lower critical solution temperature (LCST) for poly(N-isopropyl acrylamide) (PNIPAAm) in water. Polymers with temperature-responsive properties have been studied since the 1960s [38]. A temperature of 32°C was determined to be the LCST of thermosensitive poly(N-isopropyl arylamide). Phase diagrams and the entropy effect have been used to determine the thermodynamic properties of the system [39]. In this case, it is N-isopropyl-acrylamide that undergoes a free radical polymerization process, which then leads to a polymer known as a homopolymer. Azibisisobutyronitrile (AIBN) is commonly used as an initiator for radical polymerization.

### 2.3.5 Systems made from Biodegradable Materials

A class of materials known as biodegradable polymers is important in drug delivery. The difference between degradation and erosion, even though they are often used interchangeably, is that degradation involves covalent bond cleavage by chemical reactions. In noncross-linked systems, erosion occurs as a result of chain fragments dissolving without chemically altering the structure of the molecule [40]. For polymers to dissolve, they need aqueous solvents (such as acid–base interactions) and chemical interactions with water (such as hydrogen bonds) [42].

Both erosion and degradation are surface- and bulk-based processes. The polymer matrix of the surface is gradually removed, but the polymer volume fraction stays relatively unchanged during degradation. After polymer carriers have degraded or eroded almost completely, bulk degradation does not result in a significant change in the physical size of the carriers, but the fraction of polymer left in the carrier declines with time. Generally, the dominant process is determined by the rate at which the nonsolvent molecules penetrate the polymer, the diffusion of degradation products, or the degradation or dissolution of macromolecular structures [18]. Since biodegradable hydrogels are often polymerized in an aqueous solvent, rate considerations are especially relevant for their design.

The backbone or cross-linker of polymers must be hydrolytically or proteolytically labile to be chemically degradable. It is assumed that poly(lactic/glycolic acid) and poly(caprolactone) is mostly degraded by hydrolytic cleavage of ester bonds or ester derivatives. Hydrolysis can also adversely affect polyanhydrides, polyorthoesters, polyphosphate esters, and polyphosphazene derivatives [15,15,18]. When degradation and dissolution occur, they are accompanied by auto-acceleration; acid products generated by the degradation mechanisms may catalyze further degradation, while ionizing an initially hydrophobic structure encourages the matrix to absorb more water, as is the case when pendant anhydrides are hydrolyzed on poly(methyl acrylate) [43].

It is well known that the degradation products from biodegradable polymers can be hazardous. It is difficult to determine toxicity experimentally due to the size of fragments produced by degradation [45]. It should be possible to eliminate parenterally administered polymers naturally into small, nontoxic, metabolic compounds.

### 2.3.6 (Ethylenimine) Polymers

The linear poly(ethylenimine) (PEI) is soluble in hot water, at a low pH, and in ethanol or chloroform. Cold water cannot dissolve acetone, benzene, or ethyl ether. Aziridine has been polymerized with ring opening to produce branched PEI. Post-modifying polymers such as polyethylenimine or polyaziridines with N-substituted groups can be used to generate linear PEI [47]. Poly(2-ethyl-2-oxazoline) was hydrolyzed to produce linear PEI [48].

### 2.3.7 Methacrylamide Poly(N-(2-Hydroxypropyl))

RAFT copolymerization of N-(2-hydroxypropyl) methacrylamide (HPMA) copolymers followed by click reaction was used for the preparation of diblock and multiblock (tetrablock and hexablock) methacrylamide (HPMA) copolymers that were developed for preclinical development of diblock and multiblock conjugates of methacrylamide (HPMA) and gemcitabine (GEM) as well as paclitaxel (PTX copolymer of polymer HPMA–cyclabine and GDC0980 conjugates. Both conjugates and their combination exhibited potent cytotoxicity, suggesting that chemotherapeutic strategies might be possible [41]. A RAFT polymerization mediated by a bifunctional chain transfer agent and a tetrachelic water-soluble HPMA copolymer have been used to produce HPMA copolymer–doxorubicin (DOX) conjugates [42,43]. Filaments contain dendritic polymers. These polymers feature a large number of functional groups at the terminal end, are low viscous, and highly soluble. Synthetic procedures can control and adjust the size, branching, and functionality of their cells. There has been an increasing number of studies on the development and synthesis of dendrimers and their application to drug delivery and immunology [44,45], vaccines, antimicrobials, and antiviral drugs. Phylogenetic analysis has revealed that dendrimers can be divided into three groups: linear, branched, and dendritic. Dendrimer-based bioreducible polymers were also investigated for efficient gene delivery.

### 2.3.8 Biodegradable and Bioabsorbable Polymers

In cases where only a temporary presence of the implant is necessary, bioabsorbable drug-delivery systems are the better option [37]. In addition to poly(glycolic acid), poly(lactic acid), poly(caprolactone), and polydioxanone, there are many polymer molecules that are commonly used as drug-delivery systems. A number of biodegradable polymers are commonly used, including polyhydroxy acids, polyanhydrides, polyamides, polyesters, polyphosphoesters, polyalkyl cyanoacrylates, polyhydrocyanic acids, and natural sugars, such as chitosan. Compared with biodegradable polymers from natural polymers, synthetic polymers have immunogenic properties, making them ideal for drug-delivery systems [31–33].

### 2.3.9 Pharmaceutical Considerations in Drug Delivery

System delivery is primarily concerned with delivering therapeutics at the desired anatomical site and maintaining the concentration within a therapeutic range for a specified period (Figure 2.5).

Bioavailability in the bloodstream allows drugs to be delivered to virtually all tissues of the body, whether absorbed orally, parenterally, or with other methods, such as transdermal patches. When the drugs are in the blood, they diffuse to the tissues by crossing the endothelial barrier or by draining through holes in the endothelium of tissues with “leaky” vessels [20,21]. In addition, the drug, its polymer carrier, or polymer–drug conjugate may employ active targeting mechanisms to distribute itself disproportionately within the tissue of interest.

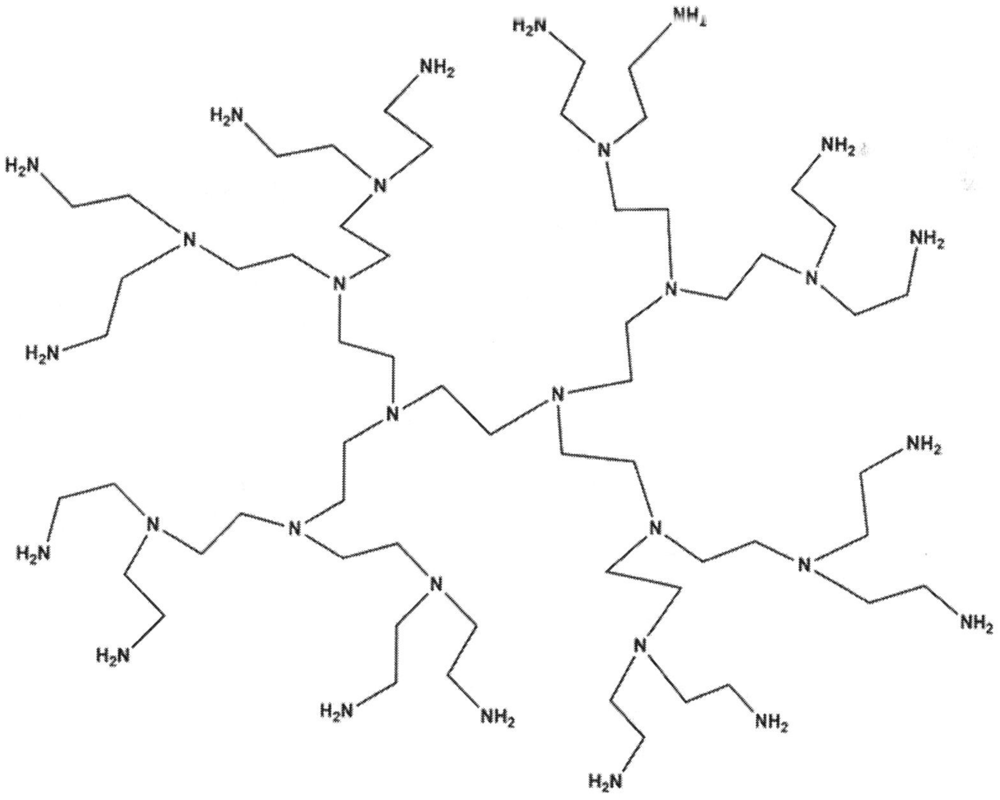

**FIGURE 2.5** The chemical structure of poly(ethylenimine)s.

### 2.3.10 Delivery Through the Oral Cavity Physiology

The most common delivery method for drugs is through oral formulations. A typical pharmaceutical formulation, including tablets and capsules, delivers relatively small molecules passively using concentration gradients based on the level of oil/water partitioning, the molecular weight, and the ionization degree. When drugs pass through transcytosis before entering intestinal capillaries or lacteal ventricles, they generally encounter the same physical barriers that limit the absorption of nutrients and ions [19,23]. After entering the lamina propria, in which substances pass either by diffusion through the endothelial cells or through the central lacteal, first-pass metabolism is avoided as a mucus-lined cavity surrounds the lumen of the intestine, the brush border (microvilli), the cytoplasm, and the basal membrane and basement membrane. Due to the substantial perfusion of blood vessels in the intestine, all substances absorbed take the capillary route unless they are exceptionally large or partition heavily into chylomicrons [25].

The bioavailability of an agent does not necessarily imply its release in the vicinity of the mucous membrane. Several second-pass metabolic pathways may remove a significant fraction of a drug that diffuses into mucous membranes [16,17]. These include refluxing back into the intestinal lumen, metabolizing in the intestinal mucosa, or being removed by the hepatic portal system.

### 2.3.11 Polymer Films made from Synthetic Polymers

An original purpose of polymer films was to cover surgical incisions. Polymers have been studied for their potential to improve wound healing using dressings or medical devices [26,33]. Since they became occlusive dressings used in wound healing, however, they have been used as wound dressings. For wound management, wound dressings have been modernized over the past few years, mainly consisting of synthetic polymers. Dressings made from synthetic polymers are classified as passive or interactive. Nonocclusive synthetic polymer-made clothing provides wound coverage and restores function underneath the coating. Tulle and gauze are examples of passive synthetic polymers [29]. Having an occlusive or semi-occlusive surface prevents bacteria from penetrating the dressing. The hydrocolloids can be in the form of foams or hydrogels.

Polymer films trap wound exudates, thereby keeping wounds moist. Because polyurethane (PU) provides a good barrier and permeability to oxygen, it is commonly used in semi-permeable dressings [31]. These films are characterized by their incompatibility with bacteria and liquids, although air and moisture vapor can pass through them. Under such semi-occlusive dressings, wound exudates may accumulate since the dressing is not absorbent, which is a major drawback. Although bacteria do not seem to thrive in this environment, fluid seepage may cause a break in the environment maintained by the occlusive dressing [31,34].

One experiment investigated polymeric wound dressings made from synthetic polyurethane films. Compared to TegadermTM, a standard wound dressing that is used commercially to treat wounds, polyurethane films form a thinner scab with lower inflammatory cell infiltration [40]. Furthermore, granulation tissue containing collagen and vascularization form earlier on the wound site. Compared to TegadermTM, which was used as a control, wound healing exhibited a better level of epithelial cell organization.

### 2.3.12 Alginates

Leg ulcers and infected postoperative wounds are the most common wounds treated with alginate dressings. Depending on the application, these can be made from porous sheets made by freeze-drying or from fibrous materials designed to be packed in the wounds [36,37]. Alginate dressings undergo an ion exchange reaction when in contact with an exuding wound. Dressings stretch by exchanging sodium ions with calcium ions in the dressing and tissue fluid. In addition to its chemical composition, dressings vary in how much swelling they produce based on their botanical origin.

It is common for alginate-containing dressings to contain a considerable amount of mannuronic or guluronic sugar groups [38].

Alginates are commonly used in wound dressings because of their hemostatic properties [43]. In comparison with nonalginate dressings, zinc and calcium alginate dressings have been found to have coagulation effects. This was the case even with alginate dressings when compared to nonalginate dressings [40]. The best hemostatic ability was observed in zinc-containing alginates. The gelling time for alginate dressings is up to a month, as opposed to hydrocolloids that degrade much more quickly.

During an experiment, it was found that removing calcium from wounds improved the gelling of alginate dressings, while introducing a specific concentration of calcium increased the degradation of the gel [36].

The bioengineering potential of alginate dressings can stimulate both tissue regeneration and bioengineering [36]. The composition of alginate salts can affect their use as cell growth substrates. Moreover, skin scaffolds can also become regenerated using this technique. Several studies have demonstrated that fetal rat chondrocytes cultured on alginate beads did not undergo dedifferentiation. Cell proliferation can be achieved using alginate thanks to its comforting properties [40].

Exudates from large leg ulcers vary from low to high, while pyoderma gangrenosum shows low to high exudates [26,30]. Besides being able to absorb high amounts of fluid, wound dressings need to prevent healthy skin from releasing fluid. For wound healing to be effective, dressings must absorb additional fluid while retaining the proper level of moisture. Alginate is derived from brown algae and is made by treating it with sodium hydroxide. After the algae have been extracted, a precipitate is produced by further treating the extract with NaCl or $CaCl_2$. In order to create alginic acid, dilute hydrochloric acid is added to a precipitate [29].

In comparison with alginate extracted from seaweed, alginate produced from bacteria may possess better chemical and physical properties. Nanoscale fibers have been developed for wound management systems as a result of the latest technological advancements [32]. Their small size and light weight also make them pore-filled with low surface-to-volume ratios. Alginate nanofibers are produced using the electrospinning technique. Synthetic polymers, along with these natural polymers, are used in the process. With an ion exchange reaction, alginate dressings interact with wound exudates. Sodium ions in wound fluids and calcium ions in wound exudates cause alginate dressings to swell and keep wounds moist. Alginate gels are capable of absorbing approximately 20 times their weight. In addition to this, alginate gels have an excellent ability to absorb exudates in low to high concentrations [42].

Modern wound dressings such as alginate-based wound dressings allow wounds to heal in moist environments. In vivo and ex vivo tests have proven that alginate is a natural polymer that is biocompatible [43]. The fact that alginate is obtained from natural sources means that it may contain some impurities. Implants and injections can be tainted with many types of impurities, including endotoxins, heavy metals, proteins, or phenolic compounds. In one experiment, however, alginate from highly purified and complex extraction procedures was implanted into animals without causing any immunologic reaction [43,44]. Another study also showed that subcutaneous injection of alginate gel did not cause any inflammatory reactions. Since calcium alginate is a natural hemostat, wounds that bleed can be dressed with alginate dressings.

Alginate dressings are characterized by gel formation, which reduces pain during dressing removal and changes. The presence of moisture promotes significant epithelialization and granulation. A controlled study showed that calcium alginate dressings were more effective than paraffin gauze dressings. In comparison to patients wearing paraffin gauze [45], patients using alginate gauze were fully healed at day 10, which was faster. By dressing split-skin graft donor sites with calcium alginate, healing rates have been significantly improved [46]. As a single treatment for split-thickness donor sites, bio-occlusive membrane dressings can cause pain, leakage, and seroma. Bio-occlusive dressings and calcium alginate can effectively treat these conditions. The delivery of low-molecular-weight drugs can also be achieved using alginate gels [48].

A nonporous alginate gel is commonly used to rapidly diffuse small drugs. Antineoplastic agents can be delivered locally and controlled using partially oxidized alginate gels.

### 2.3.13 Parenteral Delivery Physiology

This detects the possibility of a site being cleared by the tissues' capillary network or lymphatic system. Accordingly, substances with smaller hydrodynamic diameters are transported by lymphatic capillaries to the thoracic duct, which then drains them into the systemic circulation. Despite a high tissue perfusion rate, the lymphatic drainage process is dominated by molecules of less than 5 kDa [23].

The constructs usually have to be implanted surgically after being placed outside the body. However, increasing numbers of researchers are becoming interested in ways to use minimally invasive techniques like laparoscopy, perhaps even simple injections [2,6]. When applied to this type of application, the construct requires low viscosity (so it can be injected easily), but with a cohesive, gel-like consistency inside the body. When concurrent drug delivery is desired, this attribute is particularly desirable, because the rate of drug release is related to viscosity.

Many angles have been used to deliver drugs or develop tissue. It is possible to form cross-links physically by interactions such as hydrophobicity, charge, hydrogen bonds, stereo complexation, and supramolecular chemistry with certain polymers. An example would be the thermomolecular polymers, which use the difference in temperature between bodies to demonstrate their properties. There are many specific chemistries that contribute to similar goals as the polypropylene oxide and polyethylene oxide triblock polymers [20]. Polymers can also cross-link via covalent bonds. Various chemistries can be used. There has been extensive use of UV photopolymerization [21], and several cross-linker-mediated small molecules [22].

Despite this, it is important to note that small-molecule cross-linkers could have the disadvantage that they would require additional equipment, and medical professionals may be concerned about UV exposure, while UV photopolymerization in situ could have toxicity concerns from residual unreacted chemicals. Covalent cross-links can be created in situ on polymers themselves [23]. A variety of chemistries can also be employed here [24,25]. Upon mixing two polymers with complementary functional groups, a covalent bond is formed between the two polymers. Its main disadvantage, despite its relatively complex chemistry, is that it separates the prepolymers before they are used (for example, in a double-barreled syringe) and that the gelation of the materials takes place within the delivery device (for instance, a needle). By forming the construct as microspheres, injectability can also be provided.

A large, porous biodegradable polymer microsphere could, for example, be used for delivering cells [26]. A polymeric microsphere has also been surface-modified so that chondrocytes can adhere to its surface [26], and then be delivered into the articular cavity. With polymers with shape memory [28], such that they can return to a predetermined shape when exposed to a defined stimulus, it might be possible to deploy minimally invasively structured constructs.

We have access to a variety of polymers, including thermoplastics and biodegradables. These could be implanted via a small incision, then expanded to their full geometry, which could reduce the aspect of a relatively bulky object. There have been other polymers developed that can change shape in response to light [23], as well as ones that can change shape more than once in response to graded temperature increases [20]. Physical cross-links mediate the first shape change, while covalent links, or their rupture, mediate the second.

## 2.4 TISSUE ENGINEERING WITH SURFACE-MODIFIED POLYMERS

In addition to chemical cues, cells also respond to morphological aspects of the surfaces with which they come into contact [21,22]. A common synthetic polymer scaffold is not able to provide finer cues as to how degradable they are or their other properties, although their manufacturing process and reproducibility make them attractive. Depending on how they are provided, you can have their surfaces modified [3,4], adsorb compounds physically, or chemically alter them. It is possible to use this method to immobilize proteins or other compounds for biochemical effects (for example, adhesion) as well as to increase polymer hydrophilicity and hopefully resist proteins (for example, in the peritoneum). Micro- or nanopatterning is another common surface modification that creates cellular arrays [11] and influences cell behavior [13].

## 2.5 POLYMERIC MATERIALS

A polymeric material is a compound whose characteristics are organic and based on chemicals (i.e., carbon, hydrogen, oxygen, nitrogen, silicon). A second distinctive characteristic of these molecules is that their molecular structures are usually chainlike, and their backbones usually consist of carbon atoms. There have been extensive studies on polymeric biomaterials [4]. Various polymer materials have been used successfully in tissue engineering and regenerative medicine [16]. Polymers can also be used as dressings for wounds, which can aid in wound healing [17]. Other polymeric materials include foam dressings, hydrocolloids, and alginate dressings. These special polymers are used primarily for transient purposes that require them to remain in place. They are used for sutures, scaffolds for tissue regeneration, tissue adhesives, hemostats, and transient barriers for adhesion or delivery of therapy [18,19].

These materials possess unique physical, chemical, biological, and biomechanical properties for their specific uses. Metals, ceramics, and polymers comprise composites which are constructed from more than two materials. The best qualities of the constituent materials are combined in composite materials, with these qualities not present in individual materials.

Composites made of organic and inorganic polymers and fillers are known as bionanocomposites [21]. The composition and chemistry of bionanocomposites can be used for a variety of biomedical applications, including tissue engineering and bone reconstruction. Metal nanoparticles fill the gaps of polymer–metal nanocomposites, which possess various advantages including electrical, mechanical, and optical properties [22].

Nanoparticles can be added to hydrogels to make nanomaterials with varied applications, including electrical, mechanical, and optical properties [23,24]. Biocompatible materials can be constructed from nanocomposite hydrogels [24]. Graphene nanocomposites have been studied extensively, and they have exhibited biocompatibility [25] and antibacterial [26] properties, as well as improved mechanical [27] and electrochemical [28] properties for utilization in medicine and dentistry.

## 2.6 POLYMERIC MATRICES IN THREE DIMENSIONS

Three-dimensionality is becoming increasingly important in engineered tissue constructs. An application-based approach, which was motivated by the need to deliver nutrients and oxygen within large organs, as well as to remove waste (the need for microvascularization), relied largely on advanced fabrication tools for creating patterned cell structures [9–12]. Microfabrication and similar techniques have been used to pattern polymeric materials in 3-D, utilizing computational fluid dynamics designs. Researchers fabricated polymer networks [16] on micropatterned silicon [15] and stacked them into 3-D networks as early applications of micropatterned silicon [15]. Hydrogels such as calcium alginate and relatively hydrophobic polymers like PLGA have also been developed for use in such devices. Additionally, these devices can be modified to deliver drugs and alter their surfaces, as discussed here. In contrast, other approaches have focused more on the interaction of nanoscale biological structures with tightly defined ultrastructures. Molecular recognition and cell interaction can be achieved through self-assembling nanofibrillar networks incorporated with molecular determinants (e.g., peptide sequences). These networks can even guide cell differentiation. Amphiphilic peptides were used for the latter systems. However, polymer systems are used in a similar manner despite the difference in biologic approach. A synthetic hydrogel degrades in response to the proteolytic activity of migrating cells [12], allowing for the formation of three-dimensional networks that promote angiogenesis. This is an interesting approach since the cells interact with the matrix and the matrix interacts with the cells. A nanoscale architecture is required because fibrils form within the extracellular matrix at this scale. In addition to shape, proliferation, migration, and gene expression, nanoscale topography influences many aspects of cell behavior. The polymers used for these applications have been widely varied [13]. Electrospinning is widely used to produce nanofibers. In this process, natural polymers may be denatured.

## 2.7 ENGINEERING TISSUES WITH POLYMERS

It can also provide support, adhesion, and other mundane functions; as such, the underlying scaffold or matrix in tissue engineering is not inert. It is undeniable that tissue engineering matrices are highly active, both directly and indirectly, which is why they are often used to deliver drugs and achieve specific biologic goals. Through recruiting inflammatory cells (cytokines and chemokines), indirect effects can be accomplished. The introduction of particles into vaccine delivery systems can benefit from this [8]. The use of these techniques has enabled carbohydrate-based matrices to act directly on putatively inert materials. When the cross-linked chitosan created severe adhesions in the peritoneal cavity, the levels of tumor necrosis factor alpha (TNF-α) and macrophage inflammatory protein 2 (MIP-2) increased in the mesothelial cells [9]. The anti-adhesion properties of cross-linked hyaluronic acid [33] were observed in mesothelial cells when the compound was used to protect against peritoneal adhesions. The increase in tissue plasminogen activator could be related to the effectiveness of the compound. A biomaterial with a defined biologic function, such as encouraging proliferation or differentiation, has been grafted, incorporated, or delivered with that function.

Despite the fact that biomaterials with these functions are possible through rational design, this process is extremely difficult, although glycomics – the study of carbohydrates – and structural biochemistry seek to achieve this. Additional researchers have synthesized thousands of possible biomaterials at nanoliter scales and screened them on chip-like systems to see how cells and polymers interact [10]. With this approach, some polymers were more effective in converting human embryonic stem cells into epithelial cells. There was also a wide variation in the reactions that several polymers had with retinoic acid, a bioactive molecule [15]. The different cell responses could have been caused by adsorption of the polymeric molecules on surfaces or by other factors, but this does not change the fact that the shape of the polymer molecules was responsible [23]. Methods such as these may make it possible to identify biomaterials with or without the addition of ligands for tissue engineering applications [28].

## 2.8 CONCLUSION

The field of polymer science is rapidly emerging as a means for drug delivery, utilizing both natural and synthetic polymers. This strategy relies on pathogens as well as mammalian cells to evade the immune system by utilizing a variety of delivery mechanisms. In addressing a variety of challenges, biomimicry and polymer drug-delivery systems show promising results. Biocompatible and bio-related copolymers and dendrimers, which are biocompatible and bio-related, have made significant advances in the treatment of cancer due to their development as anticancer drug-delivery systems, including cisplatin and doxorubicin. Unique properties of dendrimers include their multi-valances, globular architectures, and well-defined molecular weight. Thanks to these features, dendrimers are promising polymeric drug-delivery scaffolds. Developing synthetic and natural polymers requires many micro-processes for maintenance in cells, including in vivo treatments and bioengineering.

These two systems differ substantially historically. Biological systems have been studied in order to bridge the gap between biomaterials and biological systems. Drug-delivery systems based on biopolymer carriers, such as bacteria and viruses, have the potential to produce negative immune responses in humans. The immune response from pathogen-based carriers can be beneficial as long as they have adjuvant properties. In the near future, synthetic and biological perspectives will converge to define a new paradigm for polymeric drug delivery.

## REFERENCES


[1] Langer R, and Peppas NA (2013). Advances in biomaterials, drug delivery, and bionanotechnology. *AICHE J*;49:2990–3006.

[2] Heller A (2015). Integrated medical feedback systems for drug delivery. *AICHE J*;51:1054–66.

[3] Din F, Aman W, Ullah I, Quereshi OS, Mustapha O, Shafique S, and Zeb A (2017). Effective use of nanocarriers as drug delivery systems for the treatment of selected tumors. *Int J Nanomedicine*;12:7291–309.

[4] Tiwari G, Tiwari R, Sriwastawa B, Bhati L, Pandey S, Pandey P, and Bannerjee SK (2012). Drug delivery systems: an updated review. *Int J Pharm Investig*;2:2–11.

[5] Basua A, Kundurua KR, Doppalapudib S, Domba AJ, and Khanb W (2016). Poly (lactic acid) based hydrogels. *Adv Drug Deliv Rev*;107:192–205. https://doi. org/10.1016/j.addr.2016.07.004

[6] Cacciatore I, Ciulla M, Fornasari E, Marinelli L, and Di Stefano A (2016). Solid lipid nanoparticles as a drug delivery system for the treatment of neurodegenerative diseases, Expert. *Opin Drug Deliv*;13(8):1121–31. https://doi.org/10.1080/17425247.2016.1178237.

[7] Xia WS, Liu P, and Liu J (2018). Advance in chitosan hydrolysis by non-specific celluloses. *Bioresour Technol*;99:6751–62.

[8] Bajaj G, van Alstine WG, and Yeo Y (2012). Zwitterionic chitosan derivative, a new biocompatible pharmaceutical excipient, prevents endotoxin-mediated cytokine release. *PLoS One*;7:1–10.

[9] Čalija B, Cekić N, Savić S, Daniels R, Marković B, and Milić J (2013). pH-sensitive micro-particles for oral drug delivery based on alginate/oligo-chitosan/Eudragit®L100–55 "sandwich" polyelectrolyte complex. *Colloid Surf B*;110:395–402.

[10] Pourasghar M, Koenneke A, Meiers P, and Schneider M (2019). Development of a fast and precise method for simultaneous quantification of the PLGA monomers lactic and glycolic acid by HPLC. *J Pharm Anal*;9:100–7. https://doi.org/10.1016/j.jpha.2019.01.004

[11] Katt ME, Placone AL, Wong AD, Xu ZS, and Searson PC (2016). In vitro tumor models: advantages, disadvantages, variables, and selecting the right platform. *Front Bioeng Biotechnol*;4:12–8. https://doi.org/10.3389/fbioe.2016.00012. 00012 PMCID: PMC4751256. PMID: 26904541.

[12] Hanak BW, Hsieh CY, Donaldson W, Browd SR, Lau KS, and Shain W (2018). Reduced cell attachment to poly (2-hydroxyethyl methacrylate)-coated ventricular catheters in vitro. *J Biomed Mater Res B Appl Biomater*;106:1268–79. https://doi.org/10.1002/jbm.b.33915

[13] Heskins M, and Guillet JE (2020). Solution properties of poly(N-isopropyl acrylamide). *J Macromol Sci Part A Chem*;2:1441–55.

[14] Zheng L, Qiulin L, Duanguang Y, Yong G, and Xujun L (2013). Well-defined poly(N-isopropylacrylamide) with a bifunctional end-group: synthesis, characterization, and thermoresponsive properties. *Des Monomers Polym*;16:465–74. https://doi.org/10.1080/15685551.2012.747165

[15] Weber C, Richard H, and Schubert US (2012). Temperature responsive bio-compatible polymers based on poly (ethylene oxide) and poly (2-oxazoline)s. *Prog Polym Sci*;37:686–714.

[16] Bae YH, Okano T, and Kim SW (2020). Temperature dependence of swelling of crosslinked poly(N,N′-alkyl substituted acrylamides) in water. *J Polym Sci Part B Polym Phys*;28:923–36. https://doi.org/10.1002/polb.1990.090280609

[17] Zhang Q, Ko NR, and Oh IK (2012). Recent advances in stimuli-responsive degradable block copolymer micelles: synthesis and controlled drug delivery applications. *Chem Commun*;48:7542–52. https://doi.org/10.1039/c2cc32408c

[18] Yang J, Zhang R, Pan H, Li Y, Fang Y, Zhang L, and Kopeček J (2017). Backbone degradable N-(2-hydroxypropyl) methacrylamide copolymer conjugates with gemcitabine and paclitaxel: impact of molecular weight on activity toward human ovarian carcinoma xenografts. *Mol Pharm*;14:1384–94. https://doi.org/10.1021/acs.molpharmaceut.6b01005

[19] Zhang R, Yang J, Zhou Y, Shami PJ, Kopeček J (2016). N-(2-hydroxypropyl) methacrylamide copolymer–drug conjugates for combination chemotherapy of acute myeloid leukemia. *Macromol Biosci*;16:121–8.
[20] Zhang L, Zhang R, Yang J, Wang J, and Kopecek J (2016). Indium-based and iodine-based labeling of HPMA copolymer-epirubicin conjugates: Impact of structure on the in vivo fate. *J Control Release*;240:306–18.
[21] Nam HY, Nam K, Lee M, Kim SW, and Bull DA (2012). Dendrimer type bio-reducible polymer for efficient gene delivery. *J Control Release*;160:592–600. https://doi.org/10.1016/j.jconrel.2012.04.025
[22] Törmälä P, Pohjonen T, and Rokkanen P (1998). Bio-absorbable polymers: materials technology and surgical applications. *Proc Inst Mech Eng*;212:101–11. https://doi.org/10.1243/0954411981533872
[23] Lee TS, and Bee ST (2019). *Polylactic Acid: A Practical Guide for the Processing, Manufacturing and Applications of PLA*, second edition. Plastics design library: Elsevier. pp. 53–95.
[24] Kamaly N, Yameen B, Wu J, and Farokhzad OC (2016). Degradable controlled-release polymers and polymeric nanoparticles: mechanisms of controlling drug release. *Chem Rev*;116:2602–63.
[25] Choi JW, Nam JP, Nam K, Lee YS, Yun CO, and Kim SW (2015). Oncolitic adenovirus coated with multi-degradable bio-reducible core-cross-linked poly- (ethyleneimine) for cancer gene therapy. *Biom acromolecules*;16:592–600. https://doi.org/10.1021/acs.biomac.5b00538
[26] Ray L (2019), Polymeric nanoparticle-based drug/gene delivery for lung cancer, in Keshav Moharir, Vinita Kale, Abhay Ittadwar, Manash K. Paul (eds) *Nanotechnology-Based Targeted Drug Delivery Systems for Lung Cancer*; Chap. 4, pp. 77–93. CRC Press, Taylor & Francis.
[27] Li Y, Thambi T, and Lee DS (2018). Co-delivery of drugs and genes using polymeric nanoparticles for synergistic cancer therapeutic effects. *Adv Healthcare Mater*;7:1700886. https://doi.org/10.1002/adhm.201700886
[28] Yang L, Li J, and Kopeček J (2019). Biorecognition: a key to drug-free macromolecular therapeutics. *Biomaterials*;190:11–23.
[29] Ahn C, Mulligan P, and Salcido RS (2018). Smoking-the bane of wound healing: biomedical interventions and social infuences. *Adv Skin Wound Care*;21(5):227–36. https://doi.org/10.1097/01.
[30] Akturk O, Tezcaner A, Bilgili H, Deveci MS, Gecit MR, and Keskin D (2017). Evaluation of sericin/collagen membranes as prospective wound dressing biomaterial. *J Biosci Bioeng*;112:279–88.
[31] Alrubaiy L, and Al-Rubaiy KK (2019). Skin substitutes: a brief review of types and clinical applications. *Oman Med J*;24(1):4.
[32] Arnold-Long M, Johnson R, and Reed L (2020) Negative pressure wound therapy overlay technique with collagen dressings for nonhealing wounds. *J Wound Ostomy Cont Nurs*;37(5):549–53.
[33] Atiyeh BS, Amm CA, and El Musa KA (2013a). Improved scar quality following primary and secondary healing of cutaneous wounds. *Aesthet Plast Surg*;27(5):411–17.
[34] Atiyeh BS, El-Musa KA, and Dham R (2013b). Scar quality and physiologic barrier function restoration after moist and moistexposed dressings of partial-thickness wounds. *Dermatol Surg*;29(1):14–20.
[35] Bao X, Hayashi K, Li Y, Eramoto A, and Abe K (2017). Novel agarose and agar fbers: fabrication and characterization. *Mater Lett*;64(22):2435–7.
[36] Baoyong L, Jian Z, Denglong C, and Min L (2015). Evaluation of a new type of wound dressing made from recombinant spider silk protein using rat models. *Burn*;36:891–6.
[37] Basu P, Uttamchand NK, and Inderchand M (2017). Wound healing materials – a perspective for skin tissue engineering. *Curr Sci*;112:2392–404.
[38] Baum CL, and Arpey CJ (2015). Normal cutaneous wound healing: clinical correlation with cellular and molecular events. *Dermatol Surg*;31(6):674–86 (discussion 686, Review).
[39] Bello YM, and Phillips TJ (2021). Recent advances in wound healing. *Jama*;283(6):716–18.
[40] Bello YM, Falabella AF, and Eaglstein WH (2016). Tissue-engineered skin. *Am J Clin Dermatol*;2(5):305–13.
[41] Berthet MA, Angellier-Coussy H, Guillard V, and Gontard N (2016). Vegetal fber-based biocomposites: which stakes for food packaging applications? *J Appl Polym Sci*;133:42528. https://doi.org/10.1002/app.42528
[42] Bishop Ah (2018). Role of oxygen in wound healing. *J Wound Care*;17(9):399–402 (Review).
[43] Boa O, Beaudoin Cloutier C, Genest H, Labbé R, Rodrigue B, Soucy J et al (2013). Prospective study on the treatment of lower extremity chronic venous and mixed ulcers using tissue-engineered skin substitute made by the self-assembly approach. *Adv Wound Care*;26(9):400–09.

[44] Boateng JS, Matthews KH, Stevens HNE, and Eccleston GM (2018). Wound healing dressings and drug delivery systems: a review. *J Pharm Sci*;97(8):2892–923.
[45] Bolton L, and Van Rijswijk L (2021). Wound dressings: meeting clinical and biological needs. *Dermatol Nurs Dermatol Nurses Assoc*;3(3):146.
[46] Boonkaew B, Kempf M, Kimble R, Supaphol P, and Cuttle L (2013). Antimicrobial efcacy of a novel silver hydrogel dressing compared to two common silver burn wound dressings: Acticoat™ and polymem Silver. *Burns*; 40(1):89–96. https://doi. org/10.1016/j.burns.2013.05.011
[47] Boyce ST, Kagan RJ, Meyer NA, Yakubof KP, and Warden GD (2019). THE 1999 clinical research award cultured skin substitutes combined with integra artifcial skin* to replace native skin autograft and allograft for the closure of excised full-thickness burns. *J Burn Care Res*;20(6):453–hyhen.
[48] Bramhill J, Ross S, and Ross G (2017). Bioactive nanocomposites for tissue repair and regeneration: a review. *Int J Environ Res Public Health*;14(1):66. https://doi.org/10.3390/ijerph14010066

# 3 Biopolymers and Their Applications in Biomedicine

*Kalpitha M. Kumar, Shristi Ram, Manjula N.G., and Shilpa Borehalli Mayegowda*

## CONTENTS



## 3.1 INTRODUCTION

Biopolymers are produced by living cells by possessive enzymes that link building blocks like sugars, amino acids, or hydroxy fatty acids, that help to yield high-molecular-weight molecules. Microbes' synthesize various types of biopolymers like polysaccharides (composed of sugars and/or sugar acids connected by glycosidic linkages), polyamides (composed of amino acids connected by peptide bonds), polyesters (composed of hydroxy fatty acids linked by ester bonds), and polyphosphates (polyPs; composed of inorganic phosphates linked by anhydride bonds) [1]. Biopolymers have been used as a vehicle for drug delivery for healing, skin regeneration, and skin grafting. Green polymers have been derived using hydrolytically sensitive biocellulosics, furan-based polymers,

DOI: 10.1201/9781003240884-3

polyesters and their amides, polypeptides, polysaccharides, polyphosphazenes, polyanhydrides, polyurethanes, pseudo-polyamino acids, etc. that are synthesized as hydrogels and fibers that are spun and fabricated into fibrous scaffolds to be used in medical applications [2]. In lieu of these effects, biopolymers have made tremendous progress due to their biocompatibility, biodegradation, and non-cytotoxicity, which make them a suitable candidate for use in the medical field [3]. With their varied characteristics and properties these microbial biopolymers have been used in the medical and industrial fields, etc. With recent developments in synthetic biology and genetic engineering methods, the production of innovative biopolymers has gained popularity such as hyaluronate as a biomaterial, as additives in cosmetic products, xanthan and dextran as additives in food, and as biopolyesters in packaging [4,5].

The future holds exciting promise for green polymers in the areas of biomedicine, skin grafting, wound healing, and regeneration. The rational strategy of microbial biopolymer-producing cell factories has gained increasingly large research and commercial interest. With the proven efficiency of the microbial polymers, more research needs to be carried out into their various applications in medicine. With an increasing awareness of the benefits of biologically derived compounds, the search for a good biomaterial for biomedical applications is an active area of research. Biomaterials offer multiple advantages in biomedical science due to their non-toxicity, biodegradation, and biocompatible nature, and also, they possess hemocompatible, non-immunogenic and tissue-regenerative activity [6,7]. These properties make polymers find wide applicability in biodegradable packaging, biosensors, tissue engineering, prosthetics, orthopedic materials, dentistry, wound healing, and controlled drug delivery [8]. The 3D scaffold biopolymers also are reported to speed up the healing process by providing support and restoration to the wound/damaged tissue.

### 3.1.1 Classification of Biopolymers

Many microbes (bacteria, yeast, fungi, etc.) are natural biofactories for many biopolymers, viz. starch, polysaccharides, polypeptides, polynucleotides, chitosan, collagen, silk, polyesters, polyhydroxyalkanotes, alginate, bacterial cellulose, dextran, xanthan, hyaluronic acid, succinoglycan, and polyphosphates (Figure 3.1; Tables 3.1 and 3.2) [9]. Chitosan-based biopolymer has been reported to have demonstrable anticorrosive activity that reduces corrosion [10]. According to the Market Research Report, the global market size for bioplastics and biopolymers is anticipated to grow at a compound annual growth rate of 22.7% and reach USD 29.7 billion by 2026 [11]. Thus, this chapter presents detailed reviews of biopolymers, sources, types, and applications in the biomedical sector such as tissue engineering, dentistry, prosthetics, implantation, and miniature medical devices.

Currently, the use of biopolymers such as polylactic acid (PLA), silk, and chitosan has been increasingly researched for medical applications. Properties of biopolymers, such as biocompatibility, non-toxicity, and biodegradability, provide great advantages and thus find usefulness in implantable medical applications. Biopolymers have great advantage over synthetic materials as artificially produced components may be more cheap but they do not fulfill the needs of the living system in adopting to tissue organizations. Recent studies have proven that the use of biopolymers, when combined with synthetic materials, has the possibility to revolutionize the medical field.

## 3.2 TISSUE ENGINEERING

With the advent of tissue-engineering biopolymers, the significance of microbes for the production of biomaterials has spurred increased investigation. The biocompatibility of biopolymers such as bacterial cellulose, chitosan, PGA, PLA, etc. has been successfully implemented with promising potential in wound care and healing of injured organs/tissues [28]. Additionally, the implementation of biopolymers in tissue engineering has addressed the most pressing question of wound healing, i.e. skin regeneration. A good contender for tissue engineering is marked by its biocompatibility,

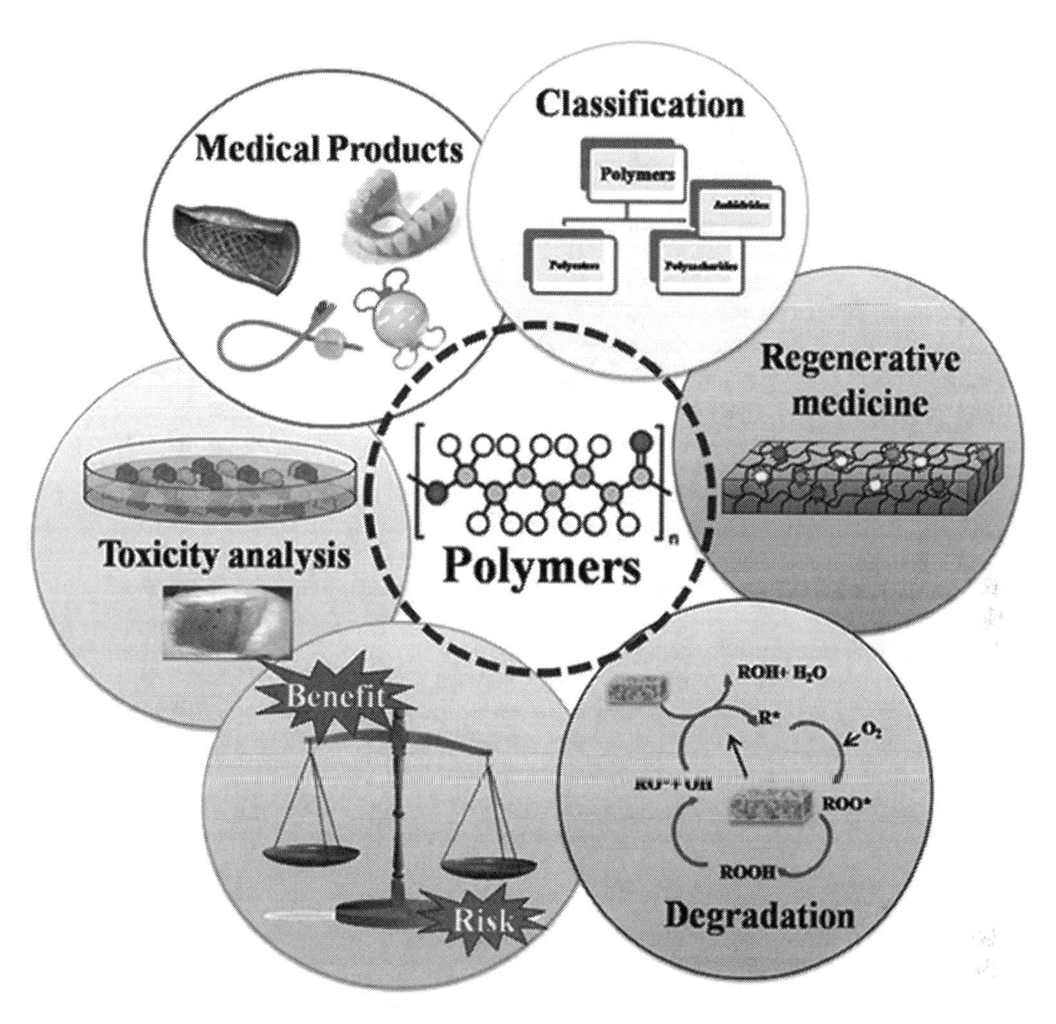


**FIGURE 3.1** Biopolymers and their applications.

hydrophilicity, non-toxicity, non-immunogenicity, good mechanical strength, biosorption, and facilitation of healing by aiding the proliferation and differentiation of underlying healthy cells [29–31]. Interest in bacterial cellulose (BC) as a useful scaffold material escalated during the 1980s after its beneficial effect on wound dressing and tissue regeneration was reported [32] (Tables 3.1 and 3.2). In addition, PGA is reported to form hydrogel along with a 3D scaffold to support tissue regeneration [33,34]. Another major advantage is the high Young's modulus, at 114 GPa for a single filament [35]. However, a challenge lies with its mechanical properties and immunogenicity which can be further optimized by attuning its culture conditions to avail higher strength and porosity of the polymer.

The adherence property of such biopolymers also depends on the water-holding capacity and water activity of these polymers as it increases its healing power [36]. The biocompatibility, adherence, healing capability, and resulting porous structure are further used for controlled drug release and drug delivery by incorporating bioactive compounds. BC has a water-holding capacity of up to 60–700 times the dry weight depending on the formulation, which increases the elastic content of the biopolymer [37] and allows charging the bioactive compound in the wound-dressing material [38].

**TABLE 3.1**
**Classification of Biopolymers – Classified According to the Monomeric Unit Used and the Structure of the Biopolymer**

| Type | Occurrence | Reference |
|---|---|---|
| **Bio-Polynucleotides** | Most of the genetic material, DNA, and RNA which are composed of long polymers with 13 or more nucleotide monomer | [12] |
| **Bio-Polypeptides** | Short polymers of amino acid | [13] |
| **Bio-Polysaccharides** | Have linear, bonded polymeric carbohydrate structures | [14] |
| **Classified depending on their origin** [15] | | |
| **Polyesters** | Polyhydroxyalkanoates, polylactic acid | [16] |
| **Proteins** | Threads of silk and collagen, gelatin, elastin, polyaminoacid, resilin, adhesive, soy, zein, wheat protein gluten, milk protein casein, serum protein albumin | [16] |
| **Bacterial Polysaccharides** | Polygalactosamine, xanthan, dextran, gellan, levan, curdlan, cellulose | [17] |
| **Fungal Polysaccharides** | Pullulan, elsinan, yeast glucans | [18] |
| **Phyto- and Phyco-Polysaccharides** | Starch, cellulose, agar, alginate, carrageenan, pectin, konjan, gums | [19] |
| **Animal Polysaccharides** | Chitin, hyaluronic acid | [20] |
| **Bio-Lipids/Surfactants** | Acetoglycerides, waxes, emulsion | [21] |
| **Bio-Polyphenols** | Lignin, tannin, humic acid | [22] |
| **Specialty Polymers** | Female lac bug secreted polymer – shellac; polymer of the amino acid – poly gamma glutamic acid; plant-based polymer – natural rubber; various synthetic polymers from natural oils and fats | [23] |

**TABLE 3.2**
**Applications of Biopolymers**

| Polymer | Microorganisms | Potential Applications | Commercial Applications | References |
|---|---|---|---|---|
| **Alginate** | *Pseudomonas aeruginosa* other pseudomonads, *Azotobacter* spp. | Hydrogels, fibers, films, and nanoparticles for various purposes, such as drug delivery, cell encapsuLation, and tissue engineering | Bacterial alginates do not have GRAS status; algal alginates are widely used as biomaterials for food, cosmetic, pharmaceutical, and biomedical purposes (for example, wound dressings and antacids) | [24] |
| **Cellulose** | *Escherichia coli*, *Salmonella enterica*, *Sarcina* spp., *Agrobacterium* spp. (NP), *Rhizobium* spp. (NP), *Pseudomonas fluorescens* (NP), *Komagataeibacter hansenii* (NP), and *Komagataeibacter rhaeticus* iGEM (NP) | Hydrogels, fibers, films, and nanoparticles for various purposes, such as drug delivery and cell encapsulation | Bacterial cellulose produced by certain bacteria (for example, *K. hansenii*) has GRAS status; widely used in food, biomedical, and packaging products (for example, wound dressings, surgical and dental implants, and textile fibers) | [25] |

**TABLE 3.2 (Continued)**
**Applications of Biopolymers**

| Polymer | Microorganisms | Potential Applications | Commercial Applications | References |
|---|---|---|---|---|
| **Hyaluronate** | Group A *Streptococcus* (P), *Pasteurella multocida* (P), *Bacillus cereus* G9241 (P); *Streptococcus equisimilis* (NP), *Lactococcus lactis*, *Bacillus subtilis*, and *E. coli* (NP), and *Corynebacterium glutamicum* (NP) | Hydrogels, surface-modified liposomes, nanoparticles, and microparticles for medical, pharmaceutical, food, and cosmetic applications | Bacterial hyaluronate produced by certain bacteria (for example, *S. equi*) has GRAS status; widely used in cosmetic, topical, ophthalmologic, and visco supplementation formulations (for example to treat osteoarthritis of the knee) | [26] |
| **γ-Poly(glycolic acid** | *Pseudomonas putida*, *Aeromonas hydrophila*, *Ralstonia eutropha*, and *Alcaligenes latus* (NP) as industrial hosts; *E. coli* and *Ralstonia eutropha* (NP) | Hydrogels or nanoparticles for biomedical applications (for example, regenerative medicine and drug delivery); for delivering high-energy phosphate for synthesis reactions | Generally, polyP has GRAS status as a direct food additive; widely used for industrial purposes such as in liquid phosphate fertilizers, water filter cartridges, and wastewater treatment | [27] |

Further modifications by making composite using alginate in BC polymer have been shown to decrease the porosity and increase the water uptake ability, increase the ease of gelation, with outstanding swelling ratios and tensile strength in comparison to pure BC [39]. The development of BC integrated with aloe vera gel increased the water vapor permeability and water-holding capacity by 1.5-fold [40]. In another experiment, a lignin-composite-derived BC polymer showed improved antibacterial action on wound dressings, thus aiding faster skin regeneration and resulting in a significant reduction in pain [41].

Some microbes naturally produce a highly anionic polymer, poly(glutamic acid) (PGA), using D- and L-glutamic acid having high hydrophilicity and biodegradability [42]. This PGA polymer has great applications in tissue engineering [43] and as an anticoagulant as it mimics heparin [44]. Further, attempts were made to integrate hydrogels with tissue plasminogen activator and fibroblast growth factor-2. This resulted in a semi-interpenetrating network of a stable hydrogel possessing growth-stimulating activity. This semi-IPN is useful in tissue restoration and sustained drug release [45,46].

Microbes known to produce γ-PGA include some members of prokaryotes belonging to bacteria such as *Bacillus licheniformis*, *Bacillus subtilis*, and *B. licheniformis* ATCC 9945A. Under archaea, a extremohilic microbe, *Natrialba aegyptiaca*, is known to produce γ-L-PGA (>1000 kDa), but they not suited to commercial production for PGA as they need highly specialized care in cultivation and are difficult to grow [47]. Even some eukaryotes, such as hydra, are known to produce γ-PGA [48]. γ-PGA synthetase from the pgsBCA gene clusters (racE, yrpC) discovered in *B. subtilis* was found to be involved in PGA production used in the making of biodegradable plastics [49]. Also, transcription factors like ComPA, DegSU, and DegQ regulated the transcription of gene activity under quorum sensing, high cell density, and an increase in osmolarity, however the mechanism of DegQ is unknown. Furthermore, the type of strain and the physicochemical parameters in the culture condition dictate the molecular weight of γ-PGA. For instance, *B. subtilis* synthesizes γ-PGA of over 10,000 kDa, whereas *Bacillus* spp. RKY3 form up to 10–50 kDa γ-PGA [50,51]. To increase the mechanical strength, a heteropolymer is synthesized by mixing chitosan and γ-PGA scaffold, which exhibited greater mechanical strength and enhanced cytocompatibility than were offered by either γ-PGA or chitosan independently [52]. Additionally, fabricated hydrogel scaffold exhibited sustained rhBMP-2 release, thereby implying its use in bone regeneration therapy [53].

### 3.2.1 Medical Adhesive

Surgical glues suffer from disadvantages of non-biodegradability and induction of chronic inflammation. Therefore, alternatively, gelatin or collagen composites of α-PGA polymer with a molecular weight of between 60–100 kDa have been explored as suitable eco-friendly alternatives [54]. Human blood fibrin-derived surgical adhesives are in use, but they suffer from the risk of viral infection due to their origin [55].

## 3.3 DENTISTRY

Pure biopolymers suffer from the disadvantage of low mechanical strength. To address this shortcoming, hybrid biopolymers such as polylactic acid, collagen, PGA, and gelatin are synthesized and indeed possesses multifarious utility in the field of dentistry [56]. For example, collagen-based polyglycolide and polylactide are popularly used in periodontal regeneration membranes [57]. Dental stabilizers are derived from γ-PGA [58]. Dental restoration and implant fixation utilize γ-PGA-based (molecular weight 115 kDa) glass ionomer cement [59]. Other examples include porcine types I and III collagen membrane-BioGide® [60], the gelatin-based sponge-Gelfoam®, resorbable collagen membrane-Biomend®, and bovine type I collagen membrane (Osseoguard™) [61].

However, the challenge associated with hybrid biopolymer is that the cross-linked polymer, when degraded, gives rise to small molecules having potential to elicit immunogenicity. Chitosan is garnering attention due to its eco-friendly nature as the degradation by-product is completely neutral. Further, the integration of bio-glass into chitosan can maintain its mechanical strength, enhancing the hardness and elasticity [62]. The application of biopolymers as a carrier for drug complexes with chitosan can furnish them as local drug-delivery systems for sustained drug release for the treatment of periodontitis, tooth caries, or root canal procedures [63]. Controlled releases of therapeutic agents maintain the drug concentration for a longer duration, thereby assisting in faster healing.

## 3.4 PROSTHESES AND IMPLANTATION

Biopolymers, whose degradation products are not immunogenic, have great potential to be used in the development of therapeutic devices such as temporary prostheses, three-dimensional porous structures as scaffolds for tissue engineering, and as controlled/sustained-release drug-delivery vehicles and also in applications such as suturing, fixation, or adhesion. Some of the developments of implantable devices made of PLA, silk, chitosan, PHA, collagen, etc. are described herein. Apart from having extensive applications, these materials sometimes are not suitable for a wide range of uses. Traditional materials like metals, ceramics, and synthetic polymers have been applied but they undergo immunological rejection by the body. Natural scaffolds made of poly(β-L-malic acid) (PMLA) and conjugated nanomaterials have been found to be beneficial in immunotherapy at the last stage of glioma, i.e., glioblastoma. This was shown in an animal study by conjugating the

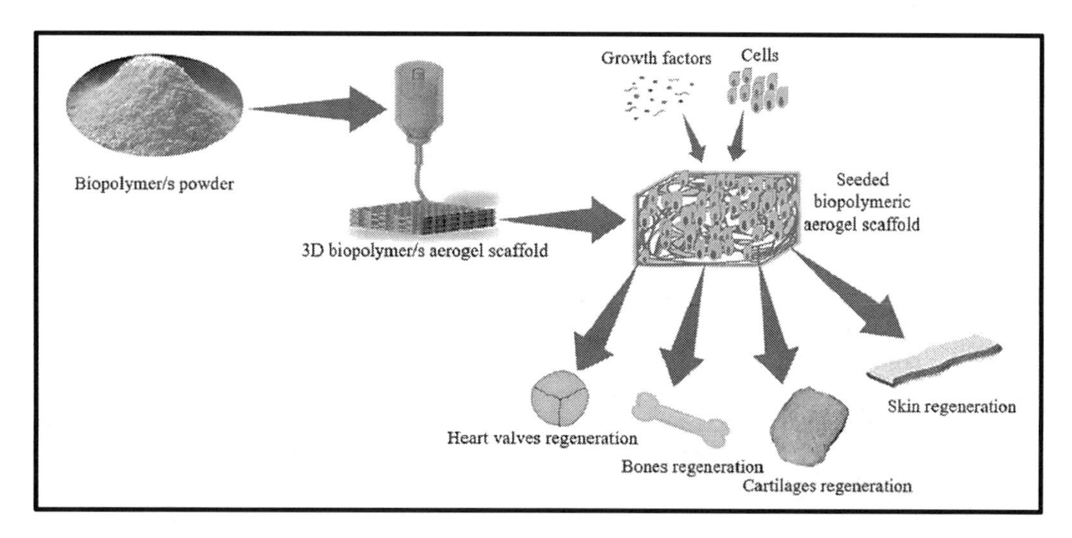


**FIGURE 3.2** Role of biopolymer-derived scaffolds in tumor therapy [65].

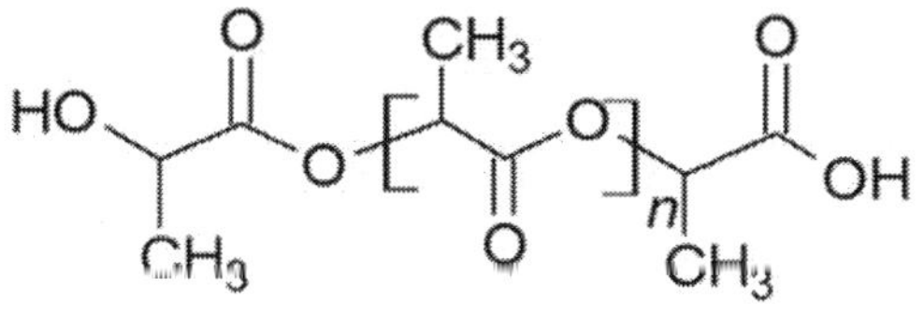

**FIGURE 3.3** Chemical structure of poly lactic acid.

biopolymer anticancer drug Temozolomide and anti-BCL2 cargo, which arrests tumor growth and prolongs the survival time [64] (Figure 3.2).

### 3.4.1 Polylactic Acid

PLA is naturally derived from potatoes, corn starch rice, and other natural sources (Figure 3.3). It is one of the most evolving biodegradable polymers. Properties of synthetic polymers and PLA are similar to those of polypropylene, and they have advantages as they are highly abundant and cost effective [66] (Table 3.1). Its characteristic of being bio-absorbable show some benefits toward their use in implantation devices. In 2007, Chang and team for the first time prepared and examined a novel porous PLA for its capacity as a carrier for the recombinant bone morphogenetic protein 2 (rhBMP 2) [67], it was also examined for the formation of bone in 2 weeks. PLA or octadecylamine is used to develop nanocomposites like nano-diamonds, which are used for tissue engineering, with excellent results as it provides good affinity between the polymer and filler in the composites [68]. Hence, this biocompatibility and the non-cytotoxicity of these properties of PLA indicate that it is applicable for tissue engineering [69].

Co-biopolymers of PLA, such as PLA–PGA devices made out of it, are used for implantability in orthopedic applications [70]. Researchers have developed and patented a technology made up of PLA and polyethylene terephthalate (PET) which is used in vascular surgery for segmenting blood vessels and reestablishment of blood flow [71]. This group has successfully produced many other biopolymers like PLA, PGA, and PDO, which are used to manufacture a wide range of ligaments because of their adaptable nature when fabricated, good rate of biodegradability, and their major biocompatibility with biomolecules in the cells and tissues. These biocompatible elements provide excellent structures when they are combined with PLA and are absorbed by the body while the aspects other provide mechanical support [72].

### 3.4.2 Silk

Silk is a traditional biopolymer commonly used in implantable devices. Recent studies suggest that core silk fibroin fibers exhibit comparable *in vitro* and *in vivo* bio-compatibility with other biopolymers such as PLA and collagen, besides possessing high mechanical resistance [73]. Nowadays, new drug-delivery platforms have been reported, where drug delivery uses microneedles and silk fibroin scaffolds to fine-tune protein drug delivery in a biomimetic fashion [74,75]. Therefore, biopolymers are used extensively for the drug-delivery/biomaterial interface which has led to further studies with relevant results for their use in musculoskeletal disease. Franck studied and determined the effects of an extracellular matrix protein coating on the ability of the silk matrix to support the cell response [148]. They reported that structural morphology and tissue-specific proteins coatings were found to be significant factors in silk biomaterials (Figures 3.4A and B). Combinations of fibronectin coatings with silk scaffolds have advantages as they have the highest level of primary smooth muscle cells. Park and co-workers reported the usefulness of tissue engineering in the production of 3D porous silk fibroin [76]. The properties of fabricated scaffolds include hydrophobicity; this provides proper conditions for the growth of human chondrocytes, and increases cell attachment and proliferation compared with other materials. This technique is highly evolving in the use of biomaterials.

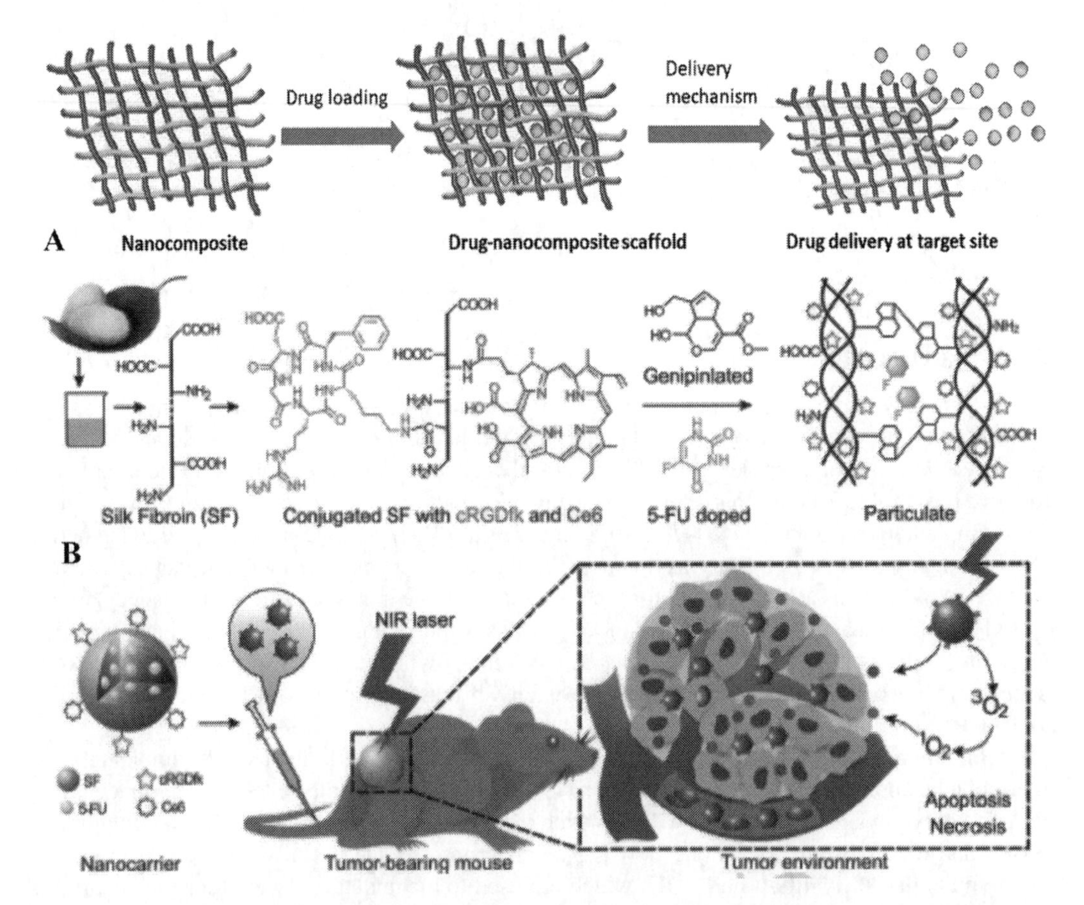


**FIGURE 3.4** (A) Schematic diagram of nanocomposite-assisted controlled drug release. (B) Silk derived nanocarrier in cancer therapeutics [77].

### 3.4.3 Chitosan

Chitosan is a cationic polysaccharide which is derived from the deacetylation of chitin which is present in the exoskeleton and is obtained by an alkalizing process at high temperature [3]. Antimicrobial, biodegradable, and biocompatible properties have enabled natural polymer chitosan to be used in different forms such as gels, films, particles, membranes, or scaffolds in a large number of applications, ranging from biomedical to industrial areas. Due to its characteristics it is used in tissue engineering, it has an important role in cell attachment and growth, and is used as a matrix in tissue engineering after the production of porous structures. The main application of chitosan-based structures is as implants, primarily with respect to bone, ligament, cartilage, tendon, liver, neural, stents, and skin regeneration. Meng proposed an idea of promoting the acceleration of re-endothelial entrenchment for healing process after coronary stent implementation in a porcine iliac artery using a coating of chitosan and heparin as this coating promotes endothelial cell compatibility and hemocompatibility to the stent surface [78]. Several applications of chitosan-based matrices in various areas are listed in Figure 3.5. The major problem with chitosan matrices is that they are quite weak, and so they need to be combined with other materials to work as compatible osteo-conductive

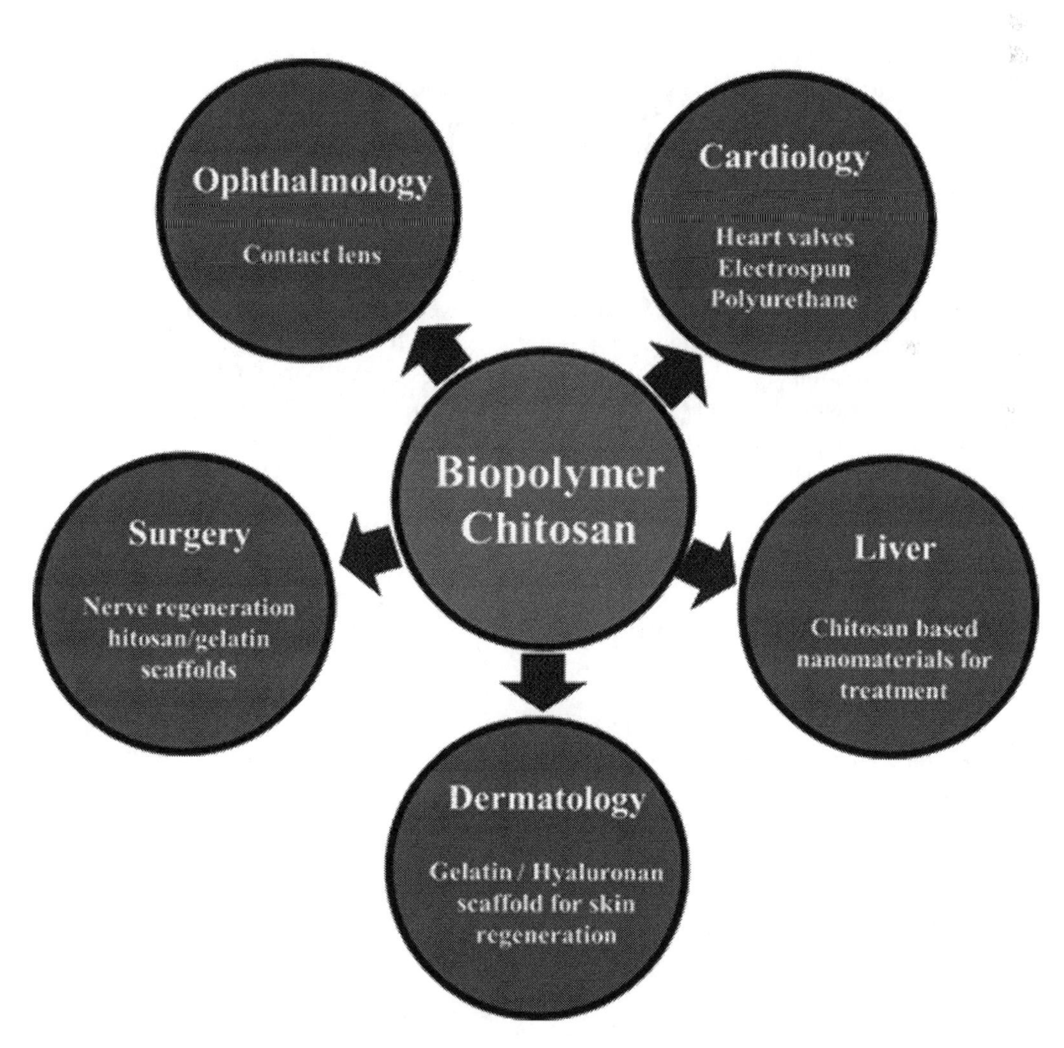


**FIGURE 3.5** Diverse medical applications of chitosan-based biopolymer.

matrices. Li and co-workers increased the mechanical strength of the scaffold by combining chitosan and alginate in 2005.

### 3.4.4 Collagen

Proteins and their polypeptide by-products are used and characterized for their applications in medical and pharmaceutical applications. For the production of micro- or nano-spheres, hydrogels, films, and scaffolds, protein-based matrices such as gelatin, albumin, elastin, casein, collagen, corn protein, and whey protein are used [79]. Collagen is the most abundant biopolymer used within biomimetic materials as it is used widely in different areas of biomedicine. Collagen is located in many connective tissues such as the skin, joints, cartilage, teeth (collagen joined to mineral crystals), tendon, bones, and others. Sources of collagen include bovine, porcine, and human, and also bovine and porcine origins. The main function of collagen is to provide mechanical stability, strength, and elasticity to native tissue, which is utilized for further biochemical applications, and it comprises 25–53% of the whole-body content [80,81]. Collagen extracted from marine sources has emerged as the most appropriate alternative. Briefly, the properties of collagen include low immunogenicity, biodegradable, biocompatibility, hydrophilicity, easy processing, and weak antigenicity; collagen has become the primary source of protein in medical applications.

However, many of them have several disadvantages as they have poor physical and chemical properties such as mechanical strength, thermo-stability, and resistance to enzymes. Due to these disadvantages, collagen has limits to its potential in biotechnological applications. Consequently, cross-linking of collagen is the solution for the improvement of its property. To minimize degradation and enhance mechanical stability, exogenous cross-links have been used. Different cross-linking methods are involved such as physical, chemical, and biochemical modifications. Physical cross-linking is carried out using UV or gamma radiation. Chemical modification – the most effective and widely used cross-linking method for collagen – uses glutaraldehyde, isocyanates, hexamethylene diisocyanate, polyepoxy compounds, as well as plant extracts or inorganic cross-linking agents. Enzymatic modification with oxidoreductases, transferase, and hydrolases is known as biochemical cross-linking. Rieu et al., using a novel process for collagen production, produced a simple collagen structure which has non-cross-linked scaffolds with uncommon mechanical properties which they applied to a 3D cell culture [82]. Also, the blend of collagen with other biomaterials and biopolymers is another alternative to prepare collagen-based biocomposites with more suitable physical and mechanical properties. Several chronic inflammations and wounds can be dressed with collagen-based biopolymers in combination with other materials such as hyaluronic acid (HA), collagen/HA/chitosan, epigallocatechin gallate (EGCG), catechin (CAT), and gallic acid (GA), and also chitosan–alginate fibroids have been found to be every effective in preventing infections, and also for enabling a more speedy recovery [83,84].

The most relevant and advanced applications of collagen in biomedicine are:

- Shielding material in ophthalmological surgeries
- Dressings burns and wounds
- Protein and drug delivery by mini-pellets
- Controlling material for transdermal delivery
- Nanoparticles for gene delivery
- Drug/gene-delivery formulations for tissue healing, used in the form of a film
- 3D scaffolds or gels for cell embedding
- Organoids or neo-organs for gene therapy

Tissue engineering includes the following procedures: skin replacement, bone substitutes, and as artificial blood vessels and valves.

## 3.4.5 Polysaccharides

### 3.4.5.1 Polysaccharides from Bacterial Sources

Microorganisms secrete exopolysaccharides (EPSs) which are of high-molecular-mass biopolymers, showing extreme diversity in terms of chemical structure and composition. EPS activity depends on their backbones, chain length, and substitution [85]. The use of these bacterial EPSs started in the middle of the 20th century as a dextran solution for clinical applications (Tables 3.1 and 3.3) Later, other bacterial EPSs were used, such as xanthan or pullulan, as pharmaceutical excipients, hyaluronic acid for surgery, arthritis treatment, or in wound healing, and bacterial cellulose was applied in wound dressings or scaffolds where it is also used in tissue engineering [86].

Production of EPSs by microbes has advantages over other plants or macroalgae-derived products which make them more suitable for industrial and commercial use, as the production time of EPSs by microbes is less than for plants, also, the area required for cultivation is also less, and there is controlled production with defined and reproducible parameters, and high quality of the final product being some of the advantages. The main disadvantage of microbial EPSs is its cost, as the expenses are directly related to the cost of the substrate as well as the cost of the bioreactor to grow the microbes. New bacterial polysaccharides are still being investigated in the fields of pharmaceuticals, cosmetics, and biomedicine [87].

### 3.4.5.2 Polysaccharides from Fungal Sources

Polysaccharides isolated from mushrooms are recognized as among the safest and most effective natural antioxidants. Mushrooms have attracted significant attention as traditional foods with a lot of medical value. Mushrooms are composed of carbohydrates, lipids, proteins, enzymes, minerals, and vitamins, and hence act as health promoters. "Mushrooms" have biochemical and medicinal applications including antitumor, antiinflammatory, immune-modulatory, and in particular, antioxidant activities. Some of the components of mushroom include pullulan, elsinan, and yeast glucans (Table 3.1).

Polymorphic fungus *Aureobasidium pullulans* is obtained from pullulan, a natural linear homopolysaccharide. It consists of three glucose units attached by an α-(1→4) glycosidic linkage, which is linked through α-(1→6) glycosidic linkage [88]. Pullulans are highly water soluble and biodegradable due to the coexistence of different glycosidic bonds [89]. Pullulan is used as a stabilizer, adhesive, as well as a coating or packaging material in many food industries. Because of its inherent non-toxic, non-immunogenic, and biodegradable characteristics, it also offers a wide range of potential applications in biomedicine, such as targeted drug/gene imaging and tissue engineering. In particular, pullulan has been used as a hydrogel for tissue engineering. Wong et al. demonstrated that pullulan hydrogels are an effective cell-delivery system and that they improve mesenchymal stem cell survival and engraftment in high oxidative-stress environments [90]. Pullulan gels have a higher water content of close to 90%, and hence they have a lot of applications as EPSs in biomedicine such as tissue engineering, specifically vascular engineering.Pullulan–collagen composite hydrogel matrices have been fabricated, resulting in a structured yet soft scaffold for skin engineering. More recently, a novel topical film prepared with verniciflua extract-loaded pullulan hydrogel was synthesized for atopic dermatitis treatment [91].

### 3.4.5.3 Polysaccharides from Plant Sources: Starch

Polysaccharides derived from herbs can be effectively used in many applications and have diverse therapeutic properties such as antioxidant, antitumor and immuno-stimulatory activities, and are effective in promoting wound healing [91]. Starch is one of the most abundant polysaccharides from plant origins, and has been used in food applications such as thickening, binding, sweetening, and also as an emulsifying agent [92]. It is mainly obtained from cereals and tubers. Chemically, starch is a polymeric carbohydrate composed of glucose units linked together, comprising two types of α-glucan: linear amylose (poly-α-1,4-D-glucopyranoside) and branched amylopectin (poly-α-1, 4-D-glucopyranoside and α-1,6-D-glucopyranoside), and so has been established as a heterogeneous material. This polysaccharide is produced from agricultural plants, mainly potatoes, rice, maize, and wheat. Its availability of hydroxyl groups makes it tremendously hydrophilic and easy to react chemically (esterification, oxidation, etherification, and cross-linking) (Table 3.1).

Several of its advantages are due to its extensive availability, low cost, and total composability without generating any hazardous residues. Starch is used for a number of biomedical applications such as tissue engineering, wound healing, bone regeneration, and drug delivery, and has also been used for adhesion, proliferation, differentiation, and regeneration of cells [93]. The employment of starch for biomedical functions is also appealing due to its similarity to the native cellular environment. In order to enable applications in tissue engineering, starch has been manipulated to improve some of its properties, including its mechanical properties and moisture sensitivity. The major problem with starch is its insufficiency to develop scaffolds. However, its mechanical stability can be improved to convert the material into an appropriate product. For instance, Waghmare et al. developed starch-based nanofiber scaffolds using polyvinyl alcohol (a non-toxic, water-soluble, biocompatible, synthetic polymer) as the plasticizer and glutaraldehyde as a cross-linking agent for application in wound healing [94]. Evaluation of the nano-fibrous scaffolds in cellular assays demonstrated their non-toxicity and their ability to promote cellular proliferation. The strategy of employing starch as a matrix not only reduced production costs, but also endowed the products with the features of biodegradation, biocompatibility, and specific interactions with biological systems. Among other biopolymers such as alginate, gelatin, and collagen, starch is also used for bone substitution to fabricate scaffolds for bone tissue engineering [95]. The ability for the growth and proliferation of bone marrow mesenchymal stem cells on the constructs confirmed the suitability of these scaffolds for bone tissue engineering applications. In regards to drug-delivery systems, starch has been used for particles and hydrogels [96].

#### 3.4.5.4 Graphene-based Substrates

Biodegradable polymers that are typically used in tissue engineering suffer from poor mechanical and electrical properties, limiting their broad implementation in regenerative sciences, in particular, for bone, muscle, nerve, and cardiac tissues [97]. Graphene is a two-dimensional (2D) nanoparticle containing a single carbon layer arranged in a honeycomb crystal lattice. Graphene-based substrates hold better opportunities in the biomedical field due to their cytocompatibility [98–100]. Integrating graphene into the matrix of polymers has been reported to drive cell differentiation and tissue regeneration [101]. Cells interact closely with nanoparticles as substrate as compared to when presented as a suspension. Some naturally derived bionanocomposites and their various applications are shown in Figure 3.6.

Graphene substrate provides a stable support and acts as a scaffold as a foundation for tissue regeneration. This ability has shown promising results in tissue engineering, drug delivery, and bioimaging, and in biomedical devices. Further, a number of studies have suggested that graphene or GO films or substrates are reported to be cytocompatible when tested on a variety of mammalian cells [98–100]. This provides a great opportunity for integrating or reinforcing graphene-derived particles into a polymer matrix to augment tissue regeneration. Additionally, the induction of cell differentiation can be evaluated by modifying graphene surfaces to enable drug encapsulation and controlled drug release.

The use of graphene as a filler at low concentration has shown improvements in mechanical, thermal, and electrical properties. A 0.1 wt.% of graphene when added to ultrahigh-molecular-weight polyethylene (UHMWPE) showed improved toughness and tensile strength by 54% and 71%, respectively [102]. Reportedly, 2.5- and 4.6-fold increases in tensile strength and Young's modulus were observed after loading 5 wt.% of GO in chitosan. GO incorporation into chitosan further was reported to provide a shift in the glass transition temperature from 118°C in pure

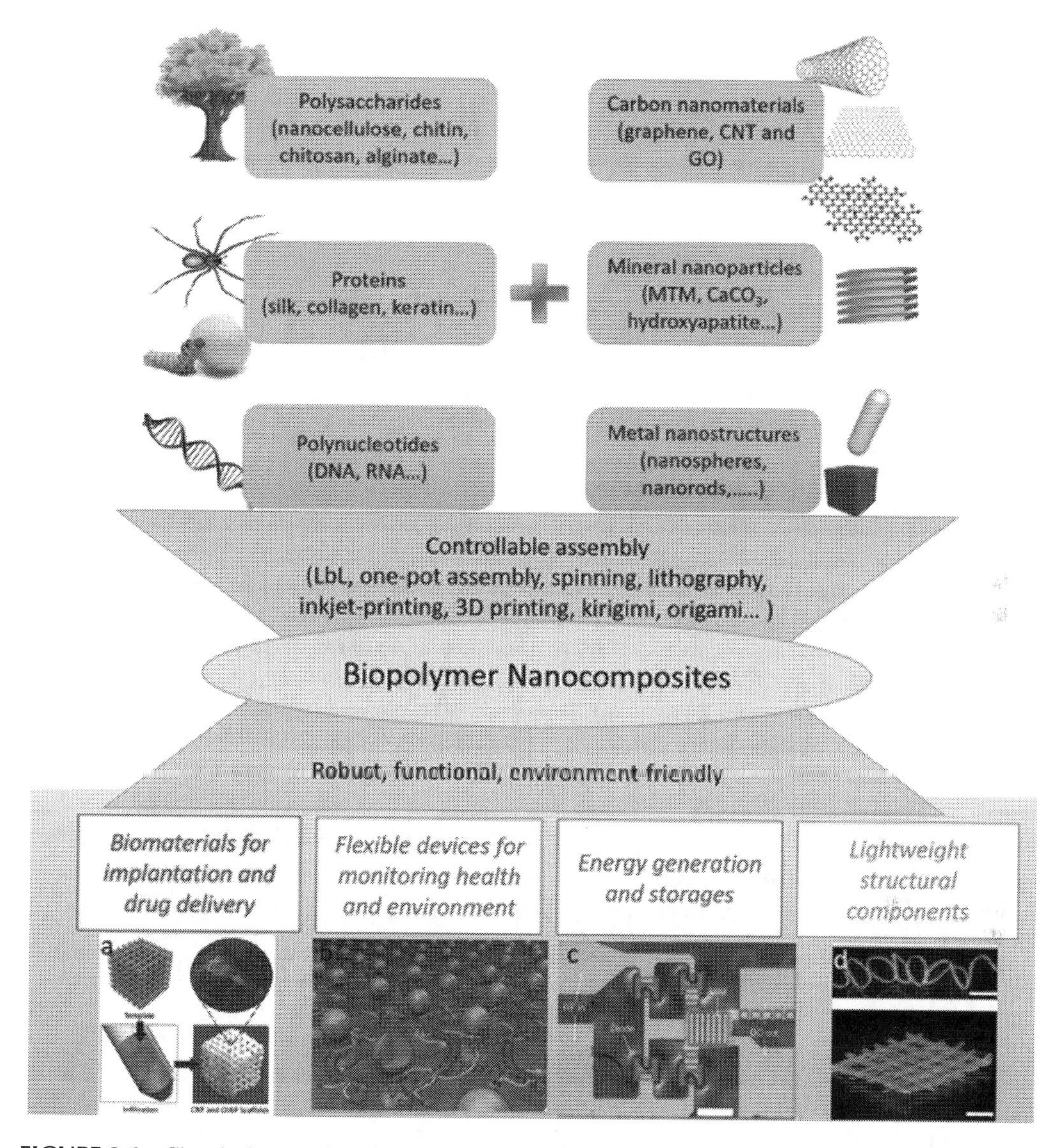


**FIGURE 3.6** Chemical properties of various nanocomposites and their applications.

chitosan to 158°C, indicating that the thermal properties of chitosan were increased as GO was added [103]. Another study showed that the incorporation of 3 wt.% of polyethylene-grafted graphene oxide sheets (PE-g-GO) in a high-density polyethylene (HDPE) matrix improved the yield strength, modulus, and elongation by 30%, 40%, and 26%, respectively [104]. Hence, graphene-derived composite enhances the mechanical and electrical properties of these biopolymers. Further, nanocomposite graphene substrate can be a useful mechanical support for load-bearing tissues and cells.

Osteoblasts showed better initial proliferation on a 2D PCL/GO composite in comparison to that in 3D porous scaffolds. Lu et al. demonstrated that chitosan–PVA polymer blend fibers reinforced with graphene particles showed complete and rapid wound healing in mice and rabbits. Furthermore, they proposed that the presence of free electrons in graphene inhibited the multiplication of prokaryotic cells, thus preventing microbial growth, without affecting the eukaryotic cells on the composite fiber, and resulting in fast wound healing [105].

Alternatively, graphene-based thermogel is an active area of research and is proving to be a promising scaffold material in 3D tissue engineering. The use of injectable GO/polypeptide thermogel as a 3D scaffold influenced adipogenic differentiation of mesenchymal stem cells by enhancing the expression of adipogenic biomarkers (PPAR-γ, CEBP-α, LPL, AP2, ELOVL3, and HSL) and aided in the adsorption of insulin and adipogenic differentiation factors [106].

A more recent effort in the use of graphene-based biomaterials is the use of a graphene particle surface decorated with metallic nanoparticles for preparing biomedical polymer composites. Such polymer nanocomposites are envisaged to show unique properties as a result of synergism between the polymer, graphene, and metallic particles. In a recent study, a multifunctional 3D macroporous PCL scaffold reinforced with strontium-decorated graphene hybrid nanoparticles has been prepared. The bioactivity of the scaffold was evaluated using mouse osteoblasts. A PCL scaffold containing the hybrid particles showed enhanced cell proliferation and mineralization due to improved wettability and release of strontium ions from the hybrid nanoparticles [97].

#### 3.4.5.5 Polysaccharides from Animal Sources

The aquatic environment as a renewable source of biocompounds leads to a positive step in the development of new systems and devices for biomedical applications, while marine polysaccharides are among the most abundant materials in the seas. Alginate, carrageenan, and fucoidan polysaccharides are extracted from algae, and chitosan and hyaluronan can be obtained from marine animal sources, which shows important biological properties such as biocompatibility, biodegradability, and anti-inflammatory activity, as well as adhesive and antimicrobial actions [107]. Among them, chitosan and its oligosaccharides have received considerable attention due to their biological activities and properties such as biocompatibility, biodegradability, and antiinflammatory activities, as well as adhesive and antimicrobial actions in commercial applications [108].

#### 3.4.5.6 N-Acetyl-D-Glucosamine and D-Glucosamine Monosaccharides

Earlier, only marine sources (shrimp, prawn, crab) were used to provide the starting chitin. Currently, new commercial chitosan, better characterized by manufacturers and with enhanced safety characteristics for certain pharmaceutical, cosmetic, and biomedical applications, has been produced at lower cost [108]. Chitosan has revealed some therapeutic activity, such as lowering of cholesterol, wound healing, antiulcer, and antimicrobial effects [100,101], however, also due to its non-toxicity (it has been approved by the US Food and Drug Administration), its biodegradability, and bacteriostatic and fungicidal characteristics [111]. Furthermore, it shows advantages in regards to its special use as a drug carrier, and thus it has been extensively exploited in the preparation of micro-/nanoparticles, beads, and capsules for controlled drug-delivery systems [112]. Ahmed and his coworkers described some of the advantages that make chitosan an appealing biopolymer for the development of polymeric particles: its mucoadhesive nature (which increases the time of attachment at the absorption site), the easy availability of free amino groups (for cross-linking), the ease of fabrication of polymeric particles without using hazardous solvents, the cationic nature that permits ionic cross-linking with multivalent anions, and its ability to control the release of an administered drug [109–113]. More recently, Bazrafshan et al. reviewed the use of chitosan to mimic fibrous assemblies [81]. Chitosan can be also mixed with other synthetic or natural polymers in order to help its processability and fine-tune its properties [114]. Dextrin- and chitosan-coated AuNPs are two important biopolymers which have been found to be the best inhibitors of fibril formation. The mechanism for the inhibition of insulin amyloid fibrils is that biopolymer-coated AuNPs strongly interact with insulin monomers and inhibit oligomer formation as well as elongation of the protofibrils. The result of cytotoxicity experiments showed that AuNP–insulin amyloid fibrils are less toxic compared to insulin amyloid fibrils alone. Therefore, as a result, both dextrin and chitosan–AuNPs could be used as therapeutic agents for the treatment of amyloid-related disorders [115].

## 3.5 MINIATURE MEDICAL DEVICES

The engagement of artificial organs, scaffolds, and other structures has enabled biopolymers to be used extensively in designing functional materials. The advantages of several stimulatory materials helped in several powerful designs of many medicinally important pregnancy, cardiovascular, and drug-delivery devices, which merged with the structural properties to match the favorable environment and physiologically functioning of a tissue or organ. The actively derived natural polymers are abundantly present in nature and basically made up of polysaccharides that include alginates, cellulose, hyaluronate, etc. Of these, alginates are most important as they have been used in various applications including biomedicine, food industries, packaging, and water purification [116] (Table 3.3).

Biopolymers have gained massive awareness in the field of biomedical applications due to their properties like non-toxicity and biodegradability. These biopolymers can be described as biocompatible as they are suitable for body and fluids exposure. These natural polymers are reliable and aid in close proximity of living cells, or work with proficiency along with the living system without causing any ill effects. In the medical field, they have been variously used in drug delivery, drug release control, tissue engineering, scaffolds in ligaments, and dentistry [130].

The main objective of implantable devices is to mimic a body part and they are used to replace a damaged organ or structure to sustain normal body function. Coatings of devices by biopolymers are considered exceedingly beneficial in biomedical applications because they offer flexibility due to chemical groups which can attach to the surface with an interaction with the tissues. Some medical

**TABLE 3.3**
**Sources of Microbial Exo-polysaccharides and Their Applications in the Biomedical Sector**

| Exopolysaccharide | Microorganism | Application | References |
|---|---|---|---|
| **Alginate** | *Azotobacter vinelandi* and *Pseudomonas aeruginosa* | Controlled drug release, encapsulation, scaffolds in ligaments, tissue engineering, and in dentistry for the preparation of forms in the presence of slow-release calcium salt, cell microencapsulation | [117,118] |
| **Xanthan Gum** | *Xanthomonas campestris* | Intra-abdominal adhesion, high thickening capacity, emulsifying, film forming, release-control agent | [119,120] |
| **Dextran** | *Leuconostoc esenteroides* | Molecule carrier or drug-delivery system, plasma volume expander, peripheral flow enhancer, antithrombotic agent, and for the rheological improvement of artificial tears | [121,122] |
| **Bacterial cellulose** | *Acetobacter*, *Gluconacetobacter xylinum* | Artificial skin, artificial blood vessels and microvessels, wound dressings, implants and scaffolds for tissue engineering, carriers for drug delivery, wound-dressing materials | [123,124] |
| **Hyaluronic acid/ hyaluronan** | *Streptococcus equisimilis*, *Streptococcus zooepidemicus*, *Bacillus subtilis* | Gelling/thickening agents, skin regenerating, collagen- and elastin-stimulating efficacy, drug release for treating tumor cells, skin regenerating, and collagen stimulating efficacy | [117,125–128] |
| **Polygalactosamine** | *Paecilomyces* spp. | Growth inhibitor of some tumor cells. With chitosan microspheres for drug-delivery system | [129] |

implants include heart, bones, eyes, ears, knees, breasts, hips, and cardiovascular system implants. Some of the uses of biopolymers are in medical materials, packaging, cosmetics, food additives, clothing fabrics, water treatment chemicals, industrial plastics, absorbents, biosensors, and even data storage elements. Furthermore, synthetic polymers may present concerns about their biodegradation products in the body, which may lead to an unwanted immunogenic response [131]. These biopolymers are used in various medical devices and also to measure, treat, or substitute any tissue, organ, or function of the body. They have gained enormous popularity due to the fact that they help in improving bodily functions without altering normal functioning or triggering any allergies or side effects. Their use has helped in further advancements in tissue culture, tissue scaffolds, implantation, and artificial grafts, wound fabrication, controlled drug delivery, bone filler materials, etc. [132]. The degradation process occurs by hydrolysis, producing carbon dioxide, which lowers the pH resulting in cell and tissue necrosis [133].

Many biopolymers like polyvinylidene fluoride (PVDF), polymethyl methacrylate (PMMA), polypropylene (PP), polyurethane (PU), and others are the most suitable biopolymers that are very used widely in making several medical tools. This is due to the fact that they possess enriched biological textiles, have piezoelectric properties, and also they have highly non-reactive thermoplastic polymers. They form structures such as meshes in surgical implants, other surgical meshes, and sutures material which is decomposable and non-reactive, with a less inflammatory reaction making them more compatible with the tissue being treated [134].

PVDF-enabled coatings and electrospinning with other copolymers with different nanomaterials are used in environmental bioremediation applications. Little is known about their use in medicine, however they showed good advantages when coated with polyaniline-coated to form nanofibers that served as self-powered piezo-organic-e-skin sensors, when used for human health monitoring. This device demonstrated incredible sensitivity through conversion of unreadable mechanical energy to readable electric energy via human finger contact at 10 V with 10 kPa. This kind of medical device is available in the market and is used to monitor actions, such as swallowing, coughing, movement, stretching, and bending, having wider implementation in hospitals and health care areas. Conjugation of PVDF with nanosilica is presented as an excellent scaffold as it possesses both hydrophilic and hydrophobic properties when used in making devices. Blending PVDF with several other polymers like polypyrrole, polyaniline, and glutamic acid-derived polyaniline showed good and increased electrical output [135]. They also exhibited high cytotoxicity toward HeLa cells. Hence, this kind of composite material is very suitable for scaffolds and could be used as a suitable tissue-engineering and wound-healing material.

PMMA is popular for its lighter weight, less cost, relatively easy alterability, and mainly its harmless subunits, and hence it is extremely apposite for use in biomedical devices. Hence, it is used in drug delivery, cementing of bones and cartilage, dentistry repairs, orthodontic retainers, and also in microsensors. Hybrids made with the combination of PMMA with poly(diallyldimethyl ammonium) chloride and chitosan–silver nanoparticles over substrates like silica, glass, polystyrene sheets, and soft rubber substrate have been confirmed to display good antibacterial effect by inhibiting the growth of pathogens like *Escherichia coli* and *Staphylococcus aureus.* Similarly, they also arrested the growth of cancerous connective tissue (L-929 fibroblast cells), proving them to have an anticancer role [136]. Numerous microorganisms belonging to the species of *Bacillus* and *Pseudomonas* are able to degrade and produce PP thermoplastic polymer, and this has been extensively used in biopolymers in medical areas, primarily in making mesh scaffolds for breast reconstruction to support soft tissue because of its firm structure and formation of a blood oxygenator membrane. Its main advantage over other cosmetic devices or implants is that it has less potential for carcinogenesis, and also low infection and inflammatory rates within the body [137]. PP cross-linked with poly(styryl bisphosphonate) has found applications in coronary treatment to protect PP films from UV radiation. Due to the high hydrophobicity of the PP surface, it is used in cross-linking with polyvinyl pyrrolidone and polyethylene glycol acrylate to make many adhesives.

Various biomedical tools such as clinical implants, tubes, contact lenses, pacemaker encapsulants, and biosensors are made up of PDMS to facilitate drug delivery and DNA sequencing. This material is suitable because of its properties such as good flexibility, excellent fabrication permeability to gases like oxygen, optical clearness, and reduced toxicity. These characteristics are helpful in studying the topography, stretching, and mechanical and electrical stimuli for planning materials. Coated PDMS serves as a good anchor for peptides fused to a cell-adhesive peptide sequence (glycine–arginine–glycine–aspartate serine) with hydrophobic interactions. This kind of coating has excellent binding performance, good fibroblast and endothelial cell attachment, and provides greater advantages in designing functionalized biomedical devices [138]. In contrast to other polymers, polyurethane is used in a small fraction when making pacemakers, lead coatings, breast implant coatings, and vascular devices. Several of modified PUs have shown noticeably reduced water permeability compared with silicon packing materials and hence they have been used in electronic implants and 3D printing [139].

### 3.5.1 Bioresorbable Electronic Patch (BEP)

BEP is a device that is used to report technologies for a flexible, sticky, and biodegradable wireless electronic device which is integrated with a bifacially designed polymer drug reservoir. Association of BEP with a mild thermic protocol leads to long drug diffusion length and drug-delivery duration [110,113]. Fabricating the implant with a material which hydrolyzes in components used in the human body reduces the risk of potential side effects from chronic neural implants and retrieval surgery is unnecessary. The flexibility of the oxidized start-based patch (OST) and its bifocal design allows for adhesion to the targeted tissue and enables local and sustained drug delivery by reducing unintended drug release to the cerebrospinal fluid (CSF). Wireless mild thermal actuation with an alternating magnetic field enhances the penetration depth of the delivered drugs [64].

### 3.5.2 Local Drug Delivery and Biocompatibility

The BEP was first designed to adhere to the curved surface of the brain for local drug delivery, for facile heat transfer during mild thermic actuation, and for prolonged drug-delivery duration by minimizing unwanted drug leakage to the CSF. This can be achieved by using oxidized start-based patch (OST) which provides strong conjugation to both brain tissues and drug molecules (e.g., doxorubicin [DOX]), and thus enables good adhesion and sustained drug release. The total amount of DOX can be increased further by using multiple BEPs. Temozolomide can also be used as an alternative drug. This was tested in a large animal (canine) model. The BEP contains MRI agents such as ferromagnetic iron oxide nanocubes which help in bioresorption and shape change in the brain can be monitored by MRI *in vivo*. The patch volume decreased significantly after 9 weeks due to its bioresorption compared to its initial stage. This method of implantation of the BEP to the canine brain did not exhibit any unexpected side effects, including brain swelling. The brain tissue reaction to the BEP was evaluated by different techniques where immunochemical staining using hematoxylin and eosin and macrophage antibodies was performed 1 week and 10 weeks after implantation. There was no observance of any significant inflammatory responses or physiological complications. Longer-term studies to observe the effects of hydrolyzed materials *in vivo* are needed in the future [110].

### 3.5.3 Wireless Mild-thermic Actuation for Accelerated Drug Delivery

The intensity of drug release by wireless mild-thermic actuation has been characterized. A radiofrequency magnetic field applied by an external coil triggered eddy current and joule heating in the heater of the implanted BEP, resulting in increased temperature of the BEP around the brain tissues.

This heater can be designed according to the sample, where the diameter, thickness of the heater, and the coil-to-heater distance can be varied to optimize heat generation. The whole array in the wireless heater slightly affects the heat generation but it is helpful for facile fabrication (transfer printing) of the device. The change in temperature depending on the coil-to-heater distance and various coil currents in 12 mm diameter were measured using our instrument to set the temperature change [113].

### 3.5.4 Evaluation of Therapeutic Efficacy in the Mouse and Canine GBM Models

Human GBM cells (U87-MG) were cultured and subcutaneously implanted near the thigh region of 6-week-old nude mice. Tumor was grown and resected, and then BEP of ~14 mm diameter was implanted. The procedure started from the tumor resection. Further studies for non-resectable tumor cases are also needed in the future. To evaluate the therapeutic effect of the BEP further, for this process two mongrel dogs were used where both the BEP and a control wafer were implanted to the brain cavity with the remaining brain tumor. Treatment with the BEP and the control wafer for 2 days resulted in apoptosis of tumor cells, which was confirmed by TUNEL assay. Both cases showed apoptosis of tumor cells within 2 mm of the cavity surface. However, the BEP induced apoptosis of tumor cells more deeply as compared to the control wafer where the apoptosis of tumor cells invaded microscopic tumor cells at 5 mm, while the control wafer could not treat tumor cells located at over 2 mm depth from the cavity surface. BEP treated more deeply seated tumor cells, as drug penetration can be extended by the mild-thermic actuation of the BEP [64,110,113].

### 3.5.5 Therapeutic Efficacy in Women's Health and Childbirth

Coating and cross-linking of polymers with biological materials has various medical applications including corrosion proofing, scratch resistance, excellent adhesion, wettability, and biocompatibility. Hence, coatings with advanced techniques and materials produce a prudent choice of material, coating methods, and production parameters in making devices and implants, as biopolymer-coated implants can serve as biomimetic surfaces in body matrices. Moreover, they have posed prodigious adaptable and mechanical properties that are comparable to those of biological tissues. Several of them exhibited reduced wettability and a stumpy surface area, which is a inheritable property of biopolymers that can be sometimes be challenging to match with biological surfaces. Hence, coated and fabricated medical devices are not only more effective, but they also result in sensible selection of material, covering methods, and manufacturing parameters. This kind of polymer-coated implants can serve as excellent biomimetics to biological processes. In addition to other applications, recently they have been shown to have great potential and are safer in several obstetrics and gynecology applications. Here, in many areas biopolymer-based materials are often used in pre- and post-delivery complications such as childbirth sutures, vaginal and abdominal care, and cesarean healing, care, and management [140]. Bio-polymeric stents used in pregnant women have been found to be more effective as they are more reliable, safe, and also avoid impediments to help in the supervision of persistent flank discomfort in women bearing children. This is mainly because they become integrated with the biological structures and also easily decompose after wound healing. Also, they have been shown to have extraordinary antibiotic roles against several pathogens, acting as a drug and preventing biofilm formation. Ethylcellulose, a biopolymer, is used as a precursor substrate in the preparation of vaginal mucoadhesive bilayer films, and has been shown to have antiviral activity which is one of the potential strategies to treat and control sexually transmitted diseases like HIV and syphilis. In addition, the adhesive material remained unharmed by the vaginal microflora, i.e., lactobacilli, and by varied vaginal pH (4–5.5) associated with vaginal drug delivery also [141]. Biopolymer stents, which act as intravascular dilators in human body fluids, may undergo biodegradation, with no foreign body remaining, thus avoiding a second surgery for stent removal

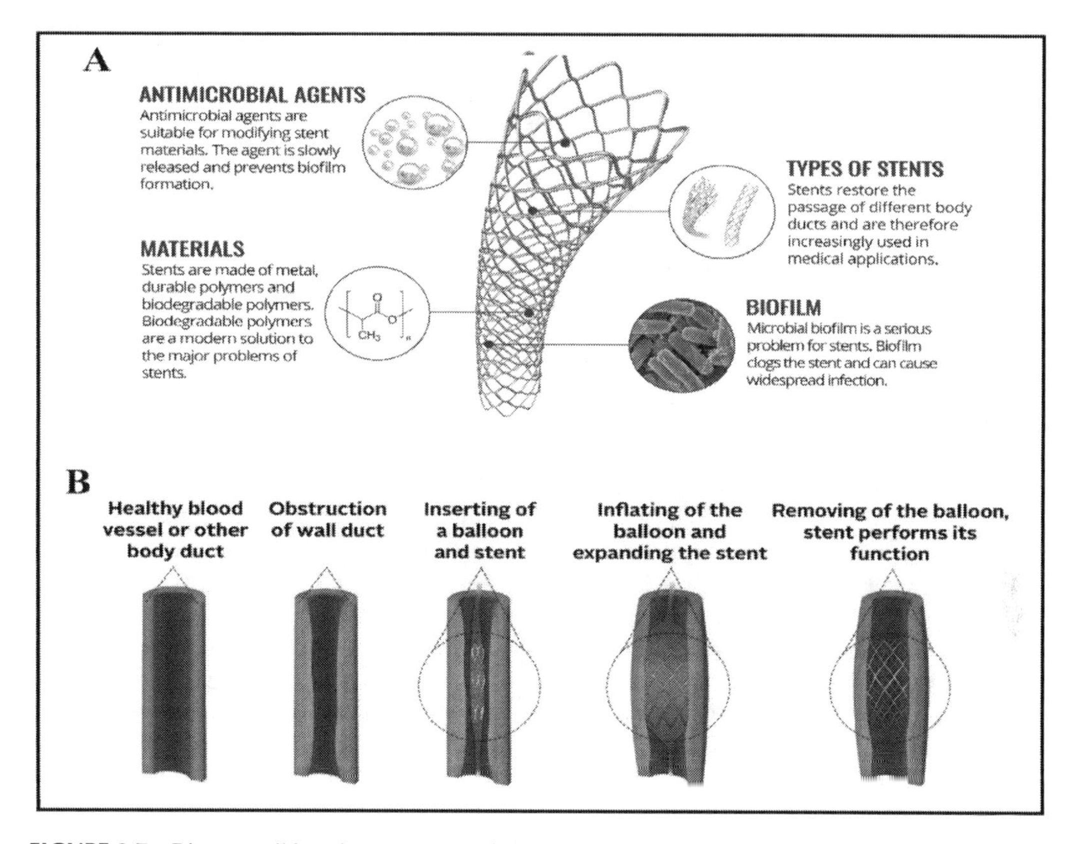


**FIGURE 3.7** Biocompatible polymers are used in producing stents [140].

(Figure 3.7). Tamai and co-workers developed PLA-based biodegradable stents which gave a promising result when applied to a human model, while it also worked well for other animals, including in rabbit aortas [142]. PLA stents are hemocompatible in comparison to stainless steel stents, and in magnetic imaging they can be used to measure luminal patency.

### 3.5.6 Miscellaneous

Compression-molded biopolymers are a type of co-polymer that are used for fracture treatment where it fills up the bone defect plates or screws for the treatment of fractures and also for scaffolding to facilitate the formation of new cartilage material in the body [143]. The advantages of co-polymer over homopolymers include that PLA-PGA co-polymer is more useful than homopolymers of PLA and PGA because the degradation rate can be changed according to the use as they are biocompatible and non-toxic [3]. Chen, in order to increase the mechanical strength, developed a polymeric stent made from chitosan-based films fixed by genipin protein which led to an improvement in the mechanical behavior, with re-endothelization of the stent-implanted vessel [144]. Lubrizol produces several medical-grade polymers which are specifically thermoplastic biopolymers, namely pellethane, isoplast, tecoflex, tecothane, and carbothane, and pathway polymers made of polyurethane-based materials. These predominantly are used for making marginally invasive catheters and silicone-based medical devices because of their superior stability, compatibility, and versatility with the biomaterial. This kind of bioresin is used in making peripheral intravenous catheters, central venous catheters, and peripherally inserted central catheters for vascular access devices. Similarly, they are used in cardiovascular devices for making short- and long-term implantable guidewires, pacemaker headers, leads, and delivery systems. In parallel, they serve effectively for creating mouth-guards for the oral cavity, dental aligners, and retainers, as well as orthodontic elastics, with the use of thermoplastic polyurethanes. Also, they are used in urology and wound care for bio-durable implants and for treatment with combination products.

Biopolymers also find usage as a packaging/coating material for implanted devices, which acts as a protective layer for the implanted device. The biopolymer hybrid obtained offers biocompatibility, and higher stability under a wide range of mechanical, electrical, chemical, and thermal behaviors. Such coverings ensure gas permeability and water permeability by the packaging material and also are able to protect the electronic circuit of the implanted device from water and ions inside the human body, thus ensuring a longer shelf-life of the implanted device. The covering of a pseudopeptide polymer on the surface of stainless steel has been further reported to improve the bioactivity and antifouling property, conferring greater resistance to corrosion in comparison to uncoated stainless steel [145]. The coating also modified the surface of a 3D vascular stent, showing potential in the prevention of late stent thrombosis and in-stent restenosis [145]. Gendenkov et al. reported that the coating of super-dispersed polytetrafluoroethylene (SPTFE) polymer on a magnesium alloy surface enhances the protective and antifriction properties of the alloy [146]. A similar observation was recorded when hydroxyapatite-polytetrafluoroethylene (PTFE) composite coatings on Mg-Mn-Ce alloys were produced, which were used in resorbable implants. Polymethyl methacrylate (PMMA) is a synthetic lightweight, cost-effective polymer used in a variety of biomedical applications. It has good mechanical capacity, low toxicity, and non-biodegradable nature, making it suitable to be used for drug delivery, as bone cement, microsensors, orthodontic retainers, rhinoplasty, intraocular lenses, and a denture base. PMMA alone does not have wider applicability due to its inability to support osseointegration. However, fibrovascular ingrowth of tissues was observed with PMMA orbital implants in surrounding tissues without any sign of inflammation [147]. Also, improved antibacterial activity against *E. coli* and *Staphylococcus aureus* by PMMA was reported when hybrid deposition of PMMA/chitosan–silver nanoparticles on soft rubber substrate was produced [136].

## 3.6 CONCLUSION

Despite the multifarious benefits of biopolymers and the high demand owing to their biocompatibility, the global market of biopolymer fields is facing challenges due to the current coronavirus pandemic. As factories are shut down and facing a lack of manpower, it will be a bumpy ride to achieve the expected target. Using composite materials coupled with innovations can benefit their use in clinical aspects that include tissue regeneration and implantable devices. Furthermore, the mechanism and biosynthetic pathways need to be completely elucidated that will open new possibilities to utilize them in various fields. With the adverse effects of synthetic polymers, these new biopolymers have great importance in medicine as they can fulfill all the requirements in living cells. Screening of such novel biopolymer-producing organisms needs to be continued and intensified with new targets and strategies that can be employed for improvising production. Another challenge is that the present market is dominated by polymers derived from petrochemical, animal, or plant origin, due to the high associated microbial production cost, particularly the downstream processing which encompasses 20–60% of total production. Therefore, further concerted efforts are needed to develop a cost-effective and economical production method for rendering large-scale biopolymer production from microbes.

## REFERENCES

1. Vroman, I., & Tighzert, L. (2009). Biodegradable polymers. *Materials*, 2(2), 307–344.
2. Iheaturu, N. C., Diwe, I. V., Chima, B., Daramola, O. O., & Sadiku, E. R. (2019). Biopolymers in Medicine. In Dhorali Gnanasekaran (ed.) *Green Biopolymers and Their Nanocomposites* (pp. 233–250). Springer, Singapore.

3. Rebelo, R., Fernandes, M., & Fangueiro, R. (2017). Biopolymers in medical implants: a brief review. *Procedia Engineering*, 200, 236–243.
4. Choi, K. R., Jang, W. D., Yang, D., Cho, J. S., Park, D., & Lee, S. Y. (2019). Systems metabolic engineering strategies: integrating systems and synthetic biology with metabolic engineering. *Trends in Biotechnology*, 37(8), 817–837.
5. Schmid, J., Sieber, V., & Rehm, B. (2015). Bacterial exopolysaccharides: biosynthesis pathways and engineering strategies. *Frontiers in Microbiology*, 6, 496.
6. Mirzaei, A., Moghadam, A. S., Abazari, M. F., Nejati, F., Torabinejad, S., Kaabi, M., ... & Saburi, E. (2019). Comparison of osteogenic differentiation potential of induced pluripotent stem cells on 2D and 3D polyvinylidene fluoride scaffolds. *Journal of Cellular Physiology*, 234(10), 17854–17862.
7. Prasanna, K., Deepthi, M. V., Ashamol, A., & Sailaja, R. R. N. (2013). Effect of nanoclay content and pH of the medium on the swelling characteristics of grafted and cross-linked mixed chitosan derivatives. *Polymer-Plastics Technology and Engineering*, 52(4), 352–357.
8. Kumar, S., Kumari, M., Dutta, P. K., & Koh, J. (2014). Chitosan biopolymer Schiff base: preparation, characterization, optical, and antibacterial activity. *International Journal of Polymeric Materials and Polymeric Biomaterials*, 63(4), 173–177.
9. Kreyenschulte, D., Krull, R., & Margaritis, A. (2014). Recent advances in microbial biopolymer production and purification. *Critical Reviews in Biotechnology*, 34(1), 1–15.
10. Mohamed, R. R., & Fekry, A. M. (2011). Antimicrobial and anticorrosive activity of adsorbents based on chitosan Schiff's base. *International Journal of Electrochemical Science*, 6, 2488–2508.
11. Market Research Report (2021). *Annual Report*. Retrieved from www.marketsandmarkets.com/Market-Reports/biopolymers-bioplastics-market-88795240.html
12. Verbeek, C. (Ed.) (2012). *Products and Applications of Biopolymers*. BoD–Books on Demand. Intechopen.
13. Numata, K. (2015). Poly (amino acid) s/polypeptides as potential functional and structural materials. *Polymer Journal*, 47(8), 537–545.
14. Younes, B. (2017). Classification, characterization, and the production processes of biopolymers used in the textiles industry. *The Journal of the Textile Institute*, 108(5), 674–682.
15. Kaplan, David L. (ed.) (1998). Introduction to biopolymers from renewable resources. In *Biopolymers from Renewable Resources* (pp. 1–29). Springer, Berlin, Heidelberg.
16. Tănase, E. E., Râpă, M., & Popa, O. (June 2014). Biopolymers based on renewable resources – A review. In *Proceedings of the International Conference Agriculture for Life, Life for Agriculture* (pp. 5–7), Bucharest, Romania.
17. Mohd Nadzir, M., Nurhayati, R. W., Idris, F. N., & Nguyen, M. H. (2021). Biomedical applications of bacterial exopolysaccharides: a review. *Polymers*, 13(4), 530.
18. Synytsya, A., & Novak, M. (2014). Structural analysis of glucans. *Annals of Translational Medicine*, 2(2), 1–14.
19. Galus, S., Arik Kibar, E. A., Gniewosz, M., & Kraśniewska, K. (2020). Novel materials in the preparation of edible films and coatings – a review. *Coatings*, 10(7), 674.
20. Mathur, N. K., & Narang, C. K. (1990). Chitin and chitosan, versatile polysaccharides from marine animals. *Journal of Chemical Education*, 67(11), 938.
21. Pérez-Gago, M. B., & Rhim, J. W. (2014). Edible coating and film materials: lipid bilayers and lipid emulsions. In *Innovations in Food Packaging* (pp. 325–350). Academic Press, Plano, TX.
22. Mukherjee, S., Basak, B., Bhunia, B., Dey, A., & Mondal, B. (2013). Potential use of polyphenol oxidases (PPO) in the bioremediation of phenolic contaminants containing industrial wastewater. *Reviews in Environmental Science and Bio/Technology*, 12(1), 61–73.
23. Yadav, P., Yadav, H., Shah, V. G., Shah, G., & Dhaka, G. (2015). Biomedical biopolymers, their origin and evolution in biomedical sciences: a systematic review. *Journal of Clinical and Diagnostic Research*, 9(9), ZE21.
24. Valentine, M. E., Kirby, B. D., Withers, T. R., Johnson, S. L., Long, T. E., Hao, Y., ... & Yu, H. D. (2020). Generation of a highly attenuated strain of Pseudomonas aeruginosa for commercial production of alginate. *Microbial Biotechnology*, 13(1), 162–175.
25. Tsouko, E., Kourmentza, C., Ladakis, D., Kopsahelis, N., Mandala, I., Papanikolaou, S., ... & Koutinas, A. (2015). Bacterial cellulose production from industrial waste and by-product streams. *International Journal of Molecular Sciences*, 16(7), 14832–14849.

26. Kang, D. H., Kim, D., Wang, S., Song, D., & Yoon, M. H. (2018). Water-insoluble, nanocrystalline, and hydrogel fibrillar scaffolds for biomedical applications. *Polymer Journal*, 50(8), 637–647.
27. Favaro, L., Basaglia, M., & Casella, S. (2019). Improving polyhydroxyalkanoate production from inexpensive carbon sources by genetic approaches: a review. *Biofuels, Bioproducts and Biorefining*, 13(1), 208–227.
28. Czaja, W. K., Young, D. J., Kawecki, M., & Brown, R. M. (2007). The future prospects of microbial cellulose in biomedical applications. *Biomacromolecules*, 8(1), 1–12.
29. Burg, K. J., Porter, S., & Kellam, J. F. (2000). Biomaterial developments for bone tissue engineering. *Biomaterials*, 21(23), 2347–2359.
30. Ogura, N., Kawada, M., Chang, W. J., Zhang, Q., Lee, S. Y., Kondoh, T., & Abiko, Y. (2004). Differentiation of the human mesenchymal stem cells derived from bone marrow and enhancement of cell attachment by fibronectin. *Journal of Oral Science*, 46, 207–213.
31. Portela, R., Leal, C. R., Almeida, P. L., & Sobral, R. G. (2019). Bacterial cellulose: a versatile biopolymer for wound dressing applications. *Microbial Biotechnology*, 12(4), 586–610.
32. Fontana, J. D., De Souza, A. M., Fontana, C. K., Torriani, I. L., Moreschi, J. C., Gallotti, B. J., De Souza, S. J., Narcisco, G. P., Bichara, J. A., & Farah, L. F. X. (1990). Acetobacter cellulose pellicle as a temporary skin substitute. *Applied Biochemistry and Biotechnology*, 24(1), 253–264.
33. Ho, G. H., Yang, T. H., & Yang, K. H. (2005). Stable biodegradable, water absorbing gamma-polyglutamic acid hydrogel. *European Patent* EP1550469A1, pp. 1–12.
34. Yang, G., Chen, J., Qu, Y. B., & Lun, S. Y. (2001). Effects of metal ions on gamma-poly (glutamic acid) synthesis by Bacillus licheniformis. Sheng wu gong cheng xue bao. *Chinese Journal of Biotechnology*, 17(6), 706–709.
35. Fernandes, S. N., Geng, Y., Vignolini, S., Glover, B. J., Trindade, A. C., Canejo, J. P., Almeida, P. L., Brogueira, P., & Godinho, M. H. (2013). Structural color and iridescence in transparent sheared cellulosic films. *Macromolecular Chemistry and Physics*, 214(1), 25–32.
36. Costa, O. Y., Raaijmakers, J. M., & Kuramae, E. E. (2018). Microbial extracellular polymeric substances: ecological function and impact on soil aggregation. *Frontiers in Microbiology*, 9, 1636.
37. Rebelo, A. L., Chevalier, M. T., Russo, L., & Pandit, A. (2021). Sweet tailoring of glyco-modulatory extracellular matrix-inspired biomaterials to target neuroinflammation. *Cell Reports Physical Science*, 100321, 1–45.
38. Shah, N., Ul-Islam, M., Khattak, W. A., & Park, J. K. (2013). Overview of bacterial cellulose composites: a multipurpose advanced material. *Carbohydrate Polymers*, 98(2), 1585–1598.
39. Phisalaphong, M., Suwanmajo, T., & Tammarate, P. (2008). Synthesis and characterization of bacterial cellulose/alginate blend membranes. *Journal of Applied Polymer Science*, 107(5), 3419–3424.
40. Saibuatong, O. A., & Phisalaphong, M. (2010). Novo aloe vera–bacterial cellulose composite film from biosynthesis. *Carbohydrate Polymers*, 79(2), 455–460.
41. Zmejkoski, D., Spasojević, D., Orlovska, I., Kozyrovska, N., Soković, M., Glamočlija, J., Dmitrović, S., Matović, B., Tasić, N., Maksimović, V., Sosnin, M., & Radotić, K. (2018). Bacterial cellulose-lignin composite hydrogel as a promising agent in chronic wound healing. *International Journal of Biological Macromolecules*, 118, 494–503.
42. Pandey, K., Pandey, A. K., Sirohi, R., Pandey, S., Srivastava, A., & Pandey, A. (2021). Production and applications of polyglutamic acid. In Pandey, A., Mohan, S.V., Chang, J.-S., Hallenbeck, P., Larroche, C. (eds) *Biomass, Biofuels, Biochemicals* (pp. 253–282). Elsevier, Amsterdam.
43. Buescher, J. M., & Margaritis, A. (2007). Microbial biosynthesis of polyglutamic acid biopolymer and applications in the biopharmaceutical, biomedical and food industries. *Critical Reviews in Biotechnology*, 27(1), 1–19.
44. Matsusaki, M., Serizawa, T., Kishida, A., Endo, T., & Akashi, M. (2002). Novel functional biodegradable polymer: synthesis and anticoagulant activity of poly (γ-glutamic acid) sulfonate (γ-PGA-sulfonate). *Bioconjugate Chemistry*, 13(1), 23–28.
45. Matsusaki, M., & Akashi, M. (2005). Novel functional biodegradable polymer IV: pH-sensitive controlled release of fibroblast growth factor-2 from a poly (γ-glutamic acid)-sulfonate matrix for tissue engineering. *Biomacromolecules*, 6(6), 3351–3356.
46. Park, Y. J., Liang, J., Yang, Z., & Yang, V. C. (2001). Controlled release of clot-dissolving tissue-type plasminogen activator from a poly (L-glutamic acid) semi-interpenetrating polymer network hydrogel. *Journal of Controlled Release*, 75(1–2), 37–44.

47. Hezayen, F. F., Rehm, B. H., Tindall, B. J., & Steinbüchel, A. (2001). Transfer of Natrialba asiatica B1T to Natrialba taiwanensis sp. nov. and description of Natrialba aegyptiaca sp. nov., a novel extremely halophilic, aerobic, non-pigmented member of the Archaea from Egypt that produces extracellular poly (glutamic acid). *International Journal of Systematic and Evolutionary Microbiology*, 51(3), 1133–1142.
48. Weber, J. (1990). Poly(gamma-glutamic acid)s are the major constituents of nematocysts in Hydra (Hydrozoa, Cnidaria). *Journal of Biological Chemistry*, 265(17), 9664–9669.
49. Kimura, K., Tran, L. S. P., Uchida, I., & Itoh, Y. (2004). Characterization of Bacillus subtilisγ-glutamyltransferase and its involvement in the degradation of capsule poly-γ-glutamate. *Microbiology*, 150(12), 4115–4123.
50. Bajaj, I., & Singhal, R. (2011). Poly (glutamic acid)–an emerging biopolymer of commercial interest. *Bioresource Technology*, 102(10), 5551–5561.
51. Sung, M. H., Park, C., Kim, C. J., Poo, H., Soda, K., & Ashiuchi, M. (2005). Natural and edible biopolymer poly-γ-glutamic acid: synthesis, production, and applications. *The Chemical Record*, 5(6), 352–366.
52. Hsieh, C. Y., Tsai, S. P., Wang, D. M., Chang, Y. N., & Hsieh, H. J. (2005). Preparation of γ-PGA/chitosan composite tissue engineering matrices. *Biomaterials*, 26(28), 5617–5623.
53. Hsieh, C. Y., Hsieh, H. J., Liu, H. C., Wang, D. M., & Hou, L. T. (2006). Fabrication and release behavior of a novel freeze-gelled chitosan/γ-PGA scaffold as a carrier for rhBMP-2. *Dental Materials*, 22(7), 622–629.
54. Sekine, T., Nakamura, T., Shimizu, Y., Ueda, H., Matsumoto, K., Takimoto, Y., & Kiyotani, T. (2001). A new type of surgical adhesive made from porcine collagen and polyglutamic acid. *Journal of Biomedical Materials Research: An Official Journal of The Society for Biomaterials and The Japanese Society for Biomaterials*, 54(2), 305–310.
55. Shih, I. L., Van, Y. T., & Shen, M. H. (2004). Biomedical applications of chemically and microbiologically synthesized poly (glutamic acid) and poly (lysine). *Mini Reviews in Medicinal Chemistry*, 4(2), 179–188.
56. Niaounakis, M. (Ed.) (2015). Chapter 1 – Introduction. In *Biopolymers: Processing and Products* (pp. 1–77). William Andrew: Oxford, UK.
57. Su, H., Fujiwara, T., Anderson, K. M., Karydis, A., Ghadri, M. N., & Bumgardner, J. D. (2021). A comparison of two types of electrospun chitosan membranes and a collagen membrane in vivo. *Dental Materials*, 37(1), 60–70.
58. Takeda, H., Howashi, G., Noda, Y., Ueki, T., & Tsukamoto, S. (2006). *Composition for Denture Stabilization.* Patent Cooperation Treaty Application Number WO2006033162.
59. Ledezma-Pérez, A. S., Romero-García, J., Vargas-Gutiérrez, G., & Arias-Marín, E. (2005). Cement formation by microbial poly (γ-glutamic acid) and fluoroalumino-silicate glass. *Materials Letters*, 59(24–25), 3188–3191.
60. Silva, E. C., Omonte, S. V., Martins, A. G. V., de Castro, H. H. O., Gomes, H. E., Zenóbio, É. G., de Oliveira, P. A. D., Horta, M. C. R., & Souza, P. E. A. (2017). Hyaluronic acid on collagen membranes: an experimental study in rats. *Archives of Oral Biology*, 73, 214–222.
61. Lee, B. S., Lee, C. C., Lin, H. P., Shih, W. A., Hsieh, W. L., Lai, C. H., Takeuchi, Y., & Chen, Y. W. (2016). A functional chitosan membrane with grafted epigallocatechin-3-gallate and lovastatin enhances periodontal tissue regeneration in dogs. *Carbohydrate Polymers*, 151, 790–802.
62. Mota, J., Yu, N., Caridade, S. G., Luz, G. M., Gomes, M. E., Reis, R. L., Jansen, J. A., Walboomers, X. F., & Mano, J. F. (2012). Chitosan/bioactive glass nanoparticle composite membranes for periodontal regeneration. *Acta Biomaterialia*, 8(11), 4173–4180.
63. Fakhri, E., Eslami, H., Maroufi, P., Pakdel, F., Taghizadeh, S., Ganbarov, K., Yousefi, M., Tanomand, A., Yousefi, B., Mahmoudi, S., & Kafil, H. S. (2020). Chitosan biomaterials application in dentistry. *International Journal of Biological Macromolecules*, 162, 956–974.
64. Peng, Y., Huang, J., Xiao, H., Wu, T., & Shuai, X. (2018). Codelivery of temozolomide and siRNA with polymeric nanocarrier for effective glioma treatment. *International Journal of Nanomedicine*, 13, 3467.
65. Yahya, E. B., Amirul, A. A., HPS, A. K., Olaiya, N. G., Iqbal, M. O., Jummaat, F., ... & Adnan, A. S. (2021). Insights into the role of biopolymer aerogel scaffolds in tissue engineering and regenerative medicine. *Polymers*, 13(10), 1612.

66. Mukherjee, T., & Kao, N. (2011). PLA based biopolymer reinforced with natural fibre: a review. *Journal of Polymers and the Environment*, 19(3), 714–725.
67. Chang, P. C., Liu, B. Y., Liu, C. M., Chou, H. H., Ho, M. H., Liu, H. C., ... & Hou, L. T. (2007). Bone tissue engineering with novel rhBMP2-PLLA composite scaffolds. *Journal of Biomedical Materials Research Part A*, 81(4), 771–780.
68. Zhang, Q., Mochalin, V. N., Neitzel, I., Knoke, I. Y., Han, J., Klug, C. A., Zhou, J. G., Lelkes, P. I., & Gogotsi, Y. (2011). Fluorescent PLLA-nanodiamond composites for bone tissue engineering. *Biomaterials*, 32(1), 87–94.
69. Rehman, A., Houshyar, S., & Wang, X. (2021). Nanodiamond-based fibrous composites: a review of fabrication methods, properties, and applications. *ACS Applied Nano Materials*, 4(3), 2317–2332.
70. DeStefano, V., Khan, S., & Tabada, A. (2020). Applications of PLA in modern medicine. *Engineered Regeneration*, 1, 76–87.
71. Klemm, D., Schumann, D., Udhardt, U., & Marsch, S. (2001). Bacterial synthesized cellulose – artificial blood vessels for microsurgery. *Progress in Polymer Science*, 26(9), 1561–1603.
72. Ravi, S., Qu, Z., & Chaikof, E. L. (2009). Polymeric materials for tissue engineering of arterial substitutes. *Vascular*, 17(1_suppl), 45–54.
73. Kundu, B., Rajkhowa, R., Kundu, S. C., & Wang, X. (2013). Silk fibroin biomaterials for tissue regenerations. *Advanced Drug Delivery Reviews*, 65(4), 457–470.
74. Bandyopadhyay, A., Chowdhury, S. K., Dey, S., Moses, J. C., & Mandal, B. B. (2019). Silk: a promising biomaterial opening new vistas towards affordable healthcare solutions. *Journal of the Indian Institute of Science*, 99(3), 445–487.
75. Holland, C., Numata, K., Rnjak-Kovacina, J., & Seib, F. P. (2019). The biomedical use of silk: past, present, future. *Advanced Healthcare Materials*, 8(1), 1800465.
76. Park, J. K., Shim, J. H., Kang, K. S., Yeom, J., Jung, H. S., Kim, J. Y., Lee, K. H., Kim, T. H., Kim, S. Y., Cho, D. W., & Hahn, S. K. (2011). Solid free-form fabrication of tissue-engineering scaffolds with a poly (lactic-co-glycolic acid) grafted hyaluronic acid conjugate encapsulating an intact bone morphogenetic protein–2/poly (ethylene glycol) complex. *Advanced Functional Materials*, 21(15), 2906–2912.
77. Jacob, J., Haponiuk, J. T., Thomas, S., & Gopi, S. (2018). Biopolymer based nanomaterials in drug delivery systems: A review. *Materials Today Chemistry*, 9, 43–55.
78. Meng, S., Liu, Z., Shen, L., Guo, Z., Chou, L. L., Zhong, W., Du, Q., & Ge, J. (2009). The effect of a layer-by-layer chitosan–heparin coating on the endothelialization and coagulation properties of a coronary stent system. *Biomaterials*, 30(12), 2276–2283.
79. MaHam, A., Tang, Z., Wu, H., Wang, J., & Lin, Y. (2009). Protein-based nanomedicine platforms for drug delivery. *Small*, 5(15), 1706–1721.
80. Bazrafshan, Z., & Stylios, G. K. (2019). A novel approach to enhance the spinnability of collagen fibers by graft polymerization. *Materials Science and Engineering: C*, 94, 108–116.
81. Bazrafshan Z., & Stylios G. K. (2019). Spinnability of collagen as a biomimetic material: A review. *International Journal of Biological Macromolecules*, 29, 693–705.
82. Rieu, C., Parisi, C., Mosser, G., Haye, B., Coradin, T., Fernandes, F. M., & Trichet, L. (2019). Topotactic fibrillogenesis of freeze-cast microridged collagen scaffolds for 3D cell culture. *ACS Applied Materials & Interfaces*, 11(16), 14672–14683.
83. Kucharska, M., Niekraszewicz, A., Wiśniewska-Wrona, M., & Brzoza-Malczewska, K. (2008). Dressing sponges made of chitosan and chitosan-alginate fibrids. *Fibres & Textiles in Eastern Europe*, 3(68), 109–113.
84. Reddy, N., Reddy, R., & Jiang, Q. (2015). Crosslinking biopolymers for biomedical applications. *Trends in Biotechnology*, 33(6), 362–369.
85. Zhou, Y., Cui, Y., & Qu, X. (2019). Exopolysaccharides of lactic acid bacteria: structure, bioactivity and associations: a review. *Carbohydrate Polymers*, 207, 317–332.
86. Ghosh, S., Lahiri, D., Nag, M., Dey, A., Sarkar, T., Pathak, S. K., ... & Ray, R. R. (2021). Bacterial biopolymer: its role in pathogenesis to effective biomaterials. *Polymers*, 13(8), 1242.
87. Rodriguez-Contreras, A. (2019). Recent advances in the use of polyhydroyalkanoates in biomedicine. *Bioengineering*, 6(3), 82.
88. Hilares, R. T., Resende, J., Orsi, C. A., Ahmed, M. A., Lacerda, T. M., da Silva, S. S., & Santos, J. C. (2019). Exopolysaccharide (pullulan) production from sugarcane bagasse hydrolysate aiming to

favor the development of biorefineries. *International Journal of Biological Macromolecules*, 127, 169–177.
89. Zhang, W., Wang, R., Sun, Z., Zhu, X., Zhao, Q., Zhang, T., ... & Lee, B. P. (2020). Catechol-functionalized hydrogels: biomimetic design, adhesion mechanism, and biomedical applications. *Chemical Society Reviews*, 49(2), 433–464.
90. Wong, V. W., Rustad, K. C., Glotzbach, J. P., Sorkin, M., Inayathullah, M., Major, M. R., ... & Gurtner, G. C. (2011). Pullulan hydrogels improve mesenchymal stem cell delivery into high-oxidative-stress wounds. *Macromolecular Bioscience*, 11(11), 1458–1466.
91. Li, T., Song, X., Weng, C., Wang, X., Wu, J., Sun, L., ... & Chen, C. (2018). Enzymatically crosslinked and mechanically tunable silk fibroin/pullulan hydrogels for mesenchymal stem cells delivery. *International Journal of Biological Macromolecules*, 115, 300–307.
92. Mahmood, K., Kamilah, H., Shang, P. L., Sulaiman, S., & Ariffin, F. (2017). A review: interaction of starch/non-starch hydrocolloid blending and the recent food applications. *Food Bioscience*, 19, 110–120.
93. Song, R., Murphy, M., Li, C., Ting, K., Soo, C., & Zheng, Z. (2018). Current development of biodegradable polymeric materials for biomedical applications. *Drug Design, Development and Therapy*, 12, 3117.
94. Waghmare, V. S., Wadke, P. R., Dyawanapelly, S., Deshpande, A., Jain, R., & Dandekar, P. (2018). Starch based nanofibrous scaffolds for wound healing applications. *Bioactive Materials*, 3(3), 255–266.
95. Shi, Y., Xu, D., Liu, M., Fu, L., Wan, Q., Mao, L., ... & Wei, Y. (2018). Room temperature preparation of fluorescent starch nanoparticles from starch-dopamine conjugates and their biological applications. *Materials Science and Engineering: C*, 82, 204–209.
96. Gholamali, I., Hosseini, S. N., Alipour, E., & Yadollahi, M. (2019). Preparation and characterization of oxidized starch/CuO nanocomposite hydrogels applicable in a drug delivery system. *Starch-Stärke*, 71(3–4), 1800118.
97. Kumar, S., & Chatterjee, K. (2015). Strontium eluting graphene hybrid nanoparticles augment osteogenesis in a 3D tissue scaffold. *Nanoscale*, 7(5), 2023–2033.
98. Chen, G. Y., Pang, D. P., Hwang, S. M., Tuan, H. Y., & Hu, Y. C. (2012). A graphene-based platform for induced pluripotent stem cells culture and differentiation. *Biomaterials*, 33(2), 418–427.
99. Li, N., Zhang, Q., Gao, S., Song, Q., Huang, R., Wang, L., ... & Cheng, G. (2013). Three-dimensional graphene foam as a biocompatible and conductive scaffold for neural stem cells. *Scientific Reports*, 3(1), 1–6.
100. Chen, H., Müller, M. B., Gilmore, K. J., Wallace, G. G., & Li, D. (2008). Mechanically strong, electrically conductive, and biocompatible graphene paper. *Advanced Materials*, 20(18), 3557–3561.
101. Hutmacher, D. W. (2000). Scaffolds in tissue engineering bone and cartilage. *Biomaterials*, 21(24), 2529–2543.
102. Lahiri, D., Dua, R., Zhang, C., de Socarraz-Novoa, I., Bhat, A., Ramaswamy, S., & Agarwal, A. (2012). Graphene nanoplatelet-induced strengthening of ultrahigh molecular weight polyethylene and biocompatibility in vitro. *ACS Applied Materials & Interfaces*, 4(4), 2234–2241.
103. Zuo, P. P., Feng, H. F., Xu, Z. Z., Zhang, L. F., Zhang, Y. L., Xia, W., & Zhang, W. Q. (2013). Fabrication of biocompatible and mechanically reinforced graphene oxide-chitosan nanocomposite films. *Chemistry Central Journal*, 7(1), 1–11.
104. Upadhyay, R., Naskar, S., Bhaskar, N., Bose, S., & Basu, B. (2016). Modulation of protein adsorption and cell proliferation on polyethylene immobilized graphene oxide reinforced HDPE bionanocomposites. *ACS Applied Materials & Interfaces*, 8(19), 11954–11968.
105. Lu, B., Li, T., Zhao, H., Li, X., Gao, C., Zhang, S., & Xie, E. (2012). Graphene-based composite materials beneficial to wound healing. *Nanoscale*, 4(9), 2978–2982.
106. Patel, M., Moon, H. J., Ko, D. Y., & Jeong, B. (2016). Composite system of graphene oxide and polypeptide thermogel as an injectable 3D scaffold for adipogenic differentiation of tonsil-derived mesenchymal stem cells. *ACS Applied Materials & Interfaces*, 8(8), 5160–5169.
107. Cardoso, M. J., Costa, R. R., & Mano, J. F. (2016). Marine origin polysaccharides in drug delivery systems. *Marine Drugs*, 14(2), 34.
108. Park, B. K., & Kim, M. M. (2010). Applications of chitin and its derivatives in biological medicine. *International Journal of Molecular Sciences*, 11(12), 5152–5164.

109. Ahmed, T. A., & Aljaeid, B. M. (2016). Preparation, characterization, and potential application of chitosan, chitosan derivatives, and chitosan metal nanoparticles in pharmaceutical drug delivery. *Drug Design, Development and Therapy*, 10, 483.
110. Lee, J., Cho, H. R., Cha, G. D., Seo, H., Lee, S., Park, C. K., ... & Kim, D. H. (2019). Flexible, sticky, and biodegradable wireless device for drug delivery to brain tumors. *Nature Communications*, 10(1), 1–9.
111. İşler, S. C., Demircan, S., Çakarer, S., Çebi, Z., Keskin, C., Soluk, M., & Yüzbaşıoğlu, E. (2010). Effects of folk medicinal plant extract Ankaferd Blood Stopper® on early bone healing. *Journal of Applied Oral Science*, 18, 409–414.
112. Huang, T., Song, X., Jing, J., Zhao, K., Shen, Y., Zhang, X., & Yue, B. (2018). Chitosan-DNA nanoparticles enhanced the immunogenicity of multivalent DNA vaccination on mice against Trueperella pyogenes infection. *Journal of Nanobiotechnology*, 16(1), 1–26.
113. Achar, A., Myers, R., & Ghosh, C. (2021). Drug Delivery Challenges in Brain Disorders across the Blood–Brain Barrier: Novel Methods and Future Considerations for Improved Therapy. *Biomedicines*, 9(12), 1834.
114. Meesaragandla, B., Karanth, S., Janke, U., & Delcea, M. (2020). Biopolymer-coated gold nanoparticles inhibit human insulin amyloid fibrillation. *Scientific Reports*, 10(1), 1–14.
115. Thakur, S., Chaudhary, J., Sharma, B., Verma, A., Tamulevicius, S., & Thakur, V. K. (2018). Sustainability of bioplastics: opportunities and challenges. *Current Opinion in Green and Sustainable Chemistry*, 13, 68–75.
116. Song, W., Su, X., Gregory, D. A., Li, W., Cai, Z., & Zhao, X. (2018). Magnetic alginate/chitosan nanoparticles for targeted delivery of curcumin into human breast cancer cells. *Nanomaterials*, 8(11), 907.
117. Hunt, N. C., Hallam, D., Chichagova, V., Steel, D. H., & Lako, M. (2018). The application of biomaterials to tissue engineering neural retina and retinal pigment epithelium. *Advanced Healthcare Materials*, 7(23), 1800226.
118. Szekalska, M., Puciłowska, A., Szymańska, E., Ciosek, P., & Winnicka, K. (2016). Alginate: current use and future perspectives in pharmaceutical and biomedical applications. *International Journal of Polymer Science*, 2016, 1–17.
119. Alhalmi, A., Alzubaidi, N., Altowairi, M., Almoiliqy, M., & Sharma, B. (2017). Xanthan gum; its biopharmaceutical applications: an overview. *World Journal of Pharmacy and Pharmaceutical Sciences*, 18, 7.
120. Song, Z., Zhang, Y., Shao, H., Ying, Y., Mei, L., Ma, X., Chen, L., Ling, P., & Liu, F. (2019). Effect of xanthan gum on the prevention of intra-abdominal adhesion in rats. *International Journal of Biological Macromolecules*, 126, 531–538.
121. Maia, J., Ribeiro, M. P., Ventura, C., Carvalho, R. A., Correia, I. J., & Gil, M. H. (2009). Ocular injectable formulation assessment for oxidized dextran-based hydrogels. *Acta Biomaterialia*, 5(6), 1948–1955.
122. Moscovici, M. (2015). Present and future medical applications of microbial exopolysaccharides. *Frontiers in Microbiology*, 6, 1012.
123. Hu, Y., Liu, H., Zhou, X., Pan, H., Wu, X., Abidi, N., Zhu, Y., & Wang, J. (2019). Surface engineering of spongy bacterial cellulose via constructing crossed groove/column micropattern by low-energy $CO_2$ laser photolithography toward scar-free wound healing. *Materials Science and Engineering: C*, 99, 333–343.
124. Moniri, M., Boroumand Moghaddam, A., Azizi, S., Abdul Rahim, R., Bin Ariff, A., Zuhainis Saad, W., Navaderi, M., & Mohamad, R. (2017). Production and status of bacterial cellulose in biomedical engineering. *Nanomaterials*, 7(9), 257.
125. Bukhari, S. N. A., Roswandi, N. L., Waqas, M., Habib, H., Hussain, F., Khan, S., ... & Hussain, Z. (2018). Hyaluronic acid, a promising skin rejuvenating biomedicine: a review of recent updates and pre-clinical and clinical investigations on cosmetic and nutricosmetic effects. *International Journal of Biological Macromolecules*, 120, 1682–1695.
126. Marengo, A., Forciniti, S., Dando, I., Dalla Pozza, E., Stella, B., Tsapis, N., Yagoubi, N., Fanelli, G., Fattal, E., Heeschen, C., & Palmieri, M. (2019). Pancreatic cancer stem cell proliferation is strongly inhibited by diethyldithiocarbamate-copper complex loaded into hyaluronic acid decorated liposomes. *Biochimica et Biophysica Acta (BBA)-General Subjects*, 1863(1), 61–72.

127. Nižić, L., Ugrina, I., Špoljarić, D., Saršon, V., Kučuk, M. S., Pepić, I., & Hafner, A. (2019). Innovative sprayable in situ gelling fluticasone suspension: development and optimization of nasal deposition. *International Journal of Pharmaceutics*, 563, 445–456.
128. Ren, Q., Liang, Z., Jiang, X., Gong, P., Zhou, L., Sun, Z., Xiang, J., Xu, Z., Peng, X., Li, S., & Li, W. (2019). Enzyme and pH dual-responsive hyaluronic acid nanoparticles mediated combination of photodynamic therapy and chemotherapy. *International Journal of Biological Macromolecules*, 130, 845–852.
129. Ishitani, K., Suzuki, S., & Suzuki, M. (1988). Antitumor activity of polygalactosamine isolated from Paecilomyces sp. I-1 strain. *Journal of Pharmacobio-Dynamics*, 11(1), 58–65.
130. Eliaz, N. (2019). Corrosion of metallic biomaterials: a review. *Materials*, 12(3), 407.
131. Poluri, K. M. (2019). Fabrication of biopolymer-based organs and tissues using 3D bioprinting. *3D Printing Technology in Nanomedicine*, 30, 43.
132. Arif, U., Haider, S., Haider, A., Khan, N., Alghyamah, A. A., Jamila, N., ... & Kang, I. K. (2019). Biocompatible polymers and their potential biomedical applications: a review. *Current Pharmaceutical Design*, 25(34), 3608–3619.
133. Kalogeris, T., Baines, C. P., Krenz, M., & Korthuis, R. J. (2012). Cell biology of ischemia/reperfusion injury. *International Review of Cell and Molecular Biology*, 298, 229–317.
134. Nathanael, A. J., & Oh, T. H. (2020). Biopolymer coatings for biomedical applications. *Polymers*, 12(12), 3061.
135. Sengupta, P., Ghosh, A., Bose, N., Mukherjee, S., Roy Chowdhury, A., & Datta, P. (2020). A comparative assessment of poly (vinylidene fluoride)/conducting polymer electrospun nanofiber membranes for biomedical applications. *Journal of Applied Polymer Science*, 137(37), 49115.
136. Suteewong, T., Wongpreecha, J., Polpanich, D., Jangpatarapongsa, K., Kaewsaneha, C., & Tangboriboonrat, P. (2019). PMMA particles coated with chitosan-silver nanoparticles as a dual antibacterial modifier for natural rubber latex films. *Colloids and Surfaces B: Biointerfaces*, 174, 544–552.
137. Teo, A. J., Mishra, A., Park, I., Kim, Y. J., Park, W. T., & Yoon, Y. J. (2016). Polymeric biomaterials for medical implants and devices. *ACS Biomaterials Science & Engineering*, 2(4), 454–472.
138. Rudy, A., Kuliasha, C., Uruena, J., Rex, J., Schulze, K. D., Stewart, D., ... & Perry, S. S. (2017). Lubricous hydrogel surface coatings on polydimethylsiloxane (PDMS). Tribology *Letters*, 65(1), 1–11.
139. Joseph, J., Patel, R. M., Wenham, A., & Smith, J. R. (2018). Biomedical applications of polyurethane materials and coatings. *Transactions of the IMF*, 96(3), 121–129.
140. Jummaat, F., Yahya, E. B., Khalil HPS, A., Adnan, A. S., Alqadhi, A. M., Abdullah, C. K., ... & Abdat, M. (2021). The role of biopolymer-based materials in obstetrics and gynecology applications: a review. *Polymers*, 13(4), 633.
141. Parvinroo, S., Eslami, M., Ebrahimi-Najafabadi, H., & Hesari, Z. (2021). Natural polymers for vaginal mucoadhesive delivery of vinegar, using design of experiment methods. *Vojnosanitetski pregled*, 79(4), 337–344.
142. Hietala, E. M. (2004). *Poly-L/D-lactide Stents as Intravascular Devices: An Experimental Study*. Presented with the permission of the Medical Faculty of the University of Helsinki, for public examination in Auditorium XII, Unioninkatu 34, University of Helsinki, on June 15, 2004, at 12 noon.
143. Sheikh, Z., Najeeb, S., Khurshid, Z., Verma, V., Rashid, H., & Glogauer, M. (2015). Biodegradable materials for bone repair and tissue engineering applications. *Materials*, 8(9), 5744–5794.
144. Chen, M. C., Tsai, H. W., Chang, Y., Lai, W. Y., Mi, F. L., Liu, C. T., Wong, H. S., & Sung, H. W. (2007). Rapidly self-expandable polymeric stents with a shape-memory property. *Biomacromolecules*, 8(9), 2774–2780.
145. Liu, S., Chen, C., Chen, L., Zhu, H., Zhang, C., & Wang, Y. (2015). Pseudopeptide polymer coating for improving biocompatibility and corrosion resistance of 316L stainless steel. *RSC Advances*, 5(119), 98456–98466.
146. Gnedenkov, S. V., Sinebryukhov, S. L., Mashtalyar, D. V., Egorkin, V. S., Sidorova, M. V., & Gnedenkov, A. S. (2014). Composite polymer-containing protective coatings on magnesium alloy MA8. *Corrosion Science*, 85, 52–59.

147. Miyashita, D., Chahud, F., Da Silva, G. E. B., De Albuquerque, V. B., Garcia, D. M., & e Cruz, A. A. V. (2013). Tissue ingrowth into perforated polymethylmethacrylate orbital implants: an experimental study. *Ophthalmic Plastic & Reconstructive Surgery*, 29(3), 160–163.
148. Franck, D., Gil, E.S., Adam, R.M., Kaplan, D.L., Chung, Y.G., Estrada Jr, C.R., & Mauney, J.R. (2013) Evaluation of silk biomaterials in combination with extracellular matrix coatings for bladder tissue engineering with primary and pluripotent cells. *PloS One* 8(2,: e56237.

# 4 pH and Thermo-responsive Systems

*Shilpa Borehalli Mayegowda, Kempahanumakkagari Sureshkumar, Sahana M., and Thippeswamy Ramakrishnappa*

**CONTENTS**



## 4.1 INTRODUCTION

Smart materials that have potential to respond to external stimuli have been proved to be the best candidates as drug carriers in designing drug-delivery systems (DDS). Nanoparticle-based external stimuli materials are extensively used in drug-delivery systems as a result of their fewer side effects, especially in diseases like cancer. The nanoparticles are used as carriers of therapeutics classified as nano-pharmaceuticals, which deliver the therapeutics to the target sites, for example, tumour sites in tissues [1]. The golden goal of a DDS system with engineered and sustainable nanomaterials as carriers is complete control over therapeutics release [2]. The nanoparticle-based smart materials are potential candidates in drug-delivery system engineering with external stimuli composed of smart engineered polymers. The nano-liposomal doxorubicin, a cancer therapeutic, finds numerous advantages over traditional drugs with respect to side effects like hair loss and damage to the heart and other organs of cancer patients [3]. The stimuli systems are broadly classified into chemical (internal) and physical (external) systems. The internal/chemical regulated stimuli systems include pH, ionic environment, redox, and enzyme-responsive systems [3]. The external/physical regulated stimuli systems include light-, thermo-, ultrasound-, electro- and magnetic-responsive systems [4–7]. pH and temperature have been proved to be the most significant external stimuli factors in DDS research recently. This is because temperature at the tumour site is higher than the average body temperature and pH is more acidic (~ 5.8) at the tumour site compared to normal cells [3].

DOI: 10.1201/9781003240884-4

Nanotechnology has an important role in the field of drug delivery, predominantly due to vital restrictions and problems that affect drug-delivery systems in old formulations and conventional pharmaceutical agents. The major drawback of conventional DDSs is the difficulty in taking out residual parts of such systems, therefore avoiding placing nonbiodegradable material within the patient's body that might be poisonous is advantageous. In most cases these conventional DDSs have a high elevated burst in drug release suddenly after the drug is administered, and the drug solubility in these conventional DDSs is very low. Therefore, the promising solution is nanopharmaceuticals. As compared with the conventional drug-delivery system, drug administration with the use of nanoparticles (NPs) has many advantages, which include the following:

- Nanoparticles are smaller than those of the basic material unit of the conventional formulated drugs. The formation of nanodrugs is by the attachment of small molecules of therapeutic agents to small nanocarriers.
- One strategy to deliver pharmaceutical agents is the nanoformulation of drugs that more accurately reach the targeted tissue and lowering of toxic side effects and overall dosage.
- The EPR (enhanced permeability and retention) effect allows the passive targeting of nano-sized drugs and accumulation at the specific sites such as pathological sites and malignant tumours.
- Nano-sized formulations lead to an increased active concentration and bioavailability compared to conventional micro-sized formulations.
- Nanoparticles have good efficacy and safety.
- Comparatively, nanodrugs are cheaper than conventional therapies.
- Drug release can occur at a constant rate over the desired timescale.

The nanopolymeric hydrogels are very sensitive to external stimuli, have high biocompatibility, and possess high stability/durability, mechanical efficiency, load-bearing capabilities and good adsorption capabilities. The above-mentioned interesting factors of smart nanopolymeric materials have enabled their use in gene vector design and drugs, toxicology, immunology and biology of the host response tissue engineering, gene delivery, and self-assembly at the nanoscale. It has been revealed that scaffolds impact the potential to proliferate, bind and differentiate cells similar to autologous tissues. Stimuli-responsive nanoparticles have the ability to enhance the delivery of therapeutics to a particular cell or region inside the body. On the other hand, there are small differences in tumour physiology that can be used to stimulate the response by using a smart polymer system, therefore enabling the design of therapeutic strategies with tumour specificity.

In aqueous medium at a particular temperature thermo-responsive polymeric materials show two critical phase transitions. The first is at the LCST (lower critical solution temperature) in which polymer is non-miscible at temperatures higher than LCST in aqueous media and the second is at the UCST (upper critical solution temperature) in which polymer is non-miscible at temperatures lower than UCST in aqueous media [8]. The dissolution of polymeric materials in aqueous media will change after heating because of a change in hydration and dehydration [9]. The temperature should be adjusted such that it is higher than the LCST compared to normal body temperature (>37°C) while using thermo-responsive nanocarriers used for cancer therapy for obtaining controlled DDS [10]. The temperature variation in vertebrates while fighting infection is through an elevation of between 1 and 5°C [11]. In order to adjust the temperature of thermo-responsive polymer nanocarrier systems, there should be more hydrophilic groups, resulting in an increased LCST [12]. Another interesting type of therapeutic carrier is nanogel, which has been used for multimode tumour therapy [13]. Recently, biomaterial-based injectable nanogels as carriers have been reported for various medical applications [14].

This chapter enumerates recently reported pH- and thermo-responsive materials as therapeutic carriers for multi-mode tumour therapy. Furthermore, synthetic strategies of these materials and their modes of action as well as their end applications in DDS are discussed briefly. Finally, the future perspective of pH- and thermo-responsive materials with respect to DDS is discussed.

## 4.2 DESIGN OF PH-RESPONSIVE BIOMATERIALS

The pH system of response makes use of specific variations in pH in order to induce localized drug delivery to various sites of the body such as the gastrointestinal tract, tumour endosomal or lysosomal compartments of cells, as well as the tumour microenvironment [15]. Examples include cancer cells, which obtain their primary source of energy from aerobic glycolysis regardless of the oxygen concentration available. This, in turn, leads to lactate accumulation in the tumour microenvironment, thereby lowering the pH in the extracellular region of the matrix. This effect is often described as the "Warburg effect" [16].

This meets the ceaseless demand for energy of a rapidly dividing cancer cell and also supplies vital precursors for the biosynthesis of various macromolecules. Responsive vectors of pH are specifically created by making use of polymers having weak acids or weak bases which are ionizable to utilize the acidic microenvironment for concise delivery of drug into tumours. The components are dependent on protonation and deprotonation to allow selective solubility in an aqueous medium. Some examples of weak acids and their derivatives are acrylic acid, methacrylic acid, maleic anhydride, 2-(methacryloyloxy)ethyl dihydrogen phosphate and N,N-dimethylaminoethyl methacrylate, whereas poly(amidoamine) (PAA or PAMAM) is an example of a polymeric weak base which is used exclusively for designing responsive vectors of pH delivery. Poly(β-amino ester), being a polymer, has high pH solubility, similarly PbAE is utilized in designing responsive vectors of pH system delivery [17].

An all-inclusive, in vitro and in vivo study using poly(ethylene oxide) PbAE (PEO-PbAE), which is a pH-responsive copolymer system, displayed a maximum level of apoptosis in MDA-MB-231 cancer cells of breast and actively clustered into the SKOV3 ovarian cancer of human xenograft model [18,19]. Various responsive vectors of pH are efficient deliverers of biologics, including gene, mi-RNA, si-RNA, proteins and peptides, showing the flexibility of system delivery [20,21]. This fusion of various quantities of polymers which are sensitive to pH, C12H25 PAA (poly acrylic acid, having a hydro-phobic group on its end) with 1-α-phosphatidyl choline, phospholipid bilayer (hydrogenated) in regard to its biological imitation and functionality is the intent. A pH sensitive poly-molecule called PAA regulates groups of –COOH and these get protonated being under an acidic pH of 4.2. A cyclo-dextrin-based pH-responsive hybrid nanosystem plays the role of a vector (nonviral) in the delivery of genes [22].

Out of the stimulus-responsive systems, pH-sensitive nanocarriers, has gained huge interest for two reasons.

1. The nanocarriers are incorporated into endosomes via endocytosis. This builds markedly acidified lumens (pH 4.5–5.5) primarily through the activity of V-type $H^+$ ATPase [23]. At low pH, by triggering drug release from nanocarriers, it provides a mechanism for the drug to escape from the endosomal compartment.
2. The acidosis in tumour tissue might be utilizable for specific targeting of tumours comparable to normal tissues, extending and complementing the selectivity achievable by the enhanced permeability and retention (EPR) effect. It is known from the expert studies of Warburg [24] that tumours have a tendency for glycolytic metabolism of glucose to lactate, which contributes to acidosis. Several studies with pH microelectrodes have exhibited lower pH values in tumours relative to normal tissues [25,26]. However, for tumour targeting it is extremely important to realize which compartments and regions in tumours have lower pH values than normal tissues – an issue that is not taken care of in the nanocarrier literature [23].

A nanovector is constructed by assimilating a cyclo-dextrin pH-responsive substance with a low molecular weight of polyethyleneimine [24]. An antisense oligo-nucleotide such as Bcl-xl was abridged with the hybrid nanosystem having maximum loading organization by a nanoemulsion procedure. This evolved pH-responsive ASON nano-therapeutic can be effectively transduced in human adenocarcinoma lung cells with precise time and dosage coverage following constructive inhibition of cell growth, appreciable repression of Bcl-xl mRNA or protein expression and efficacious apoptosis of cells [22]. Notably, the recently developed nanovector displayed exceptionally maximum potency and minimum cytotoxicity when juxtaposed by a PLGA-based counterpart often utilized cation vectors, namely Lipofectamine 2000 and a branched PEI (25,000 Da). This pH-responsive hybrid nanosystem could be used as a secure and potent non-viral vector which can be applied in gene therapy studies [25,26].

In recent years, nanoparticle-based pH-responsive drug-delivery systems have become more efficient and the most studied among other DDS. This may be due to their efficacy over trivial systems in terms of target-based delivery and other related factors found in DDS [27]. Among the pH-responsive nanoparticle-based DDS systems, first-generation nanoparticles have been proved for their drug tolerability and circulation half-life as pharmaceutical carriers. In this regard, researchers tend to develop second-generation nanoparticles for DDS in order to exhibit sustained release, targeting the specific molecules and able to respond to environmental factors [27]. Organs, tissues and cellular compartments have different pathophysiological conditions, which results in different pH gradients in these areas (Figure 4.1). When pH-responsive nanoparticle-based DDSs are exposed to these conditions, the pH-responsive nanoparticles induce physicochemical changes such as the structural and surface characteristics of DDSs. These structural and surface characteristics include swelling, dissociation of the structures and reversing the surface charge of the material [27]. The above-mentioned change triggered by the different pH environment of the tumour site made it possible to release drugs in that specific area in preference to surrounding healthy tissues. The pH-triggered nanoparticle-based drug-delivery systems can be differentiated into organ, tissue and subcellular levels [27]. At each level the drug administration paths will be different and different possible available paths include oral drug delivery, delivering at specific tumour sites and finally intracellular delivery. For organ-level drug delivery, oral-based nanoparticle DDSs are formulated in order to achieve differential drug delivery along the gastrointestinal tract [27]. Meanwhile, for tissue-level drug delivery, the nanoparticle-based DDS systems are designed such that they can distinguish the pH gradient of tumour sites over surrounding healthy sites and accordingly tumour-specific drug delivery is achieved [28]. Finally, the intracellular DDS systems are designed such that the nanoparticle DDS systems are protected in the high acidic environments of the endolysosomal compartments and allow release of drug in the cytoplasm of cells [29]. The gastrointestinal tract has different pH gradients at different parts, as illustrated in Figure 4.1. The low cost, patient comfort and convenience make the oral mode of drug administration very popular. However, it suffers a great

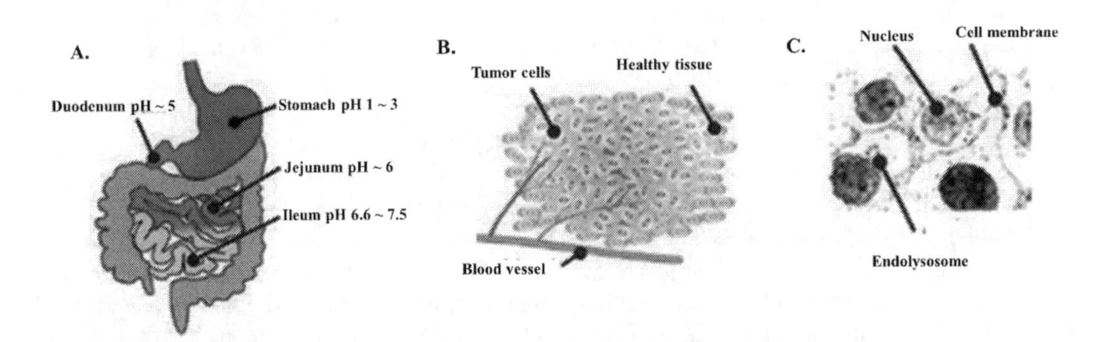


**FIGURE 4.1** Demonstration of different pH conditions at (A) organ; (B) tissue; and (C) cell levels [27].

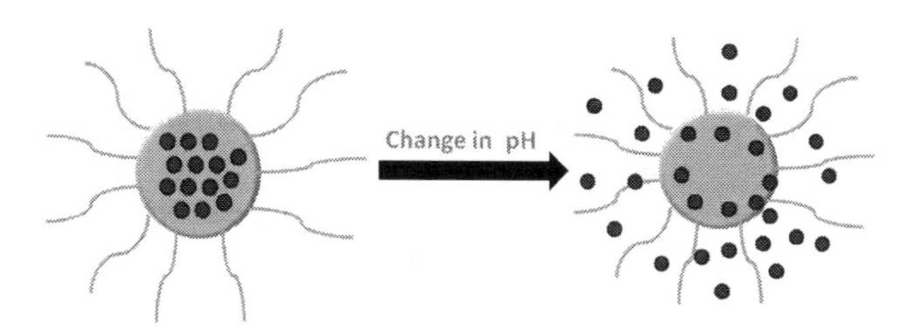


**FIGURE 4.2** A schematic representation of drug release by the nanoparticle-based pH-sensitive DDS system.

deal because of the degradation of drugs in the drastic pH conditions of the gastrointestinal tract as well as by enzymatic degradation. In contrast, the pH-responsive DDS based on surface-functionalized nanoparticles exhibited good drug-delivery efficiency in terms of gastric retention, transepithelial transport and cellular targeting [30]. Organ-specific drug delivery can be achieved through formulating DDS with pH-dependent swellable nanoparticles (Figure 4.2) [27]. The above-described DDS formulation includes use of acrylic acid-based polymers like poly(methacrylic acid) (PMMA) along with the nanoparticles [31]. In this DDS, the nanoparticles are in a hydrophobic collapsible state in the stomach due to protonation of acid groups. However, after passage of nanoparticle-based DDS from the gastric phase, the swelling of nanoparticles increases due to deprotonation of the carboxylic groups due to higher pH as well as breakdown of hydrogen bonds [31]. The pH-triggered swelling of nanoparticles was due to the presence of PMMA–poly (ethylene glycol) (PEG) diblock polymers. These polymers can achieve swelling ratios of up to 40–90-fold (mass of swollen polymer/mass of swollen polymer). The swelling rate of this PMMA–PEG system depends on the copolymer composition along with the PEG chain length [31]. When insulin-releasing studies were conducted with DDS based on the above-mentioned polymer nanoparticle systems, ~90% of insulin was delivered at neutral pH (7.4) conditions in a nanoparticle swollen state, whereas only about 10% insulin was delivered at acidic pH (1.4) conditions in a nanoparticle collapsed state [31]. The other copolymers of PMAA like polymethacrylate (PMAA-PEA) and polymethacrylate (PMAA-PAA) have been found to undergo dissolution with pH as an external stimuli. These polymer composites can be tailored such that they can undergo dissolution in the different pH conditions of the intestine [32]. Eduragit L100-55 is the commercial formulation of PMAA-PEA which is suitable to release drugs to duodenum regions of pH greater than 5.5. This is because the above-mentioned formulation undergoes dissolution at pH >5.5. The commercial formulation of PMAA-PAA is Eduragit L100, which is suitable to release a drug to the ileum region of the intestine with a pH above 7.5. This is because Eduragit L100 will undergo dissolution at pH >7 [33]. Some of the nanoparticles are engineered such that their surface charge will be reversed after passage from the acidic gastric part to the more alkaline intestinal portion [34]. This unique property of these nanoparticles is useful in delivering the drug to the alkaline intestinal part using the surface charge reversal behavioural nanoparticles. The mesoporous $TiO_2$ surface was used for functionalizing the nanoparticles with positive charges to various degrees [34]. The positive charges were induced to nanoparticles by treating with trimethylammonium salt. The positive surface charge on nanoparticles with trimethyl ammonium species helps in collecting anionic drug (sulfasalazine) in acidic environments [34]. The sulfasalazine drug has been found to be an anti-inflammatory prodrug for bowel disease. When the above-described drug-loaded nanoparticles are placed in physiological pH (7.4), the nanoparticles acquire a partial negative charge due to deprotonation of silanol groups of $TiO_2$. The electrostatic repulsive forces between the nanoparticles and surroundings triggered the sustained release of drug molecules from the nanoparticles [34]. The pH-responsive nanomaterials will specifically deliver the drug at tumour sites because of the pH gradient between tumour sites and healthy surrounding areas. Heparin–chitosan nanoparticles were formulated in order to cure *Helicobacter pylori* infections. This was based on the pH gradient between the overall acid environment (~2) of the stomach and the comparatively higher pH conditions (~7.5) of the mucus layer and epithelium of the gastric lumen [35]. The self-assembled nanoparticles of size 130–300 nm were obtained by mixing the heparin and chitosan at pH 1.2–2.5. The above nanoparticle formulation was quite stable in the gastric lumen with an acidic pH of 1.2 due to electrostatic interactions inside the structure. When nanoparticles get in contact with *Helicobacter pylori* in the gastric epithelium of pH 7.4, the deprotonation of chitosan starts and electrostatic interactions inside the structures weaken, resulting in the collapse of nanoparticles which releases the heparin [35]. A similar pH gradient-induced release of insulin was achieved by formulation of multi-ion cross-linked nanoparticles. The multi-ion cross-linked nanoparticles were formulated by using chitosan poly-γ-glutamic acid, tripolyphosphate and $MgSO_4$ [36]. The above nanoparticle formulation was encapsulated with insulin at pH<6 and the insulin was released by the same nanoparticle encapsulation at higher pH conditions compared to their encapsulated pH. This is because of the deprotonation of chitosan which resulted in weakening of electrostatic interactions within the structure at these pH conditions [36]. The nanoparticles were surface modified with molecules like lectin, small peptides and vitamins [37–39]. The surface functionalization of nanoparticles with these compounds helps in retention in the gastrointestinal tract. This is because of the stability of the surface-modified species in strong acidic conditions. The chitosan which exhibits transcytosis in the intestinal epithelium was used in formulation of the PMAA-PEG nanoparticles [40]. The vitamin B12 enhances apical to basal transport of nanoparticles in Coca-2 cells [41]. These were surface functionalized on to dextran nanoparticles and used for insulin delivery in vivo [42]. The RGD peptides were used to target the β1 integrins expressed on the apical side of M cells in vivo as well as in vitro. In order to achieve the targeted nanoparticle delivery to M cells and follicle-associated epithelium of the intestinal tract the novel peptides were identified through in vivo phage display [30].

The nanoparticles described in the above section exhibited excellent drug-delivery properties, and improved cell membrane permeability. These nanoparticles were modified with some ligands to achieve targeted delivery. Therefore, nanoparticles have been proved to be excellent delivery vehicles that can have differential retention and pass through the gastrointestinal tract. In summary, the nanoparticle-based DDS formulations discussed in the above section showed good efficacy towards nano-medicine administered orally.

Next, we focus on nanoparticle formulation for controlled drug delivery at tissue levels. Human tumour cells are associated with acidic microenvironments in the range from pH 5.7 to 7.8 [43]. One of the reasons among many for these acidic microenvironments of the tumour cells is the accumulation of lactic acid. This lactic acid accumulation occurs in rapidly multiplying tumour cells due to their increased uptake of glucose, along with the very slow rates of oxidative phosphorylation [44]. The Warburg effect, which is nothing but the production of more lactic acid in the presence of oxygen, helps in the growth of tumour cells in vivo [45]. Along with higher lactate production, other reasons for acidic microenvironments of tumour cells are poor blood supply as well as poor lymphatic drainage [46]. Researchers have exploited the acidic tumour microenvironments extensively in order to achieve targeted drug delivery/localized drug delivery in contrast to the traditional untargeted DDS which results in cell toxicity along with a greater decrease in the efficacy of the medicine [47]. The pH-dependent drug-delivery systems based on nanoparticle-modified polymers exhibit changed physical and chemical properties in tumour environments such as swelling and an increase in their solubility rates. The driving forces for these changes to nanoparticle-modified polymers in tumour microenvironments is the differential compared to the tumour soundings. Griset et al. reported acralyte–nanoparticle composites which can transform from hydrophobic to hydrophilic when exposed to acidic tumour microenvironments [48]. The formulation is based on cross-linking the nanoparticles with acralyte-hydrophobic polymers with the hydroxyl groups and later masking with labile protecting groups (2,4,6-trimethoxybenzaldehyde) towards the pH. The above-mentioned nano-formulations were stable in neutral environments. However, when exposed to slightly acidic environments (pH 5) the protecting groups are cleaved, resulting in the exposure of hydroxyl groups. This induces the switching from hydrophobic to hydrophilic nature for nanoparticle composites, which leads to swelling of the nanoparticles followed by drug release [48]. Only about <10% paclitaxel was released at basic pH conditions (7.4), whereas nearly complete release of the same was obtained in slightly acidic conditions (pH 5.0) within 24 hours. The above-discussed nanoparticle formulation, i.e. acralyte-based polymer-modified nanoparticles, inhibited growth of LLC tumour cells in C57B1/6 mice compared to non-pH-responsive nanoparticles along with free suspended paclitaxel in the solution [48]. The above interesting results of the polymer-modified nanoparticles towards targeted drug release with differential pH of tumour and healthy surrounding environment as a driving force proved to be efficient materials. The pH triggered switching from the hydrophobic to the hydrophilic state by nanoparticle–polymer composites was also used for controlled dissolution of the polymer matrix, which leads to the release of drug present inside the polymer. The nanoparticle polymer composite is PEG-poly(β-amino ester). Furthermore, pH-responsive drug delivery can be obtained by conjugating the drug molecules with the polymers via pH-labile cross-linkers. The release of cisplatin was achieved in pH<6 conditions by conjugating it to polymer nanoparticles via a hydrazone cross-linker [49]. This was due to hydrolysis of hydrazone at pH<6 conditions against polylactic acid degradation. The cisplatin-conjugated polymer with nanoparticles via hydrozone cross-linker exhibited enhanced cellular uptake and toxicity compared to free cisplatin studied in vitro [49]. Some of the acid-labile cross-linkers used for conjugating drug molecules with polymer nanoparticles and their hydrolytic products are summarized in Table 4.1. pH-responsive nanoparticle-based drug delivery is designed for specific intracellular drug delivery as well. After endocytosis, there is a rapid decrease in the endosomal pH (pH<6) due to vacuolar protein ATPase-induced protein influx [50]. These highly acidic environments of endosomes are not safe for therapeutic molecules, especially for macromolecules such as DNA, small RNA and proteins. The acidic environment of endosomes can be used as a pH trigger for drug delivery by endosomal escape followed by drug delivery. This is via the mechanism called the sponge effect [51]. This involves absorption of protons in the acidic environment of endosomes resulting in increased osmotic pressure of the endosomal compartments followed by disruption of the plasma membrane, which results in the release of nanoparticles into the cytoplasm [51]. Intracellular drug delivery was achieved through the pH-sensitive polymers grafted with other functional units which buffer the endosomal pH [50,52]. These kinds of nanoparticle platforms are described as dynamic polyconjugates. The above platforms consist of an amphipathic endosomolytic poly(vinyl ether) backbone with butyl and amino vinyl ethers. The nanoparticles are conjugated with siRNA and used to deliver through reversible disulphide linkage and additionally with functional units like PEG and targeting ligands [50,52]. The dynamic polyconjugates were used for delivering two endogenous liver genes, apolipoprotein B and peroxisome proliferator, activated receptor alpha (PPARα) in vivo [50,52].

## 4.3 DESIGN OF THERMO-RESPONSIVE BIOMATERIALS

At higher than or less than a specific temperature, a change of phase is experienced by thermo-responsive materials. Th upper and lower critical solution temperatures are the changes they go through. The materials are indissoluble below or above the critical temperature and turn fully dissoluble after exceeding the transition temperature.
The major two thermo-responsive types are:

1. As the poly(N-isopropylacrylamide) (PNIPAAm) is nearer to body temperature 37°C, it has a lower critical solution temperature, i.e., ~32°C.
2. Giving the all-round hydrophilicity of the polymer being lower or higher, copolymerization with hydrophilic or hydrophobic monomers results in refinement of the LCST of PNIPAAm.

**TABLE 4.1**
**pH-labile Substrates Used for DDS Formulations and Corresponding Hydrolytic Products after Reaction**

| Sl. No. | Reactive Group of Cross-linker | Structure of Cross-linker and Related Hydrolytic Products | References |
|---|---|---|---|
| 1 | Ester | 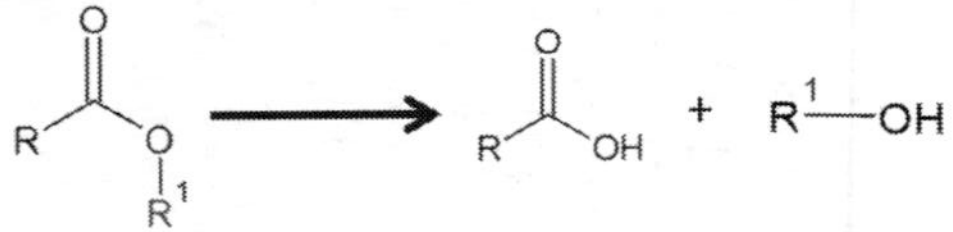 | [95,96] |
| 2 | Hydrazone | 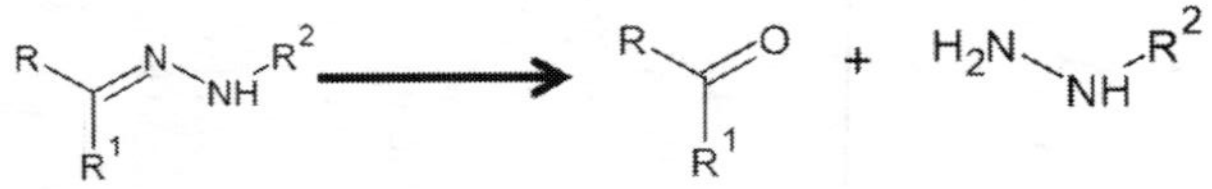 | [49,97] |
| 3 | Anhydride | 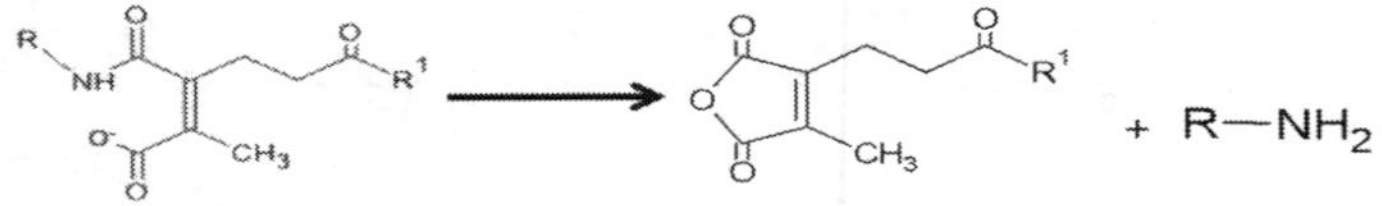 | [52] |
| 4 | Anhydride | 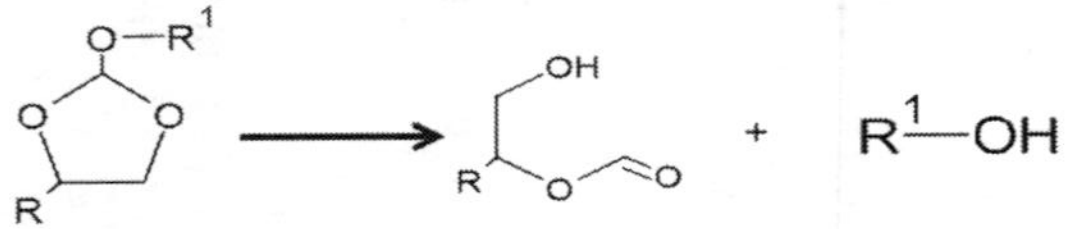 | [52] |
| 5 | Orthoester | 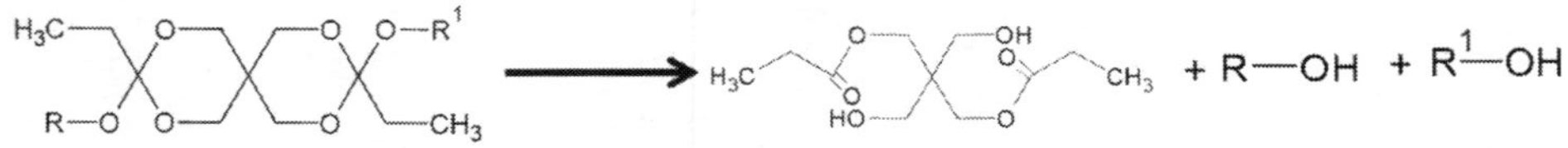 | [98] |
| 6 | Imine |  | [100] |
| 7 | β-Thiopropionate |  | [101] |
| 8 | Vinyl ether |  | [102] |
| 9 | Phosphoramidate | 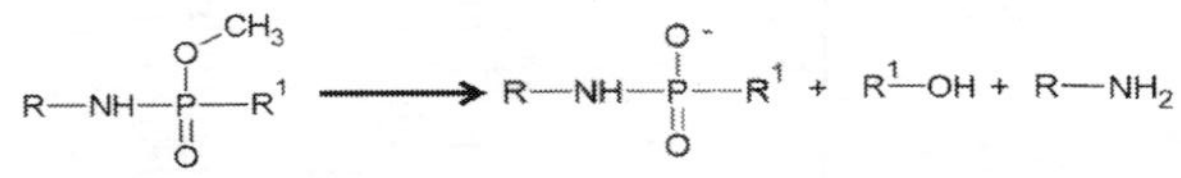 | [103] |

3. Polymers of poly[2-(dimethylamino)ethyl methacrylate] (PDMAEMA) along with 50°C of LCST poly(N-vinylcaprolactam) (PVCL) with between 25°C and 35°C of LCST
   - Poly(ethylene glycol) (PEG) of 85°C of LCST.
   - N, N-diethyl acrylamide) (PDEAAm) with the range of 25–30°C of LCST.
4. The LCST is influenced by the side group size particularly.
5. Lower critical solution temperature LCST(51) that depends on the length of the EO chain is presented by PEG methacrylate polymers (PEGMA) that have a side PEG chain of 2–10 ethylene oxide units (EO) $< 10$.
6. Architecture and molecular weight influence the LCST of a polymer.

The characteristic features of upper critical solution temperature (UCST) are described below.
The polymers in tissue are often applied in two conditions. Firstly, cell multiplication and development that are enabled by the substrates and, secondly, for the in situ scaffold they are used as injectable gels.

- The cell attachment and detachment from a surface are due to the regulation of the thermo-responsive ability of the polymers.
- The application includes cell encapsulation in 3D structures in the body [53]. For flaws of any shape, the in vivo formation of a structure enables delivery of growth factor, nutrients and encapsulated cells using slightly protruding methods in comparison with the in situ formation of the scaffold/cell. The thermo-responsive polymers are induced inside the body after mixing at room temperature (RT) along with polymers such as 2011 and 1220 cells. The polymer moulds into a physical gel at room temperature, when it is injected into the body it rises to 37°C which is higher than the polymer's LCST. Hence the cells are encapsulated in the gel's 3D structure. In the following segments that classify the polymer 3D structure, many researchers specify thermo-responsive polymers in tissue engineering [53].

Biomedical applications have high demands and are advantageous towards bioinspired thermo-responsive polymeric materials that are associated with tuneable phase's transition behaviours. The concoction of the two forms of thermo-responsive polypeptoids associated with the tuneable phase transition temperature of between 29–55°C, which are the UCST and LCST have been outlined as it is a simplified approach. Under appropriately moderate conditions, introducing ethylene glycol (EG) and alkyl groups occurs to result in a curbed phase transition behaviour. It is noted that by just tuning pH and the alkyl chain length an acute transition ($\Delta T \leq 1.5$ °C) can be brought about. A depictable UCST behaviour that can be controlled and regulated in both methanol and water is displayed by the carboxyl-containing polypeptides in particular. For practical implementations, these characteristics make the procured polymers more advantageous and favourable. Although there is an excellent capacity to control and adjust the LCST behaviour of the order, we illustrate the aspect that the hydrophilic EG group acts as a remarkable regulator to regulate the UCST behaviour, expecting that the systematic order-property studies will fulfil the particular requirements of future applications by allowing the design of smart polymer materials to do so [54].

## 4.4 RESPONSIVE MATERIALS IN DRUG-DELIVERY SYSTEMS

The cutting-edge advancement in materials science revolutionized the drug-delivery systems, particularly in cancer treatment, so that physicochemical parameters of drug-delivery systems can be optimized in order to advance intelligent or smart systems that can deliver the medicine to the targeted sites specifically. Understanding the clear microenvironment of tumour sites along with the differences between the bio-environment of the tumour and healthy sites helps in developing intelligent drug-delivery materials. Cancer is associated with a complex cellular environment with diverse cell types, extracellular matrix, signalling molecules, immune-related factors, etc. The listed complex chemical environment with the complex physiology of cancer disease makes it difficult to develop efficient drug molecules. In addition, acquired multidrug resistance to developed chemotherapeutics further complicates cures, resulting in poor clinical outcomes. The tumour site in the cell is associated with a low pH gradient, high protease activity, hypoxia, no lymphatic drainage, and a redox environment [55]. Along with cancerous cells, tumour sites also have stromal cells, immune cells, endothelial cells, and cancer stem cells (CSCs) [56]. The heterogeneous environment of tumour site increases the tumour survival chances, along with promoting the growth of cancer cells and also helps in spreading the cancerous cells to the surrounding healthy cells at a faster rate. However, the distinct tumour physiology enables the development of stimuli-responsive intelligent/smart materials in order to develop a drug-delivery system with extreme specificity to tumour sites.

Biomaterials have been the topic of several studies to pursue potential therapeutic interferences for a broad variety of diseases and disorders. The chemical and physical properties of numerous materials have been discovered to develop semi-synthetic, synthetic, or natural materials with dissimilar advantages for practice as drug-delivery systems for non-central nervous system (non-CNS) and central nervous system (CNS) diseases. In this analysis, an entire overview of well-known biomaterials as drug-delivery systems for neurodegenerative diseases is provided, equalizing the challenges and potential connected with CNS drug delivery. Being a suitable drug-delivery system, favoured properties of biomaterials are discussed, focusing on the persistent encounters, for instance, stimuli responsiveness, targeted drug delivery, and controlled drug release in vivo. Finally, we discuss the views and boundaries of incorporating extracellular vesicles (EVs) as a drug-delivery system and their usage for stable, biocompatible and targeted delivery with circumscribed immunogenicity, along with their capacity to be carried through a non-invasive approach for the treatment of neurodegenerative diseases [57].

### 4.4.1 Internal Regulated Systems

The internal regulated systems are also described as self-regulated or closed-loop systems triggered by internal factors such as pH, the presence of protease, redox environment, or other internal factors which stimulate and regulate drug delivery. The difference in the environment at the tumour site compared to normal areas changes the physical/chemical environment of DDS, which results in drug release. The degree of the drug-releasing property of DDS is entirely dependent upon the degree of internal stimuli at the tumour site. However, the degree of drug release of DDS cannot be regulated by any external factors. The DDSs have taken advantage of diverse pH conditions of the body such as the gastrointestinal tract, tumour site environment and lysosomal or endosomal compartments of cells (Figure 4.1).

Aerobic glycolysis occurs in cancerous cells as their primary energy source is independent of oxygen. This results in a dramatic increase in lactate concentrations at the microenvironment of the tumour site resulting in lowering of the pH of the extracellular matrix. The above condition is referred to as the "Warburg effect". This enables it to supply the high energy needs of the rapidly dividing cancerous cells along with the necessary precursor molecules in order to build essential macromolecules [58]. The pH-responsive DDSs for cancer therapy materials are generally polymer molecules with easily ionizable weak bases which will recognize the acidic environment of cancerous tumour cells and can deliver the drug at these sites specifically. These polymer materials containing weak bases will undergo deprotonation specifically at acidic sites depending upon their preferential solubility in aqueous media. Some of these polymer materials are discussed in Section 4.2. The pH-responsive materials-based delivery systems have also been demonstrated for delivering many biologics such as mi-RNA, si-RNA, proteins, peptides and genes [27,59]. The type of internally regulated drug-delivery system is based on redox-responsive materials. There will be redox potential between the intracellular and extracellular environments of tumour sites

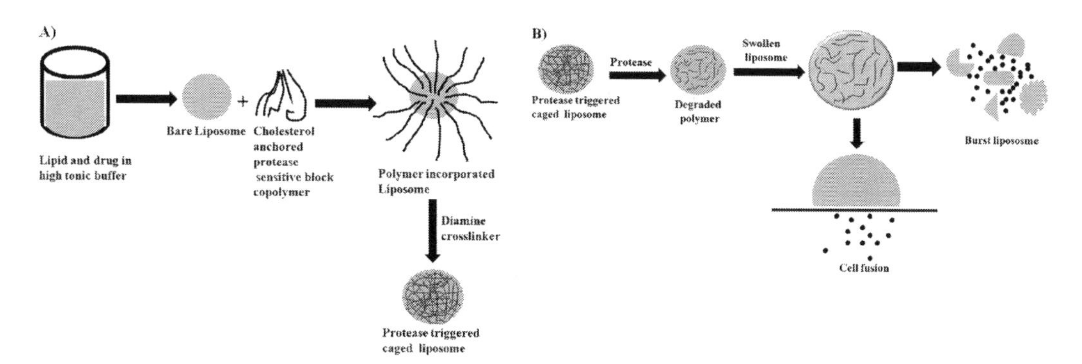


**FIGURE 4.3** (A) Demonstrating the steps involved in the synthesis of uPA-sensitive polymer caged liposomes. (B) Enzymatic action of polymer-caged liposomes [60].

because of the reducing nature of the intracellular and oxidizing nature of extracellular environments. For the development of redox-responsive materials-based delivering systems the redox potential will be the driving force. The structural configurations of the redox-based vectors will be changed inside the cytoplasm and nucleus due to their excess concentrations (2–10 mM) of tripeptide glutathione in comparison with the extracellular matrix (2–20 μM). Based on the disulphide bond, redox-responsive polymers were extensively studied for protein-based delivery systems. The third type of stimuli-responsive system is the enzyme-responsive vectors. Proteases play a vital role in tumour physiology. The cancer-associated proteases include urokinase plasminogen activators, cathepsin, and matrix metalloproteinases (MMPs). These proteases play an important role in cancer tumour cells, such as invasion, dissemination and disease progression. The matrix metalloproteinases are known to be overexpressed in most cancer types and these proteases are responsible for tumour progression as well as invasiveness. The enzyme-responsive system is made-up of enzyme-specific peptides which are triggered to deliver when substrate is broken down by enzymatic activity inside the tumour cells. Recently, enzyme-responsive DDS, based on cholesterol-anchored graft polymer containing peptide GSGRSAGK and acrylic acid was integrated in liposomes made of cholesterol, 1,2-dipalmitoyl-sn-glycero-3-phosphocholine and 1,2-dioleoyl-sn-glycero-3-phosphocholine (Figure 4.3A) [60]. These liposomes that had cross-linked polymers showed good stability along with the resistance to swelling or osmotic leaking. In the presence of enzymes the cross-linking was rapidly degraded, which caused the swelling of liposomes by drug release (Figure 4.3B) [60].

### 4.4.2 External Regulated Systems

The externally regulated DDS systems can be termed as open-loop systems where the drug delivery is induced by an external stimulus which is from outside the body. The strength and duration can be controlled by external stimuli, so that drug release by DDS made up of external stimuli vectors can have control, and can be released in a controlled manner at tumour sites precisely. Different external stimuli-responsive vectors reported in DDS are heat (thermal), light, sound (ultrasound), magnetic and electric based. Thermo- or heat-responsive materials undergo phase change at particular temperatures referred to as upper or lower critical solution temperatures. These heat-sensitive materials by crossing the transition temperature are completely soluble but they are insoluble below or above the transition temperature. More discussion regarding these materials has been done in Section 4.3. Besides the use of thermo-sensitive materials in DDS, efforts are in progress to use these materials as scaffolds and surfaces for tissue culture and imaging, engineering and diagnostics [61].

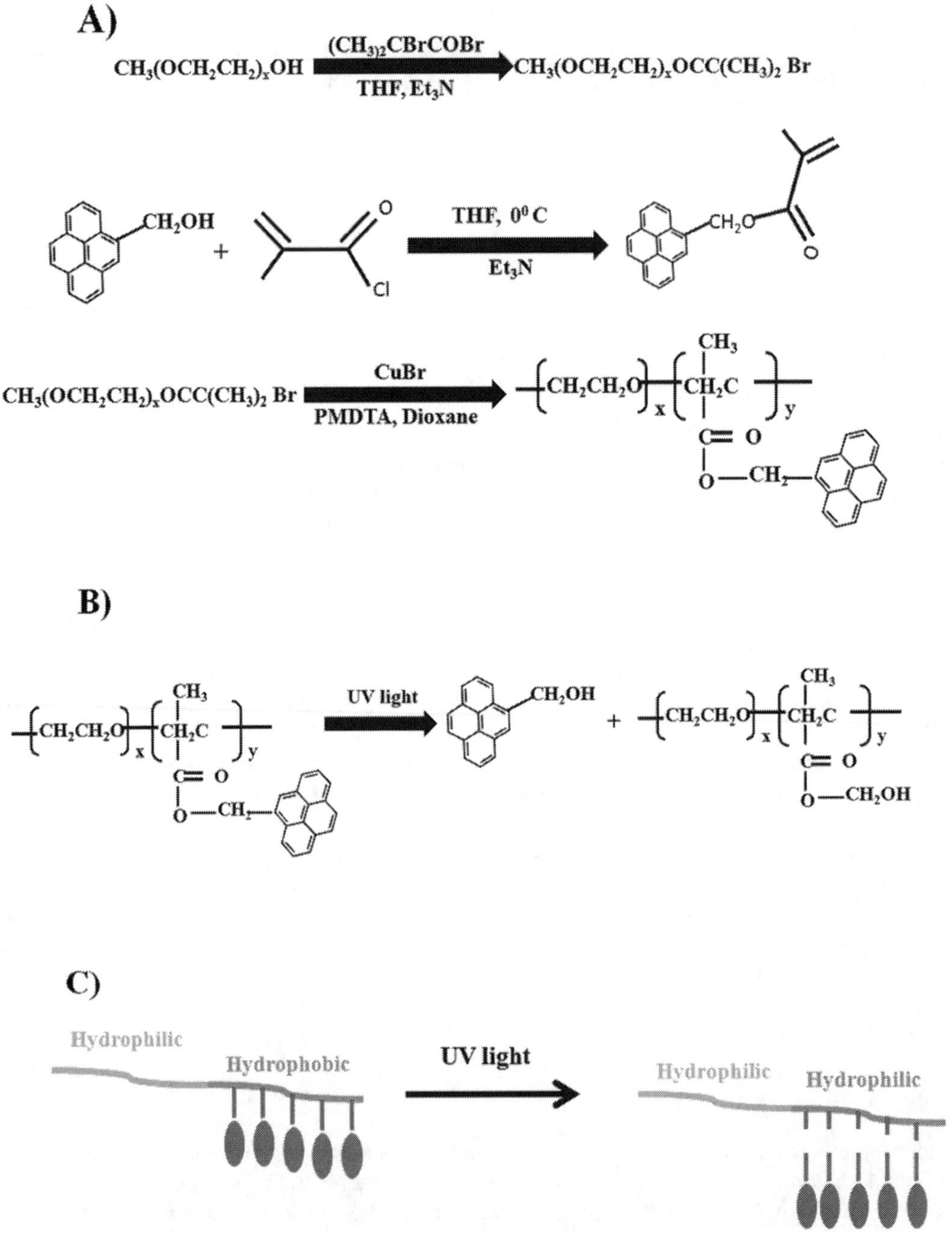

**FIGURE 4.4** (A) Steps involved in the synthesis of thermo-responsive block copolymer; (B) UV light-induced dissociation of the polymer material; (C) conversion of a hydrophobic to a hydrophilic group by photo solvolysis using UV light irradiation [61].

The other interesting polymer for DDS is light-sensitive material. Light can be used as a stimulus in DDS because of its precise control over the intensity by choosing the appropriate source along with the wavelength, and also the penetration capacity of light can be controlled. The two parameters of light make it a suitable choice to use as an external stimulus in DDS. Generally used light-sensitive materials are azobenzene, o-nitrobenzene, coumarin and pyrene moieties. The amphiphilic block copolymers have been designed and synthesized where their micellar aggregates can be dissociated by light irradiation [61]. The diblock copolymer consists of polyethylene oxide as a hydrophilic candidate and polymethacrylate with pyrene pendant units (PPY) as hydrophobic moieties. This system was synthesized using the atom transfer radical polymerization process as illustrated in Figure 4.4A and light-induced photolysis occurs as illustrated in Figure 4.4B [61].

Upon irradiation of UV light on the micelle solution of the polymer, light-induced solvolysis of pyrene moieties takes place, which results in their detachment from the polymer. This changes hydrophobic PPY to hydrophilic poly methacrylic acid. This results in complete dissociation of the micelle network of the polymer (Figure 4.4C) [61]. Another important externally regulated stimuli system is the electro-responsive system. The conducting properties of some of the inorganic and organic compounds will be used in DDS where the drug can be delivered by the applying external electric field. The small pulses of electric energy (~1 V) are sufficient to trigger the release of drugs with these kinds of systems and poses several advantages over other external stimuli systems [62]. These include easy to control and apply, fever instrumentation requirements and finally they can be integrated into chip-based devices. The electro-active materials reported in DDS based on electro-responsive systems are ferrocene, polypyrrole and carbon nanotubes. Other possible external responsive materials are ultrasound- and magnetic-responsive sheets [62].

## 4.5 TISSUE REGENERATION AFTER INJURY

### 4.5.1 Bioglasses for Cardiac Tissue Regeneration

Due to the incapacity of cardiomyocytes to regenerate and the difficulty of medical intervention, other methods of cardiac tissue restoration are needed in fatal cardiac conditions, hence biomaterials have been researched [63]. The development of polymeric heart patches to repair damaged tissue includes cell-delivery systems and heart mechanical support [64]. To achieve their objective, the team created a range of elastomeric nanocomposites made up of soft elastomer PGS (poly glycerol sebacate) and 45S5. The bioglass nanoparticles had concentrations varying from 0 to 10% by weight. PGS provides mechanical flexibility to polymer cardiac patches, whereas nano-sized bioglasses provide mechanical characteristics and also anchor points for releasing the cells [65,66]. In vitro experiments on human ESC (endometrial stromal cell)-derived cardiomyocytes show that PGS-nano-bioglass is more biocompatible than PGS without glass nanoparticles. In this regard, it is worth remembering Barabadi's experiments, in which he generated gelatine–collagen hydrogels (Gel/Col) incorporating bioglasses (Gel/Col/BG) to address the insufficiency of functional vasculature in tissue engineering. The expression of VEGF (vascular endothelial growth factor) was found to be higher [67,68].

Heart failure is a very common and major indication after myocardial infarction. In spite of their clinical significance, current treatments for myocardial infarction (MI) still have various restrictions. For myocardial infarction-induced damage, revascularization has been shown to have positive effects. For MI treatment biomaterial-based angiogenesis strategies currently represent potential candidates. Bioglasses are bioactive glasses and they are commercially available. Bioglass has angiogenic properties and it may be a substitute for MI treatments [69].

### 4.5.2 Composite Biomaterials for Nervous Tissue Regeneration

In the course of nerve regeneration, the biomaterial used in nerve tissue repair has to be flexible, tubular and mechanically stable [70,71]. A new method for fabricating electroactive nanofibres has been reported. By using vapour phase polymerization, electrospinning robust nanofibres, electrically conducting together with both biodegradable components by the use of PCL and conducting component, PPY has been established. The results indicated that the conducting materials had high potential levels for neural tissue repair [70,72].

The successful method for nerve regeneration and neuronal function is shown by electrical stimulation. These PC12 cells are seeded on conductive PPY/PDLLA conduits and exposed to electrical stimulation at 100 mV for 2 hours, and they exhibit a higher percentage of neurite-bearing cells [73,74]. Furthermore, as PPY content is increased, median neurite length increased also. The PPY/PDLLA nerve in vivo had a conduit with 5% PPY and this was used to fill a 10 mm defect in the sciatic nerve of a rat [75,76]. The findings revealed that it demonstrated effective recovery similar to an autologous nerve graft and considerable advancement in comparison to PDLLA conduits, indicating its enormous capability for nerve tissue regeneration [77]. Three different types of scaffolds were investigated by Mazzoni et al. [72] which included pristine collagen, another form was a purely oxidized one which was regenerated by cellulose-Ca (ORC-Ca) and last one was a collagen/ORC-Ca composite scaffold. On the other hand, to produce an acceptable scaffold for the Schwann cell neural supportive components in nerve tissue engineering these scaffolds were a great help. Additionally, collagen/ORC-Ca composite is a potential scaffold for the outermost regeneration and nerve repair, according to experimental findings [78].

For the progress of therapies of the next generation in restorative medicines and tissue engineering, developing and surveying of multifaceted intelligent biomaterials is necessary. New research displays how the various biomaterials could be applied to copy well-defined features of the in situ microenvironment. The in vivo tissue structure is classified by the interlink between the particular elements of the extracellular matrix (ECM). The importance of the composition of the ECM in the growth of cellular and molecular techniques, to extract the ecologically biodegradable and bioactive biomaterial adoptable to human physiology, is proved in the final substantiation. Through the progress of biomaterial and scar healing integration, those biomaterials send bioactive indications to accommodate the adjacent biological habitat. Influencing the old processes such as differentiation, self-renewal and cell adhesion, results in particular alterations to the properties of cells and fate restoring by the linkage between biomaterials and stem cells. Recent studies conducted in the field of tissue engineering and restorative medicine have been dealing with a wide range of implementations and usages, providing appropriate and effective strategies for neural restoration and repair, pointing to the potential application of biomaterials in the form of 3D tools for the in vitro neural growth of tissue models, in both physiological and pathological situations. In this route, various methods in terms of characteristics of biomaterials, for multiplied operationalization of scaffolds and in promoting neural multiplication and cell–ECM interactions, and many tools supporting cell restoration, which is subservient with cytokines and several other soluble aspects delivery through scaffold. As a matter of fact, 3D scaffolds should ensure developmental and uniform delivery of development aspects, cytokines or biomolecules, along with the need to help and provide guidance for the injured tissue. There are different layers that can be created in scaffolds, where each has its own variant physical and biochemical factors, capable of delivering and simultaneously assisting the specific cell phenotype and various ECM morphogenesis also with organization maintenance. The evidence summarizes the current advancements in operational materials which are difficult to present, while brushing off the current demerits and limits to nervous tissue regeneration and tissue engineering [79].

### 4.5.3 Treating the Hard-Tissue Interface in Tendons and Ligaments of the Bone Junction by Scaffold

The healing process of tendons or ligaments to the bone junction remains a challenging task in orthopaedic medicine, with the major issues to be resolved being the junction's low vascularity and multi-tissue transitional nature. Bioactive substances, synthetic materials and stem cells have been used in in vivo studies and clinical trials [80]. These methods, however, failed to restore the intricate structure–function relationships that exist between various tissues. Improvement in Random-Aligned-Random material enhanced fibro and bone cartilage development in the interface in vivo tests in rabbits. This outcome appears to be superior to that of the ECM of the unaltered tendon [81].

The healing of a tendon that has been injured is a difficult task. A high rate of recurrence has been reported as a result of the tension caused by scar development which is due to the use of sutures during surgery [82]. A recent study used a gradient material generated by a one-station electrospinning technique, together with osteosarcoma-like cells and fibroblasts cultured on a PLC nanofibre scaffold, to regenerate the tendon-to-bone contact [83]. The biomaterial's microstructure appears to be a promising area for simulating the transitional contact between various tissues. Additional research into this model, which represents the encouragement of the regeneration of the multi-tissue complex system by the heterogeneous cellular composition platform, is needed [84].

### 4.5.4 Biomaterials Used in Pulmonary Tissue Regeneration

The lungs can be affected by a variety of diseases, and lung tissues have poor regenerative capabilities because of which they have been highly affected in people suffering from COVID-19 in the recent pandemic. These features illustrate a grave problem for patients affected by serious COPD (chronic obstructive pulmonary disease), pulmonary hypertension along with cystic fibrosis to undergo lung transplants [85]. Bioactive sol–gel coating 58S has been used for biocompatibilty studies against murine epithelial cells of the lungs. The final outcome considered biocompatibility [65] and productivity for cell improvement in adhesion and proliferation. Another study [86] dealt with immortal cells derived from human lung that was put to test based on porous poly D,L-lactic acid scaffold blended with 45S5 bioglass, and showed an enhancement of adhesive cells and its growth based on its foam BG capacity. Therefore, further studies are required to understand lung tissue with regard to BGs in tissue engineering of lungs despite these positive results [87].

### 4.5.5 Biomaterials Used in Neurodegenerative Disorders

A huge problem worldwide is presented by neurodegenerative disorders such as Parkinson's disease, spinal cord injury, amyotrophic lateral sclerosis and Alzheimer's disease. Due to the rise in the average global age, these disorders are becoming increasingly common and prominent. The precise reasons behind these ailments are not known yet, despite several studies that have concentrated on their aetiopathology, and there remain only treatments based on symptoms. Biomaterials have become equally important for both the application of disease pathogenesis in revitalizing medicines as well as the disease pathogenesis itself. To promote the development of variant types of cells and to copy the extracellular matrix environment are among the best advantages enabled by biomaterials. Biomaterials can be used as secondary helping materials for cell multiplication to be transferred and as vectors to produce several active molecules for the treatment of neurodegenerative disorders. Biomaterials lead as a useful tool for the treatment of neural disorders, and they remain a future medical challenge. Biomaterials can be comprised of either artificial or synthetic compounds, as mentioned earlier. The studies into the biomaterials implemented above mentioned various characteristics, but they were described as bioinert, sterilizable, biocompatible, biofunctional and biodegradable. Due to their specific features, some of them are utilized in treatment and regeneration in neurological diseases (NDs). Hydrogels are outlined as optimal scaffolds for the culture and differentiation of stem cells. For the delivery of particular development factors to promote differentiation and cell growth, they are thought to be an optimal candidate. Several protective and active molecules for curing NDs have been announced as remarkable vectors. For the regeneration of nerve injuries, self-assembled peptides (SAPs), carbon-based nanomaterials, and nanofibres have been described as helpful. For the correct direction in helping and promoting axon development, both nanofibres and SAPs with a fibre-like structure have been used.

Finally, due to the capacity to increase electrical conductivity, carbon-based nanomaterials are specifically used for the restoration of axons. In-depth research needs to be made prior to implicating the materials for human neural restoration, although biomaterials were assessed in both in vivo and in vitro animal models. While biomaterials have several known beneficial results, they also have some challenges related to protection of health in humans for instance, an oxidative stress effect or triggering inflammation. By enabling developments in the treatments of human neural disorders, good advances are hoped for in the future in the biometrics field [88].

## 4.6 TAILORED STRUCTURE AND FUNCTIONS

Polymers are a typical scaffolding biomaterial because of their mass production stability, compatibility and degradability. Naturally derived polymers such as chitin-derived chitosan, gelatine, collagen and anionic polymer alginate also are utilized as polymers based on polyester such as polylactide, polycaprolactone, poly(lactic co-glycolic acid), polyurethane, and poly glycerol sebacate for both soft and hard tissue repair [89]. PCL, for example, is a low-cost linear synthetic biodegradable aliphatic polyester that, due to its flexibility, may be used to create a variety of scaffolds [90]. Conducting polymers, a new class of materials containing organic compounds, are distinguished with respect to their innate electric conductivity as well as their electrical and mechanical properties, and structural stability with compatibility of biomaterials [91]. Polypyrrole, polythiophene and similar compounds are biocompatible, and easy to synthesize and modify, making them attractive components in applications of biomedical studies such as drug-system delivery, artificial muscle structures, bio-actuators and bio-sensors [92]. Polyaniline's biological applications, on the other hand, are questioned due to its potential cytotoxicity, it does however offer certain intriguing qualities, including the following:

1. ease of manufacture from ordinary chemicals
2. excellent electrical conductivity
3. maximum electric and optic features
4. effective ion exchange and redox potential
5. environmental friendliness and stability
6. relatively minimal cost.

CPs may be used to enhance a variety of cellular functions, such as attachment of cells, growth, migratory abilities and cell differentiation with or without electrical stimulation. CP-based biomaterials are gaining prominence for developing electrically sensitive tissue, including heart cardiac muscles, nerve cells, skin and bone, and they may also be used to make scaffolds with electro-active and biodegradable properties when combined with biodegradable polymers [93].

Nowadays, these nanostructures have found worthwhile applications as advanced materials in renewable energy, biomedicine, nanotechnology, environmental science, tissue engineering and material science, to name just a few fields. In all the above-mentioned applications, the ultimate function relies on:

1. The structure of hierarchy of the peptide amyloids and protein from the atomistic to mesoscopic length scales
2. The specific mechanisms of protein aggregation
3. The physical properties of the amyloids in the surrounding environment (artificial and biological) [94].

## 4.7 CONCLUSION

The available quantum of literature regarding the stimuli-responsive systems makes it clear that they have significant importance in DDS and other applications. However, the developed systems have not passed the preclinical stage and only fever stimuli-responsive DDSs entered the clinical trial stage. The development of stimuli-responsive systems with accurate control over response to applied stimuli is crucial and further complicates the clinical translational challenges. The development of these stimuli-responsive systems involves multistep synthetic processes and multicomponent conversions that make the process more complicated. The stimuli-responsive systems are still in the developing stages, i.e., many synthetic procedures need to be optimized before translating these systems to clinical trials followed by real-world applications. Furthermore, precise control over stimuli in respect of the response applied for stimuli-responsive systems still needs to be improved by several orders in order to move the system from the preclinical to the clinical stage. External regulated systems can be controlled which are physical in nature, whereas internal regulated systems are very difficult to control and are biological triggers inside the system. Finally, the stimuli-responsive systems are promising delivery systems among other strategies available as delivery vectors after solving the existing hurdles or finding alternative pathways.

## REFERENCES


1. Farjadian F, Ghasemi A, Gohari O, et al. Nanopharmaceuticals and nanomedicines currently on the market: challenges and opportunities. *Nanomedicine (Lond)*. 2019 Jan;14(1):93–126.
2. Farjadian F, Roointan A, Mohammadi-Samani S, et al. Mesoporous silica nanoparticles: synthesis, pharmaceutical applications, biodistribution, and biosafety assessment. *Chemical Engineering Journal*. 2019 Mar 1;359:684–705.
3. Weissig V, Pettinger TK, and Murdock N. Nanopharmaceuticals (part 1): products on the market. *International Journal of Nanomedicine*. 2014;9:4357–4373.
4. Karimi M, Sahandi Zangabad P, Baghaee-Ravari S, et al. Smart nanostructures for cargo delivery: uncaging and activating by light. *Journal of the American Chemical Society*. 2017 Apr 5;139(13):4584–4610.
5. Guo B, Zhao J, Wu C, et al. One-pot synthesis of polypyrrole nanoparticles with tunable photothermal conversion and drug loading capacity. *Colloids Surf B Biointerfaces*. 2019 May 1;177:346–355.
6. Karimi M, Sahandi Zangabad P, Ghasemi A, et al. Temperature-responsive smart nanocarriers for delivery of therapeutic agents: applications and recent advances. *ACS Applied Materials & Interfaces*. 2016 Aug 24;8(33):21107–21133.
7. Pitt WG, Husseini GA, and Staples BJ. Ultrasonic drug delivery–a general review. *Expert Opinion on Drug Delivery* 2004;1(1):37–56.
8. Hoogenboom R. Chapter 2 – Temperature-responsive polymers: properties, synthesis, and applications. In: Aguilar MR, and San Román J, editors. *Smart Polymers and Their Applications* (Second Edition). Woodhead; 2019. Pp. 13–44.
9. Zhang Q, Weber C, Schubert US, et al. Thermoresponsive polymers with lower critical solution temperature: from fundamental aspects and measuring techniques to recommended turbidimetry conditions [10.1039/C7MH00016B]. *Materials Horizons*. 2017;4(2):109–116.
10. Meeussen F, Bauwens Y, Moerkerke R, Nies ELF, and Berghmans H. Molecular complex formation in the system poly(vinyl methyl ether)/water. *Polymer*. 2000;41(10):3737–3743.
11. Repasky EA, Evans SS, and Dewhirst MW. Temperature matters! And why it should matter to tumor immunologists. *Cancer Immunology Research*. 2013 Oct;1(4):210–216.
12. Ramos J, Imaz A, and Forcada J. Temperature-sensitive nanogels: poly(N-vinylcaprolactam) versus poly(N-isopropylacrylamide) [10.1039/C2PY00485B]. *Polymer Chemistry*. 2012;3(4):852–856.
13. Lee JH. Injectable hydrogels delivering therapeutic agents for disease treatment and tissue engineering. *Biomaterials Research*. 2018;22(27):1–14.
14. Boyer C, Bulmus V, Davis TP, et al. Bioapplications of RAFT polymerization. *Chemical ReviewsChemical Reviews*. 2009 Nov 11;109(11):5402–5436.

15. Ward C, Meehan J, Gray ME, et al. The impact of tumour pH on cancer progression: strategies for clinical intervention. *Exploration of Targeted Anti-tumor Therapy*. 2020;1(2):71–100.
16. Prado-Garcia H, Campa-Higareda A, and Romero-Garcia S. Lactic acidosis in the presence of glucose diminishes Warburg effect in lung adenocarcinoma cells [original research]. *Frontiers in Oncology*. 2020 June 12;10(807):1–13.
17. Iqbal S, Qu Y, Dong Z, et al. Poly(β-amino esters) based potential drug delivery and targeting polymer: an overview and perspectives (review). *European Polymer Journal*. 2020 Dec 5;141:110097.
18. Shenoy D, Little S, Langer R, et al. Poly(ethylene oxide)-modified poly(beta-amino ester) nanoparticles as a pH-sensitive system for tumor-targeted delivery of hydrophobic drugs. 1. In vitro evaluations. *Molecular Pharmaceutics*. 2005 Sep–Oct;2(5):357–366.
19. Devalapally H, Shenoy D, Little S, et al. Poly(ethylene oxide)-modified poly(beta-amino ester) nanoparticles as a pH-sensitive system for tumor-targeted delivery of hydrophobic drugs: part 3. Therapeutic efficacy and safety studies in ovarian cancer xenograft model. *Cancer Chemotherapy and Pharmacology*. 2007 Mar 1;59(4):477–484.
20. Lam JKW, Chow MYT, Zhang Y, et al. siRNA versus miRNA as therapeutics for gene silencing. *Molecular Therapy – Nucleic Acids*. 2015 Jan 1;4:e252.
21. Tarvirdipour S, Skowicki M, Schoenenberger C-A, et al. Peptide-assisted nucleic acid delivery systems on the rise. *International Journal of Molecular Sciences*. 2021;22(16):9092.
22. Chen H, Liu X, Dou Y, et al. A pH-responsive cyclodextrin-based hybrid nanosystem as a nonviral vector for gene delivery. *Biomaterials*. 2013 May;34(16):4159–4172.
23. Kanamala M, Wilson WR, Yang M, et al. Mechanisms and biomaterials in pH-responsive tumour targeted drug delivery: a review. *Biomaterials*. 2016 Apr;85:152–167.
24. Li X, Guo X, Cheng Y, et al. pH-responsive cross-linked low molecular weight polyethylenimine as an efficient gene vector for delivery of plasmid DNA encoding anti-VEGF-shRNA for tumor treatment [original research]. *Frontiers in Oncology*. 2018 Sep 25;8(354).
25. Kafil V, and Omidi Y. Cytotoxic impacts of linear and branched polyethylenimine nanostructures in a431 cells. *Bioimpacts*. 2011;1(1):23–30.
26. Zakeri A, Kouhbanani MAJ, Beheshtkhoo N, et al. Polyethylenimine-based nanocarriers in co-delivery of drug and gene: a developing horizon. *Nano Reviews and Experiments*. 2018;9(1):1488497–1488497.
27. Gao W, Chan JM, and Farokhzad OC. pH-responsive nanoparticles for drug delivery. *Molecular Pharmaceutics*. 2010 Dec 6;7(6):1913–1920.
28. Asokan A, and Cho MJ. Exploitation of intracellular pH gradients in the cellular delivery of macromolecules. *Journal of Pharmaceutical Sciences*. 2002 Apr;91(4):903–913.
29. Whitehead KA, Langer R, and Anderson DG. Knocking down barriers: advances in siRNA delivery. *Nature Reviews Drug Discovery*. 2009 Feb;8(2):129–138.
30. Roger E, Lagarce F, Garcion E, et al. Biopharmaceutical parameters to consider in order to alter the fate of nanocarriers after oral delivery. *Nanomedicine (Lond)*. 2010 Feb;5(2):287–306.
31. Colombo P, Sonvico F, Colombo G, et al. Novel platforms for oral drug delivery. *Pharmaceutical Research*. 2009 Mar;26(3):601–611.
32. Dai S, Tam KC, and Jenkins RD. Aggregation behavior of methacrylic acid/ethyl acrylate copolymer in dilute solutions. *European Polymer Journal*. 2000 Dec 1;36(12):2671–2677.
33. Dai J, Nagai T, Wang X, et al. pH-sensitive nanoparticles for improving the oral bioavailability of cyclosporine A. *International Journal of Pharmaceutics*. 2004 Aug 6;280(1–2):229–40.
34. Lee C-H, Lo L-W, Mou C-Y, et al. Synthesis and characterization of positive-charge functionalized mesoporous silica nanoparticles for oral drug delivery of an anti-inflammatory drug. *Advanced Functional Materials*. 2008;18(20):3283–3292.
35. Lin YH, Chang CH, Wu YS, et al. Development of pH-responsive chitosan/heparin nanoparticles for stomach-specific anti-Helicobacter pylori therapy. *Biomaterials*. 2009 Jul;30(19):3332–3342.
36. Lin YH, Sonaje K, Lin KM, et al. Multi-ion-crosslinked nanoparticles with pH-responsive characteristics for oral delivery of protein drugs. *Journal of Controlled Release*. 2008 Dec 8;132(2):141–149.
37. Akande J, Yeboah KG, Addo RT, et al. Targeted delivery of antigens to the gut-associated lymphoid tissues: 2. Ex vivo evaluation of lectin-labelled albumin microspheres for targeted delivery of antigens to the M-cells of the Peyer's patches. *Journal of Microencapsulation*. 2010 Jan 1;27(4):325–336.
38. Higgins L, Lambkin I, Donnelly G, et al. In vivo phage display to identify M cell-targeting ligands. *Pharmaceutical Research*. 2004;21:695–705.

39. Zhang Z, and Feng S-S. Self-assembled nanoparticles of poly(lactide)–Vitamin E TPGS copolymers for oral chemotherapy. *International Journal of Pharmaceutics*. 2006 Nov;324(2):191–198.
40. Sajeesh S, and Sharma CP. Novel pH responsive polymethacrylic acid-chitosan-polyethylene glycol nanoparticles for oral peptide delivery. *Journal of Biomedical Materials Research Part B: Applied Biomaterials*. 2006 Feb;76(2):298–305.
41. Russell-Jones GJ, Arthur L, and Walker H. Vitamin B12-mediated transport of nanoparticles across Caco-2 cells. *International Journal of Pharmaceutics*. 1999 Mar 15;179(2):247–255.
42. Chalasani KB, Russell-Jones GJ, Jain AK, et al. Effective oral delivery of insulin in animal models using vitamin B12-coated dextran nanoparticles. *Journal of Controlled Release*. 2007 Sep 26;122(2):141–150.
43. Vaupel P. Tumor microenvironmental physiology and its implications for radiation oncology. *Seminars in Radiation Oncology*. 2004 Jul;14(3):198–206.
44. Kim JW, and Dang CV. Cancer's molecular sweet tooth and the Warburg effect. *Cancer Research*. 2006 Sep 15;66(18):8927–8930.
45. Christofk HR, Vander Heiden MG, Harris MH, et al. The M2 splice isoform of pyruvate kinase is important for cancer metabolism and tumour growth. *Nature*. 2008 Mar 13;452(7184):230–233.
46. Brahimi-Horn MC, and Pouysségur J. Oxygen, a source of life and stress. *FEBS Letters*. 2007 Jul 31;581(19):3582–3591.
47. Lee ES, Gao Z, and Bae YH. Recent progress in tumor pH targeting nanotechnology. *Journal of Controlled Release: Official Journal of the Controlled Release Society*. 2008;132(3):164–170.
48. Griset AP, Walpole J, Liu R, et al. Expansile nanoparticles: synthesis, characterization, and in vivo efficacy of an acid-responsive polymeric drug delivery system. *Journal of the American Chemical Society*. 2009 Feb 25;131(7):2469–2471.
49. Aryal S, Hu C-MJ, and Zhang L. Polymer–cisplatin conjugate nanoparticles for acid-responsive drug delivery. *ACS Nano*. 2010 Jan 26;4(1):251–258.
50. Murphy RF, Powers S, and Cantor CR. Endosome pH measured in single cells by dual fluorescence flow cytometry: rapid acidification of insulin to pH 6. *Journal of Cell Biology*. 1984 May;98(5):1757–1762.
51. Freeman EC, Weiland LM, and Meng WS. Modeling the proton sponge hypothesis: examining proton sponge effectiveness for enhancing intracellular gene delivery through multiscale odelling. *Journal of Biomaterials Science, Polymer Edition*. 2013;24(4):398–416.
52. Rozema DB, Lewis DL, Wakefield DH, et al. Dynamic polyconjugates for targeted in vivo delivery of siRNA to hepatocytes. *Proceedings of the National Academy of Sciences of the United States of America*. 2007 Aug 7;104(32):12982–7.
53. Ward MA, and Georgiou TK. Thermoresponsive polymers for biomedical applications. *Polymers*. 2011;3(3):1215–1242.
54. Fu X, Xing C, and Sun J. Tunable LCST/UCST-type polypeptoids and their structure–property relationship. *Biomacromolecules*. 2020 Dec 14;21(12):4980–4988.
55. Iyer AK, Singh A, Ganta S, et al. Role of integrated cancer nanomedicine in overcoming drug resistance. *Advanced Drug Delivery Reviews*. 2013 Nov;65(13–14):1784–1802.
56. Meacham CE, and Morrison SJ. Tumour heterogeneity and cancer cell plasticity. *Nature*. 2013 Sep 19;501(7467):328–337.
57. Dugger BN, and Dickson DW. Pathology of neurodegenerative diseases. *Cold Spring Harbor Perspectives in Biology*. 2017 Jul 5;9(7):1–22.
58. Talekar M, Boreddy SR, Singh A, et al. Tumor aerobic glycolysis: new insights into therapeutic strategies with targeted delivery. *Expert Opinion on Biological Therapy*. 2014 Aug;14(8):1145–1159.
59. Convertine AJ, Diab C, Prieve M, et al. pH-responsive polymeric micelle carriers for siRNA drugs. *Biomacromolecules*. 2010 Nov 8;11(11):2904–2911.
60. Basel MT, Shrestha TB, Troyer DL, et al. Protease-sensitive, polymer-caged liposomes: a method for making highly targeted liposomes using triggered release. *ACS Nano*. 2011 Mar 22;5(3):2162–2175.
61. Jiang J, Tong X, and Zhao Y. A new design for light-breakable polymer micelles. *Journal of the American Chemical Society*. 2005 June 1;127(23):8290–8291.
62. Ge J, Neofytou E, Cahill TJ, et al. Drug release from electric-field-responsive nanoparticles. *ACS Nano*. 2012 Jan 24;6(1):227–233.
63. Isomi M, Sadahiro T, and Ieda M. Progress and challenge of cardiac regeneration to treat heart failure. *Journal of Cardiology*. 2019 Feb;73(2):97–101.

64. McMahan S, Taylor A, Copeland KM, et al. Current advances in biodegradable synthetic polymer based cardiac patches. *Journal of Biomedical Materials Research Part A*. 2020 Apr;108(4):972–983.
65. Kargozar S, Hamzehlou S, and Baino F. Potential of bioactive glasses for cardiac and pulmonary tissue engineering. *Materials (Basel)*. 2017 Dec 15;10(12):1–17.
66. Kerativitayanan P, and Gaharwar AK. Elastomeric and mechanically stiff nanocomposites from poly(glycerol sebacate) and bioactive nanosilicates. *Acta Biomaterialia*. 2015 Oct;26:34–44.
67. Barabadi Z, Azami M, Sharifi E, et al. Fabrication of hydrogel based nanocomposite scaffold containing bioactive glass nanoparticles for myocardial tissue engineering. *Materials Science and Engineering C-Materials for Biological Applications*. 2016 Dec 1;69:1137–1146.
68. Laflamme MA, Chen KY, Naumova AV, et al. Cardiomyocytes derived from human embryonic stem cells in pro-survival factors enhance function of infarcted rat hearts. *Nature Biotechnology*. 2007 Sep;25(9):1015–1024.
69. Qi Q, Zhu Y, Liu G, et al. Local intramyocardial delivery of bioglass with alginate hydrogels for post-infarct myocardial regeneration. *Biomedicine and Pharmacotherapy*. 2020 Sep;129:110382.
70. Fornasari BE, Carta G, Gambarotta G, et al. Natural-based biomaterials for peripheral nerve injury repair [review]. *Frontiers in Bioengineering and Biotechnology*. 2020 Oct 16;8(1209):1–26.
71. Subramanian A, Krishnan UM, and Sethuraman S. Development of biomaterial scaffold for nerve tissue engineering: biomaterial mediated neural regeneration. *Journal of Biomedical Science*. 2009 Nov 25;16(1):108.
72. Mazzoni E, Iaquinta MR, Lanzillotti C, et al. Bioactive materials for soft tissue repair [review]. *Frontiers in Bioengineering and Biotechnology*. 2021 Feb 19;9(94):1–17.
73. Yu Q-z, Dai Z-w, and Lan P. Fabrication of high conductivity dual multi-porous poly(l-lactic acid)/polypyrrole composite micro/nanofiber film. *Materials Science and Engineering B-Advanced Functional Solid-State Materials*. 2011;176:913–920.
74. Chang YJ, Hsu CM, Lin CH, et al. Electrical stimulation promotes nerve growth factor-induced neurite outgrowth and signaling. *Biochimica et Biophysica Acta*. 2013 Aug;1830(8):4130–4136.
75. Duffy P, McMahon S, Wang X, et al. Synthetic bioresorbable poly-α-hydroxyesters as peripheral nerve guidance conduits; a review of material properties, design strategies and their efficacy to date [10.1039/C9BM00246D]. *Biomaterials Science*. 2019;7(12):4912–4943.
76. Xu H, Holzwarth JM, Yan Y, et al. Conductive PPY/PDLLA conduit for peripheral nerve regeneration. *Biomaterials*. 2014 Jan;35(1):225–235.
77. Sinis N, Kraus A, Tselis N, et al. Functional recovery after implantation of artificial nerve grafts in the rat: a systematic review. *Journal Brachial Plexus Peripheral Nerve Injury*. 2009 Oct 25;4:19.
78. Chan BP, and Leong KW. Scaffolding in tissue engineering: general approaches and tissue-specific considerations. *European Spine Journal.* 2008;17(Suppl 4):467–479.
79. Abud EM, Ramirez RN, Martinez ES, et al. iPSC-derived human microglia-like cells to study neurological diseases. *Neuron*. 2017 Apr 19;94(2):278–293.e9.
80. Baldino L, Cardea S, Maffulli N, et al. Regeneration techniques for bone-to-tendon and muscle-to-tendon interfaces reconstruction. *British Medical Bulletin*. 2016 Mar;117(1):25–37.
81. Liu W, Thomopoulos S, and Xia Y. Electrospun nanofibers for regenerative medicine. *Advanced Healthcare Materials*. 2012;1(1):10–25.
82. Yang G, Rothrauff BB, and Tuan RS. Tendon and ligament regeneration and repair: clinical relevance and developmental paradigm. *Birth Defects Research Part C: Embryo Today*. 2013 Sep;99(3):203–222.
83. Nowlin J, Bismi MA, Delpech B, et al. Engineering the hard-soft tissue interface with random-to-aligned nanofiber scaffolds. *Nanobiomedicine (Rij)*. 2018 Jan–Dec;5:1849543518803538.
84. Mantha S, Pillai S, Khayambashi P, et al. Smart hydrogels in tissue engineering and regenerative medicine. *Materials (Basel)*. 2019;12(20):3323.
85. Barnes PJ, and Celli BR. Systemic manifestations and comorbidities of COPD. *European Respiratory Journal*. 2009 May;33(5):1165–1185.
86. Verrier S, Blaker JJ, Maquet V, et al. PDLLA/Bioglass composites for soft-tissue and hard-tissue engineering: an in vitro cell biology assessment. *Biomaterials*. 2004 Jul;25(15):3013–3021.
87. Fiume E, Barberi J, Verné E, et al. Bioactive glasses: from parent 45S5 composition to scaffold-assisted tissue-healing therapies. *Journal of Functional Biomaterials*. 2018 Mar 16;9(1):1–33.
88. Bordoni M, Scarian E, Rey F, et al. Biomaterials in neurodegenerative disorders: a promising therapeutic approach. *International Journal of Molecular Sciences* 2020 May 4;21(9).

89. Thavornyutikarn B, Chantarapanich N, Sitthiseripratip K, et al. Bone tissue engineering scaffolding: computer-aided scaffolding techniques. *Progress in Biomaterials*. 2014;3:61–102.
90. BaoLin G, and Ma PX. Synthetic biodegradable functional polymers for tissue engineering: a brief review. *Science China Chemistry*. 2014 Apr 1;57(4):490–500.
91. K N, and Rout CS. Conducting polymers: a comprehensive review on recent advances in synthesis, properties and applications [10.1039/D0RA07800J]. *RSC Advances*. 2021;11(10):5659–5697.
92. Balint R, Cassidy NJ, and Cartmell SH. Conductive polymers: towards a smart biomaterial for tissue engineering. *Acta Biomaterialia*. 2014 June 1;10(6):2341–2353.
93. Dong R, Ma PX, and Guo B. Conductive biomaterials for muscle tissue engineering. *Biomaterials*. 2020 Jan 1;229:119584.
94. Wei G, Su Z, Reynolds NP, et al. Self-assembling peptide and protein amyloids: from structure to tailored function in nanotechnology [10.1039/C6CS00542J]. *Chemical Society Reviews*. 2017;46(15):4661–4708.
95. Sengupta S, Eavarone D, Capila I, et al. Temporal targeting of tumour cells and neovasculature with a nanoscale delivery system. *Nature*. 2005 July 1;436(7050):568–572.
96. Tong R, and Cheng J. Ring-opening polymerization-mediated controlled formulation of polylactide–drug nanoparticles. *Journal of the American Chemical Society*. 2009 Apr 8;131(13):4744–4754.
97. Banerjee SS, and Chen DH. Multifunctional pH-sensitive magnetic nanoparticles for simultaneous imaging, sensing and targeted intracellular anticancer drug delivery. *Nanotechnology*. 2008 Dec 17;19(50):505104.
98. Mok H, Park JW, and Park TG. Enhanced intracellular delivery of quantum dot and adenovirus nanoparticles triggered by acidic pH via surface charge reversal. *Bioconjug Chem*. 2008 Apr;19(4):797–801.
99. Liu R, Zhang Y, Zhao X, et al. pH-Responsive nanogated ensemble based on gold-capped mesoporous silica through an acid-labile acetal linker. *Journal of the American Chemical Society*. 2010 Feb 10;132(5):1500–1501.
100. Wang B, Xu C, Xie J, et al. pH controlled release of chromone from chromone-$Fe_3O_4$ nanoparticles. *Journal of the American Chemical Society*. 2008 Nov 5;130(44):14436–14437.
101. Ali MM, Oishi M, Nagatsugi F, et al. Intracellular inducible alkylation system that exhibits antisense effects with greater potency and selectivity than the natural oligonucleotide. *Angewandte Chemie International Edition in English*. 2006 May 5;45(19):3136–3140.
102. Li Y, and Lee PI. A new bioerodible system for sustained local drug delivery based on hydrolytically activated in situ macromolecular association. *International Journal of Pharmaceutics*. 2010 Jan 4;383(1–2):45–52.
103. Jeong JH, Kim SW, and Park TG. Novel intracellular delivery system of antisense oligonucleotide by self-assembled hybrid micelles composed of DNA/PEG conjugate and cationic fusogenic peptide. *Bioconjugate Chemistry*. 2003 Mar 1;14(2):473–479.

# 5 A Road to Future Sensors

## *Polymeric Biomaterials*

*Jyoti Tyagi and Shahzad Ahmad*

**CONTENTS**



## 5.1 INTRODUCTION

Sensors play a primary role in improving human quality of life, especially sensors used in the biomedical field and personal healthcare. At present, wearable and flexible electronics that possess remarkable stretchability, good mechanical properties, and biocompatibility are being designed and employed for various applications [1]. Thus, the demand for sensors with high sensitivity and selectivity prevails in the industry, leading to frequent updates in electronic devices using sensors [2]. Over the past few decades, sensors have been developed mainly based on non-biodegradable materials that negatively affect the environment after disposal. The need for sustainable development leads to intensive studies on the fabrication of biomaterial-based sensors that can be combined with electronic devices [3]. Of the three biomaterials, metals, ceramics, and polymers, polymeric biomaterials are flexible, which is an unrivalled property compared to their counterparts. Besides flexibility, polymeric biomaterials are also lightweight and available in desired shapes, such as latex, film, and fibres. In addition, polymeric biomaterials are easy to synthesize, cost-effective, readily available, and can also form composites [4]. Because of their remarkable physical, chemical, and mechanical properties, polymeric biomaterials find applications in almost every field, including in sensors. Furthermore, the advancements in manufacturing technology and development have opened the door to the design and synthesis of sensors based on polymeric biomaterials. Sensors based on naturally derived, synthetic non-biodegradable, and biodegradable polymeric biomaterials are synthesized for numerous uses [2]. For example, there are sensors for sensing pressure, temperature, sound, light, humidity, pH, glucose, DNA, pathogen, and many more [3,5]. Among various precursors employed in the fabrication of sensors, conducting biopolymers offer an excellent choice owing to their efficient electrochemical activity, good electrical conductivity, optimum mechanical elasticity, and biocompatibility. These polymers also possess the unique capability of transferring electric charge from redox enzymes to electrodes [6]. The electrochemical sensors based on biopolymers have applications in various fields, especially in the detection of biologically active molecules in the aqueous environment [7]. Furthermore, optical sensors based on biopolymers and conducting polymers are employed for detecting heavy metal ions [8]. Recently, conducting polymer composites based on renewable resources have been attracting

DOI: 10.1201/9781003240884-5

the attention of researchers owing to their easy availability, low cost, biodegradability, and biocompatibility [7]. Among the various categories of sensors, optical and electrochemical sensors are more frequently fabricated and utilized for practical applications due to their phenomenal detectability, cost-effectiveness, and comparative experimental simplicity [2,9]. Recently, layered double hydroxides (LDHs), or simply anionic clays, have been also classified as biomaterials. The LDHs have attracted the attention of many researchers from various areas owing to their versatile properties and potential applications in the biomedical field, cosmetics, sensing, and biosensing. Although LDHs belong to a class of inorganic materials, they exhibit high biocompatibility along with highly controllable synthesis processes [10,11].

The use of plasma polymers as a different type of sensor has been exemplified in many research articles [5]. The treatment of polymers with $O_2$ and/or $N_2$ plasma to obtain a polymer plasma is a simple, dry, one-step process which will play a vital role in developing various types of sensors such as stimuli-responsive, temperature, glucose, DNA, etc. [5]. Plasma polymers possess reactive surfaces due to the development of new functional groups such as carbonyl, ether, carboxyl, amine, and hydroxyl on exposure to plasma. These plasma polymer-based sensors have also exhibited potential in practical biomedical applications. A variety of sensors based on polymeric biomaterials or composites of biomaterials are utilized in biomedical systems for various applications. The use of these sensors makes the early diagnosis of some diseases possible, which helps in timely and better treatment, thus improving quality of life and health [12].

In this chapter, sensors developed based on conducting biopolymers, layered double hydroxides, and plasma polymers are discussed in detail. Optical and electrochemical sensors are discussed separately because of their utilization in various sophisticated sensors. Furthermore, sensors in biomedical systems are also discussed due to their importance in human life.

## 5.2 CONDUCTING BIOPOLYMERS

Over the last two decades, conducting biopolymers have attracted significant attention due to their efficient electrical conductivity and biodegradability. The conducting biopolymers are increasingly utilized materials that find significant applications in the biomedical field, especially for diagnostic applications [13]. Polypyrrole, polyaniline, and poly(3,4-ethylenedioxythiophene) (PEDOT) are the most commonly exploited conducting polymers for the fabrication of electrochemical and optical sensors owing to their remarkable electrical and optical properties [9]. However, these conducting polymers are non-biodegradable and biocompatibility is also a challenge that limits their use as biomaterials. In contrast, biopolymers are mostly biodegradable, biocompatible, and eco-friendly, which is a desirable property for biomedical applications. Furthermore, depending on the nature of biopolymers, they are classified as biodegradable or non-biodegradable. Natural biopolymers such as cellulose, chitosan, silk fibroin (SF), and synthetic biodegradable biopolymers such as polylactic acid, polycaprolactone, and poly(glycolic acid) are promising biomaterials that can be combined with conducting polymers via ester linkages for fabricating conducting biopolymers [14]. A conducting biopolymer can be prepared using processes such as blending, composite formation, and copolymerization [13]. To synthesize wearable and flexible sensors, as presented in Figure 5.1, conductive hydrogels based on natural biopolymers have been summarized by Yang and co-workers in a review article [1].

Silk is a natural biopolymer, mainly available in two forms, which are SF and sericin. SF film is an efficient and promising option in the fabrication of biosensors owing to their high biocompatibility, biodegradability, robust mechanical properties, adaptive structure, and remarkable optical properties. The conductive and mechanical properties of the silk can be further improved by adding graphene oxide, nanotubes, and poly(3,4-ethylenedioxythiophene): poly(styrene sulfonate) (PEDOT: PSS) [3]. Figure 5.2 shows a biocomposite of conducting polymer, PEDOT:PSS, and silk fabricated via photolithography, which acts as a flexible biosensor [15].

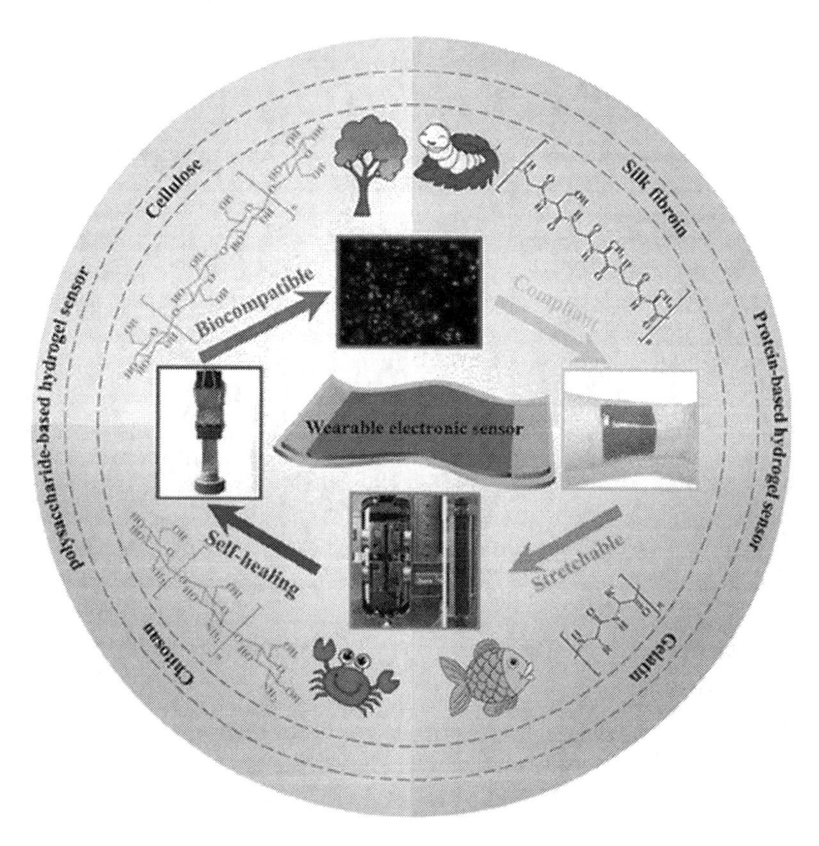


**FIGURE 5.1** Wearable and flexible sensors based on natural biopolymers (cellulose, chitosan, SF gelatine). (Reprinted with permission from Ref. [1]. Copyright 2021, American Chemical Society.)

To be specific, the PEDOT:PSS microstructures with high resolution and reliability were prepared over a large area using a benchtop water-based photolithographic setup. The synthesized biosensor exhibited high sensitivity to detect dopamine and ascorbic acid, non-specifically. In addition, the biosensor displayed exceptional electrochemical activity and stability for approximately one month. Besides being mechanically flexible, the microfabricated device was fully biocompatible, biodegradable, and also optically active. Furthermore, the biosensor was also specifically able to detect glucose with remarkable selectivity and sensitivity due to the encapsulation of the glucose oxidase enzyme at ambient temperature and biological pH. The biosensor retained its properties even after repeated mechanical deformations, and was fully degradable after nearly one month with enzymatic actions. Because of the phenomenal mechanical properties and controllable biodegradability, the developed devices are promising candidates for the 'implant and forget' type of applications. The reported sensitive, low-cost, and robust biosensor can be used practically to fabricate devices in a biomedical field such as for smart skin or implantation applications because of the scalable technique along with controllable biodegradability.

A novel wearable strain sensor was fabricated using carbonized silk fabrics (CSFs) [16]. The synthesized robust strain sensor was highly sensitive and ultra-stretchable, and displayed potential applications for sensing human motion, either subtle or large, such as jumping, facial expressions, etc. For making sensors, SF was converted into graphitic nanocarbon via thermal treatment, which exhibited outstanding electrical conductivity. The developed CSFs with plain-weave structure exhibited phenomenal stretchability (up to 500%), a broad strain sensing range from 0% to

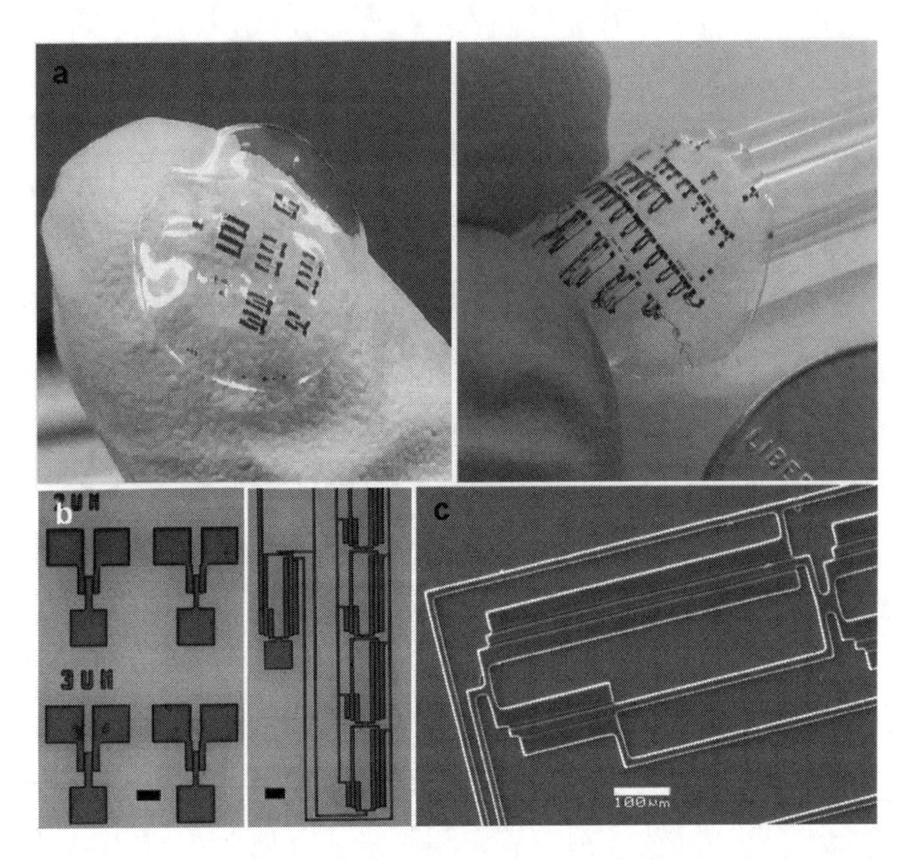


**FIGURE 5.2** Formation of microstructures by photolithography. (a) Large area micropatterns of PEDOT:PSS can be formed on flexible and conformable SF sheets; (b) optical micrographs; and (c) SEM images of PEDOT:PSS micropatterns on glass. Scale bars =100 μm. (Reprinted with permission from Ref. [15]. Copyright 2016 Elsevier.)

>500%, high sensitivity, high durability (10,000 cycles), and short response time (<70 ms), which suggests its significant applications for fabricating smart robots and wearable devices. The excellent characteristics of the sensor were due to the unique hierarchical structure of the very first time exploited carbonized woven fabrics. Besides applications in monitoring human motion, this sensor can also be utilized for personal healthcare. This strategy of carbonizing natural biomaterials could also apply to other fabrics, such as cotton and modal, thus opening new doors for inexpensive and extendable fabrication of wearable sensors.

Cellulose, a natural biopolymer, has been utilized by many researchers to fabricate tuneable sensors owing to its large surface areas, easy transformation to films, environment-friendly, remarkably conductive, and mechanical properties. To fabricate the variety of flexible biosensors based on cellulose, its conductive and mechanical properties are altered by combining it with some conductive materials such as conducting polymers, nanotubes (single-, double-, and multi-walled), graphene oxide, metals, etc. [3]. A flexible strain sensor was synthesized from bacterial nanocellulose (BNC) after modification with double-walled carbon nanotubes (DWCNTs) and multi-walled carbon nanotubes (MWCNTs) [17]. The electrical conductivity of the fabricated sensor depended on the modifying agent (DWCNTs and MWCNTs) and its dispersion process. MWCNT-modified BNC displayed higher conductivity than the DWCNT-modified BNC. The highest conductivity of the former was 1.6 S $cm^{-1}$, while the latter showed the highest conductivity of 0.39 S $cm^{-1}$. Furthermore, the strain-induced electrochemical behaviour of the modified BNC films was studied, wherein MWCNT-treated BNC exhibited higher sensitivity than DWCNT-treated BNC. The MWCNT-treated sample of BNC films produced a gauge factor of 252. The strain sensitivity comparison of the carbon nanotube-treated samples of BNC with different conductivity revealed that conductivity and strain sensitivity are correlated. This observation will be useful for the development of improved strain sensors in the future. Another light-weight flexible strain sensor based on conductive bacterial cellulose (BC)/MWCNT nanocomposite aerogel was *in situ* biosynthesized from BC/MWCNT nanocomposite hydrogels using the supercritical $CO_2$ process [18]. A small percolation threshold value of 0.0041 (volume concentration) was estimated for the synthesized nanocomposite aerogel. The strain-sensing response of the nanocomposite aerogel was also studied under tensile loading. The results of the study suggested that, on introducing nanotubes into culture media, the average pore size of aerogels decreased to 8.6 nm and volume shrunk by 2%. Furthermore, a response time of 390 ms and gauge factor of 21, obtained for nanocomposite aerogel of BC/MWCNTs, suggested that the developed strain sensor can be employed for monitoring human motion.

Chitosan is again a natural biopolymer possessing properties similar to silk and cellulose, which has been used to prepare sensors or biosensors after integrating with some conductive materials such as MWCNTs, graphene, metal nanoparticles, etc. An amperometric biosensor based on a composite of MWCNTs and biopolymer chitosan (MWCNTs-CHIT) for detecting lactate was fabricated on a glassy carbon electrode [19]. A homogeneous nanobiocomposite film of MWCNTs-CHIT and lactate dehydrogenase displayed the ability to increase the current responses, decrease the electro-oxidation potential, and prevent surface fouling of electrodes. The electroanalytical capability of the biosensor was optimized after studying the effect of various experimental factors, such as the pH of a solution, enzyme loading, applied potential, and $NAD^+$ concentration. The optimized biosensor exhibited a response time of 3 s and a sensitivity of 0.0083 $AM^{-1} cm^{-2}$. Furthermore, 65% of the original response of the biosensor was retained after seven days. Another chitosan-based glucose biosensor was prepared at the gold electrode using thin films of chitosan, containing nanocomposites of graphene and gold nanoparticles (AuNPs) to make the nanocomposite conductive [20]. The fabricated biosensor immobilized the glucose oxidase (GOx) enzyme. The resulting nanocomposite film showed good electro-catalytical activity toward $H_2O_2$ and $O_2$ owing to the synergistic effect between graphene and AuNPs. In addition, with GOx as a model, the graphene/AuNPs/GOx/CHIT composite-modified electrode was synthesized using a simple casting process. The biosensor displayed a good amperometric response for glucose at a range of 2–10 mM and 2–14 mM at –0.2 V and 0.5 V, respectively. Moreover, good reproducibility and a sensing limit of 180 μM were also observed. The developed nanocomposite film has shown an excellent electrochemical response to glucose, making the biosensor a promising candidate for electrochemical glucose detection.

## 5.3 OPTICAL AND ELECTROCHEMICAL SENSORS

In an electrochemical sensor, the information linked with electrochemical reactions is transformed into a detectable signal. The electrode acts as a transducer element in these sensors. The electrochemical sensors mainly fall into three categories: potentiometric, conductometric, and amperometric/voltametric [21]. On the other hand, there are four types of optical sensors: colorimetric, electrochemiluminescence, fluorescence, and surface plasmon resonance (SPR). The optical and electrochemical sensors/biosensors are the most fabricated and pragmatic because of their outstanding detection characteristics for analyses with high selectivity and high sensitivity. In addition, these sensors are important due to various other factors such as fast response time, low cost, low limit of detection, easy to fabricate, and user friendliness [8]. To date, silica-based materials are the prime materials utilized in optical sensors owing to their outstanding optical properties. However, the absence of biocompatibility and flexibility limits the use of these materials for

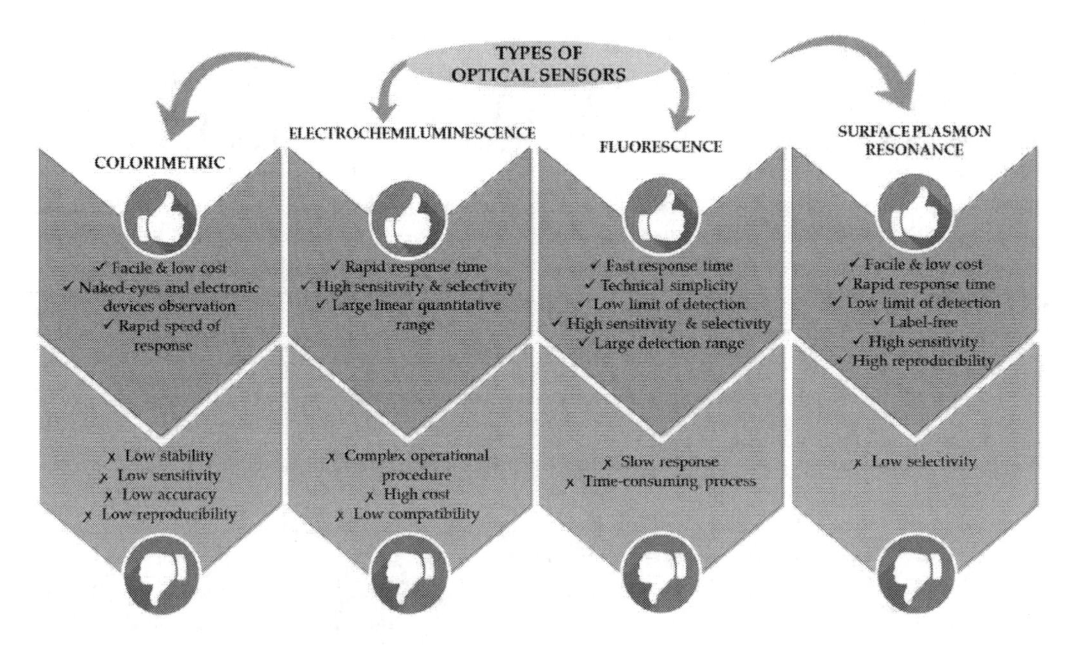


**FIGURE 5.3** The pros and cons of different optical sensors for heavy metal ion detection. (Reprinted from Ref. [8] with License under the Creative Commons Attribution.)

biological applications. The use of polymeric biomaterials to fabricate optical and electrochemical sensors provides a solution to many challenges such as biocompatibility, flexible mechanics, controllable biodegradability, easy synthesis process, cost-effectiveness, easy availability, and design versatility [2]. These sensors have found their applications in diverse fields such as biomedical, sensors, biosensors, agriculture, defence settings, and environmental science.

Heavy metal ions present in water have damaging effects on the environment and human health. The concentration of these non-biodegradable and toxic substances is not very high. Therefore, sensors with high sensitivity, selectivity, and capability for detecting metal ions in very low concentrations are in high demand. Polymeric-biomaterial-based optical sensors have emerged as promising sensors for detecting heavy metal ions in aqueous media. For example, chitosan, silk, cellulose, and nanocrystalline cellulose are incorporated in optical sensors owing to their versatile properties and potential applications, especially in the field of sensors [8]. The various pros and cons of different types of optical sensors are presented in Figure 5.3, especially for the detection of heavy metal ions.

Of the four types of optical sensors, the SPR sensors are becoming most popular due to their excellent performance in biological and environmental detection. In addition, these sensors are inexpensive, highly sensitive, and based on simple procedural techniques. The different types of optical sensors based on biopolymers exhibited different performances in sensing an analyte with respect to sensitivity and detection limit, as summarized in Table 5.1. Although the detection limit of SPR sensors based on biopolymers is relatively high, it could be further improved using a better composition of biopolymer and conducting polymer in the future. For further insight into optical sensors being utilized as environmental sensors, a review article by Fen and co-authors can be considered [8].

In the field of optical sensors, chitosan is a promising biopolymer to sense gases also, because of its superior capability to form thin films and tuneable optical properties such as the refractive index. An optical waveguide $H_2S$ gas sensor based on nanocomposites of chitosan with Au and Ag nanoparticles was synthesized using spin-coated nanofilms (150 nm thick) of chitosan by

**TABLE 5.1**
**A Comparison of the Limit of Detection of Different Biopolymer-Based Optical Sensors**

| Optical Sensor | Biopolymer Composite | Lowest Limit of Detection (nM) | Heavy Metal Ion |
|---|---|---|---|
| Electrochemi-luminescence | Chitosan/Ru$(bpy)_3^{2+}$/silica nanoparticle-modified glass carbon electrode | 0.3 | $K^+$ |
| Colorimetric | Chitosan functionalized gold nanoparticles assembled on sulphur-doped graphitic carbon nitride | 0.275 | $Hg^{2+}$ |
| Fluorescence | Three-dimensional fluorescent chitosan hydrogel fluorescence | 0.9 | $Hg^{2+}$ |
| Surface plasmon resonance | PPy–chitosan/ITO/silver | 1.29 | $Cd^{2+}$ |

*Source:* Based on Ref. [8].

adsorption and reduction of $[AuCl_4]^-$ and $Ag^+$ ions [22]. The resulting nanocomposites, chitosan/Au and chitosan/Ag, were deposited as sensitive coatings onto planar optical waveguides, which allowed $H_2S$ detection at a concentration of between 5 to 300 ppm, and 0.1 to 100 ppm, respectively. In addition, the sensor effectiveness was investigated by considering the various parameters such as sensitivity, the kinetics of the sensor, cross-sensitivity, and the effect of humidity. The developed sensor displayed a highly sensitive, selective, and reproducible response signal. Interestingly, the composite films can be a potential candidate to indicate freshness in the food packaging arena due to no humidity dependence, irreversibility, and capability to sense the colorimetric response by the naked eye. Chitosan-based optical sensors also have been reported to sense relative humidity in several studies [23,24]. Similarly to chitosan, other natural biopolymers as well as synthetic biopolymers are employed to design and fabricate optical sensors [2]. Besides being applied in various fields, optical sensors are also utilized for biophotonic applications where optical techniques are integrated into the domains of biology and medicine to improve the sensing, imaging, and therapeutic approaches. Specifically, optical biosensors enable the early detection and diagnosis of various types of cancers such as cervical, oral, epithelial, etc., and some other diseases such as celiac and Alzheimer's diseases [2]. Optical biosensors are also utilized for monitoring wound healing. For example, a wearable biosensor using polyacrylamide hydrogels, a synthetic polymeric biomaterial, was fabricated for detecting multi-signals [25]. The resulting sensor was based on changes in pH and concentration of inflammatory C-reactive protein (CRP). For sensing the pH, the polymer hydrogel was responsive to swelling or shrinking when the pH of the surroundings changed, and also the pH measuring range can be accustomed using different chemical compositions of the hydrogels. For sensing the CRP concentration, a functional surface using phosphorylcholine receptors was designed, capable of explicitly recognizing the targeted protein. Furthermore, changes in the pH and CRP concentration were detected using an optical signal that was based on probing the changes in refractive index, an optical property. When the developed sensor was tested with blood serum, it successfully demonstrated reversible pH measurements between a range of 6–8 and CRP concentration detection range between 1–100 µg/mL. The portable sensing device enables remote and real-time *in situ* monitoring of wound healing, and this technology can be further extended to monitor skin grafts and ulcer medication. The biopolymer-based optical sensors also find their application in sensing biological active molecules such as glucose. An optical sensor using natural biopolymer silk and a gold nanostructure was synthesized as shown in Figure 5.4, which can be utilized for real-time and *in vivo* supervision of analytes in biological fluid [26].

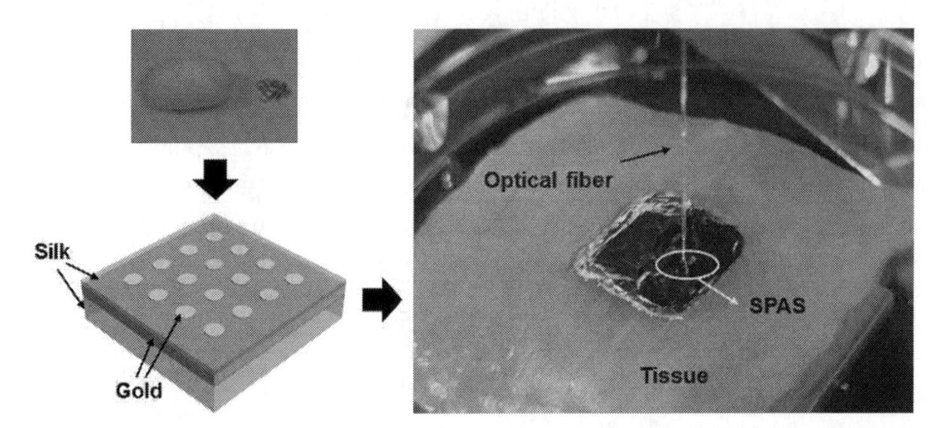


**FIGURE 5.4** A schematic representation of a SPAS device from silk protein and gold nanostructures to fabricated model, and then the device (diffracted colours) on a chicken breast tissue, attached with an optical fibre for measurements. (Reprinted with permission from Ref. [26]. Copyright 2015, American Chemical Society.)

This sensor was developed using the concept of manipulative plasmonic resonances. In the sensor, the silk protein acted as an insulating spacer, and analytes were absorbed in the spacer. Because of the hydrogel properties of silk protein, it swelled controllably on exposure to water–alcohol mixtures which were employed as stimuli to modulate the swelling ratios of the silk spacer. Furthermore, radical shifts in reflectance spectra were observed via numerical simulations and experiments owing to the changes in some physical quantities of the silk spacer, such as volume and refractive index, after absorbing the analytes. The resulting biocompatible silk plasmonic absorber sensor (SPAS) acts as a bio/chemical sensor with high sensitivity and high tunability. Moreover, the fabricated SPAS device was investigated as a glucose sensor and displayed a sensitivity of 1200 nm/RIU (refractive index units) which showed the significant potential of this device to be utilized practically.

In the domain of electrochemical sensors/biosensors, chitosan which is derived from chitin via deacetylation is the most studied natural biopolymer owing to its excellent ability to form thin films, non-toxicity, abundant functional groups (amine and hydroxyl), low cost, and easy availability. In addition, its characteristics of forming hydrogel and higher solubility in an aqueous solution of mild acidic pH compared with other natural biopolymers further make it more suitable for synthesizing electrochemical biosensors. Various studies have been reported to synthesize electrochemical sensors based on chitosan such as nucleic acid biosensors, immunosensors, enzyme biosensors, environmental sensors, and many more [4,21]. To detect various environmental pollutants such as organic pollutants and heavy metals ions, chitosan-based nanocomposites with other materials like MWCNTs, graphene oxide, nanoparticles of metals, carbon dots, and graphene have been investigated in different researches, as summarized in Table 5.2.

Besides the potential use in environmental science, chitosan-based electrochemical sensors are also utilized in the medical field to detect different classes of molecules such as pharmaceuticals, neurotransmitters, and antioxidants [4]. The synthesis of the glucose sensor is also realized using nanocomposites of chitosan. For example, recently a ternary bionanocomposite was fabricated using polyaniline, cadmium stannate nanoparticles, and chitosan by the chemical oxidative polymerization method [35]. The resulting polyaniline-cadmium stannate-chitosan composite exhibited remarkable electrochemical non-enzymatic detection of glucose and ammonia solution. Specifically, the weight percentage of chitosan was varied in the bionanocomposites to optimize the electrocatalytic and electrochemical activities, and polyaniline/cadmium stannate/chitosan (10%) showed the best

**TABLE 5.2**
**Chitosan-Based Electrochemical Sensors to Detect Organic Pollutants and Heavy Metal Ions**

| Target Analyte | Composite Composition | Real Sample | Reference |
|---|---|---|---|
| **Organic Pollutants** | | | |
| o- and p-Nitrophenols | Reduced graphene oxide-cyclodextrin-chitosan | River water | [27] |
| 2, 4-Dichlorophenol | Carbon dots – hexadecyltrimethyl ammonium bromide-chitosan | Lake water | [28] |
| 4-Nitroaniline | Copper nanoparticles – chitosan | Milk | [29] |
| 17α-Ethinylestradiol | Chitosan-MWCNTs | Human urine | [30] |
| **Heavy Metals** | | | |
| $Cd^{2+}$ and $Pb^{2+}$ | $Fe_3O_4$/MWCNTs/ laser scribed graphene /Chitosan | Tap water | [31] |
| $As^{3+}$ | Chitosan-$Fe(OH)_3$ | Drinking water | [32] |
| $Hg^{2+}$ | Chitosan-AuNPs | Drinking water, fruit juice | [33] |
| $Cr^{3+}$ | Chitosan/MWCNTs/$MnO_x$ | Drinking water | [34] |

**TABLE 5.3**
**Chitosan-Based Electrochemical Sensors to Detect Glucose**

| Nanocomposite Composition | Limit of Detection | Reference |
|---|---|---|
| Chit/PPy nanotubes-AuNPs | 3.10 mM | [36] |
| $GO_x$-Chit-CNT85 | – | [37] |
| FTO/Nano-CuO/Chit/Gox | 27 μM | [38] |
| Gold, magnetic $Fe_3O_4$ & $Fe_3O_4$-Chit NPs | 0.43 μM | [39] |
| AgNPs/NSC | 0.046 mM | [40] |
| AuNPs-MWCNTs-Chit cryogel | 0.5 μM | [41] |

*Source:* Adapted from Ref. [4].

results. In addition, a short response time and a low limit of detection were observed for a wide range of concentrations of the target analytes. The fabricated sensor was highly sensitive, selective, stable, and reproducible for the target analytes. Thus, this novel sensor can be efficiently employed on a large scale with no further modification. Apart from the discussed sensor, there are various other electrochemical sensors based on chitosan that have been designed and synthesized for glucose detection, as presented in Table 5.3.

Other than chitosan, electrochemical sensors based on cellulose, silk, and other synthetic biopolymers also have been reported, however a discussion of these studies is beyond the scope of this chapter. For illustrating cellulose application in electrochemical sensors, one example is discussed. An electrochemical biosensor to detect lactate was fabricated using BC as a substrate due to its excellent mechanical resistance [42]. The biosensor successfully demonstrated lactate detection in artificial sweat via direct immobilization of the enzyme (lactate oxidase) on the surface of BC, after chemical functionalization. The developed biosensor displayed an excellent detection range between 1–24 mM for lactate in artificial sweat with 1.31 mM as a limit of detection, and

4.38 mM as the limit of quantification. The proposed BC substrate for biosensors can be employed in wearable devices because of its versatile properties, such as biocompatibility, easy large-scale production of screen-printed electrodes on its surface, and remarkable flexibility. Another electrochemical novel sensor based on cellulose was reported to detect paracetamol in commercial drugs, such as Doliprane® 500, which can be beneficial for laboratories to control the quality of drugs [43].

## 5.4 LAYERED DOUBLE HYDROXIDES

Layered double hydroxides (LDHs) have emerged as potential biomaterials owing to their remarkable features such as layered structure, high compositional diversity, large surface area, easy synthesis process, cost-effectiveness, excellent adsorptive capacity, ion-exchange capability, good electrical conductivity, low toxicity, high stability, and biocompatibility. The LDHs are known as anionic clays which belong to the class of inorganic materials as well as hydrotalcite nanocomposites. They are represented by the general formula $[M^{2+}_{1-x}M^{3+}_{x}(OH)_2]^{x+}.[A_{\frac{x}{n}}]^{n-}.mH_2O$, where $M^{2+}$ is the divalent metal ion, such as $Mg^{2+}$, $Cu^{2+}$, $Fe^{2+}$, $Ca^{2+}$, etc., $M^{3+}$ is the trivalent metal ion such as $Fe^{3+}$, $Al^{3+}$, $Cr^{3+}$, etc., $A^{n-}$ (present in the interlayer) is an inorganic or organic anion with charge n– such as hydroxyl, carbonate, nitrate, m is the number of water molecules present in the interlayer, and x is the layer charge density whose value varies between 0.22 and 0.33 and it is equal to $M^{3+}/(M^{2+} + M^{3+})$ [44].

The increasing demand for easily operable sensors and biosensors with high selectivity and sensitivity in industry has led to the assessment of different types of materials that possess intrinsic qualities desirable for sensor synthesis. The LDHs fulfil these requirements to a large extent with no extra cost or compromise in quality for sensitivity and selectivity. Moreover, the composition and structure of the LDHs can also be easily modified depending upon the need of an application, which further enhances their utilization in the field of sensors and biosensors. Several LDH-based electrochemical sensors, as mentioned in a review article [45], find their application for sensing different types of environmental pollutants such as heavy metal ions ($Cd^{2+}$, $Pb^{2+}$, $Hg^{2+}$), $H_2O_2$, pesticides (thiourea, methyl parathion, glyphosate), phenols (2-chlorophenol, pentachlorophenol, bisphenol A, catechol, hydroquinone, p-cresol, phenol, m-cresol), and some other environmental toxins such as nitrite and iodate. Furthermore, LDHs, when combined with natural or synthetic biopolymers, result in bionanocomposites which find their application as a catalyst, adsorbent, flame retardant, biopolymer reservoir, sensor, in drug and gene delivery, and UV protection [44]. Moreover, various LDH-based optical sensors utilizing colorimetric, chemiluminescence, and fluorometric sensing methods also have been fabricated by several researchers [46].

In the biosensor-designing domain, two types of LDH-based sensors are fabricated: enzymatic and non-enzymatic. Nowadays, the fabrication of non-enzymatic sensors is significantly promoted due to the low enzyme stability, high cost, and pH and temperature dependence of enzymatic sensors. However, non-enzymatic sensors' selectivity is not as high as that of enzymatic sensors but their sensitivity is excellent and the synthesis process is simple [11]. At present, the need for glucose sensors around the world is constantly increasing due to the increase in the numbers of diabetic patients. Therefore, LDHs are also exploited for glucose detection by various researchers to see their applicability in this growing area. A non-enzymatic glucose sensor based on CuAl-LDH composites was fabricated that detected glucose with a detection limit of 0.02 μM and a short response time of less than 5 s [47]. In this highly sensitive sensor, glucose was electrocatalytically oxidized on a glassy carbon electrode, modified by the synthesized composites of CuAl-LDHs. The developed sensor also exhibited good anti-interference capability and stability in the presence of various biological molecules, such as uric acid, vitamin C, acetaminophen, dopamine, and some carbohydrates. The sensor was successfully demonstrated for glucose detection in human blood samples. Due to the low-cost precursors and easy synthesis process, this sensor can be used practically as a disposable glucose sensor chip. Moreover, the same sensor might detect glucose in other samples (tear, saliva,

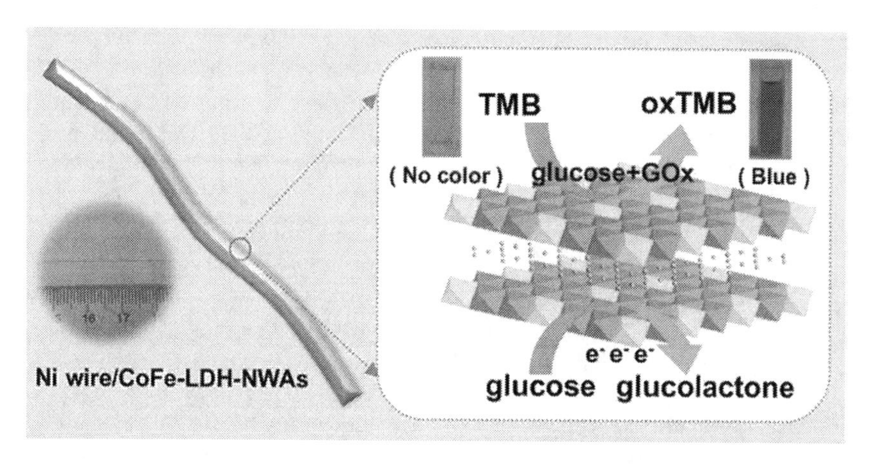


**FIGURE 5.5** Schematic representation of the Ni wire-CoFe-LDH-NSA and the electrochemical and colorimetric detection of glucose. (Reproduced from Ref. [48] with permission from the Royal Society of Chemistry. The article is licensed under a Creative Commons Attribution 3.0 Unported Licence.)

**TABLE 5.4**
**Some Representatives of Non-enzymatic Electrochemical Sensors Based on LDH Nanocomposites**

| Target Analyte | LDH Nanocomposite | Limit of Detection | Reference |
|---|---|---|---|
| Glucose | CuAl-LDHs | 0.02 μM | [47] |
| | Ni wire-CoFe-LDH-NSA | 0.27 μM | [48] |
| | Graphene quantum dots-CoNiAl-LDH | 6.0 μM | [49] |
| | NiCo-LDH/CC | 0.12 μM | [50] |
| | Au/NiAl-LDH-CNTs-Graphene | 1.0 μM | [51] |
| Lactate | NiCo-LDH | 0.53 mM | [52] |

*Note:* CC-carbon cloth.

and sweat) and can be used as a non-invasive sensor. Another, bifunctional enzyme-free glucose microsensor with high flexibility was synthesized using CoFe-LDH nanosheet array (NSA) on Ni wire, as shown in Figure 5.5 [48].

The Ni wire-CoFe-LDH-NSA fabricated sensor successfully detected the glucose electrochemically and colorimetrically with limits of detection of 0.27 μM and 0.47 μM, respectively. The sensor was highly sensitive and selective and displayed a good detection limit for glucose in a human urine sample also. This study provides an effective approach to the synthesis of multi-functional non-enzymatic glucose sensors. Some non-enzymatic electrochemical sensors based on LDHs nanocomposites are summarized in Table 5.4.

LDH-based electrochemical sensors are also utilized to detect the drug diazepam, which is commonly used as medication for anxiety, acute alcohol withdrawal, and some other ailments [53]. For making this sensor, CoAl-LDH nanoparticles were embedded in a poly(tyrosine) biopolymer, a novel sensitive platform for detecting diazepam electro-analytically. The sensor successfully detected the drug with a detection limit of 0.078 μM. The developed sensor displayed good stability, reproducibility, and also the anti-interference ability to various species present in a biological system such as ascorbic acid, uric acid, glucose, tryptophan, $K^+$, $Na^+$, $Mg^{2+}$, $Cl^-$,$NO_3^-$,$CO_3^{2-}$, and $SO_4^{2-}$. In addition, the sensor was investigated for real sample analysis such as commercial tablets and plasma, and reasonably good results were obtained, suggesting the utility of the sensor for accurate and consistent detection. Apart from the mentioned applications of the LDH-based electrochemical sensors, they are also exploited for sensing volatile organic compounds such as acetone, ammonia, chlorine vapours, and ethanol [54]. In this study, four different LDHs, namely, ZnAl-LDHs, ZnFe-LDHs with $Cl^-$ as the anion, ZnAl-LDHs, and MgAl-LDHs with $NO_3^-$ as the anion were considered to explore the sensing performances of the synthesized sensors. The sensors were able to detect volatile organic compounds, selectively and reversibly with up to 6% sensing response values at room temperature (21°C). The results suggested that the sensitivity and selectivity of the sensor can be further improved by altering the composition of the LDHs. The proposed microsensor could be potentially utilized for monitoring the indoor and outdoor quality of air and thus help in improving everyday life. The LDHs are also applied to detect other environmentally hazardous pollutants. For example, a NiAl-LDH was fabricated via a co-precipitation process and used in an electrochemical sensor to detect isoproturon, a pollutant reaching water bodies via surface run-off and leaching from agricultural soil [55]. In the sensor, the carbon paste electrode was modified by the prepared LDHs and better sensing results were obtained compared to an unmodified electrode. The proposed sensor was optimized with respect to various factors that can influence the stripping response of the sensor, and after optimization the sensor displayed a detection limit of 1 nM. Finally, the sensor was also tested to detect isoproturon in real spiked water samples. The results suggested that the fabricated sensor could be employed to sense the analyte in a real environment. Another recent study has reported an interesting application of a thin film of Li-Al-$CO_3$-LDH which is calcined at 300°C as a luminescence sensor for reversibly sensing $CO_3^{2-}$ in an aqueous medium [56]. In particular, when the $CO_3^{2-}$ anion was re-intercalated in the prepared sample of calcined LDH at 300°C, the photoluminescence (blue light) of the sample was quenched. In addition, when other anions ($F^-$, $Cl^-$, $Br^-$, $I^-$, $NO_3^-$, $SO_4^{2-}$ and $PO_4^{3-}$) were introduced in the LDH interlayer, no quenching of luminescence was observed, suggesting that the proposed thin film of LDH can be used effectively to sense the $CO_3^{2-}$ anion selectively in solution with a concentration range between 0 to 2500 ppm.

## 5.5 PLASMA POLYMERS

Unlike conventional polymers, plasma polymers' properties are not completely dependent on the type and structure of monomer used in the polymerization process. Plasma polymer characteristics are a function of the monomer, plasma used for polymerization, whether it is hot or cold plasma, plasma reactor, power, process time, gas composition, pressure, and distance between the plasma and substrate surface [5]. Plasma polymer synthesis also involves three stages, similar to conventional polymers: initiation, propagation, and termination. These polymers are highly branched and possess a high degree of cross-linkage. Furthermore, these polymers are highly selective and sensitive in detecting various molecules. The use of air or nitrogen ($N_2$) as plasma is popular at the industrial level owing to their low cost, but inert gas use along with some reactive gases such as $O_2$, $NH_3$, and steam is also mentioned in some studies. The properties of these polymers can be easily tuned and modified, leading to their application in various fields including sensors. The sensing properties of these polymers have been exploited for biomedical applications such as neurotransmitter, DNA, glucose, humidity, temperature, pressure, and pH sensors [5].

A plasma polymer using the selective-patterning method was synthesized via an $O_2$ plasma treatment at 500 W for 5 minutes, acting as a capacitive touch sensor [57]. This sensor is stretch-unresponsive stretchable and transparent, an advantage in the category of the capacitive touch sensor, which normally changes its capacitance upon dimensional changes in the dielectric material. The choice of a dielectric layer and substrate material, both being low responsive to strain, are the key parameters in the development of stretch-unresponsive touch sensors. This plasma polymer sensor consists of silver nanowires/reduced graphene oxide (stretchable and transparent) as electrodes, polyurethane (PU) as elastomeric dielectric, and polydimethylsiloxane (PDMS) as substrate. The Poisson's ratio of dielectric PU is lower than for PDMS (substrate), which makes this sensor immune to stretching. This capacitive touch sensor based on plasma polymer will be of great importance in the future wearable stretchable electronics and the interface between humans and machines. Another plasma polymer using $N_2$ plasma at 50 W for 3 minutes was prepared by Wang and co-workers [58], which exhibited the possibility of making a pressure sensor with high performance in the future. The tactile sensor was synthesized by combining $N_2$ plasma-modified PEDOT:PSS film with a patterned interdigitated electrode of indium-tin-oxide (ITO) as a layer on the flexible poly(ethylene terephthalate) (PET) substrates (Figure 5.6).

After plasma treatment for 3 min, the polymer film showed enhanced piezoresistive sensitivity and response owing to an increase in its water contact angle from 88° to 95°. After the 3 min exposure of $N_2$ plasma, the chemical structure of the PEDOT:PSS film changed, due to which an increment in the hydrophobicity of the polymer film was observed. The cleavage of an electrostatic bond between PEDOT & PSS and alteration in conjugation present at PEDOT chains was reported owing to the formation of new sulfamate and thiocyanate bonds, respectively. This plasma polymer possesses great potential for its application in making future pressure sensors with improved piezoresistive characteristics.

The application of plasma-treated polymers in temperature and humidity sensors has been reported in a few studies. These sensors can be utilized for biomedical applications, as demonstrated by Aliane and his co-workers [59]. The enhanced printed temperature sensor on flexible polyethylene naphthalate substrate having a large area was developed which showed enhanced sensitivity when treated with an $O_2$ plasma at 120 W for 1 min. The electrical sensitivity of the temperature sensor was found to be increased up to 0.025 V °C$^{-1}$ at normal human body temperature after plasma treatment.

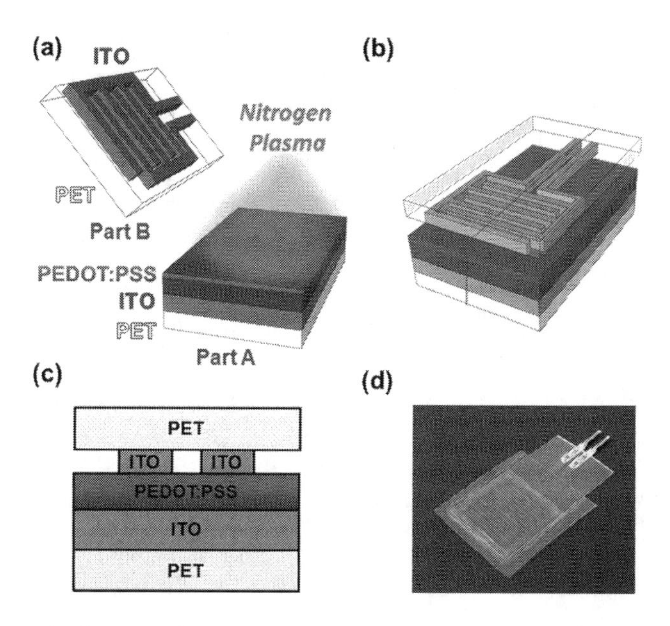


**FIGURE 5.6** Schematic diagrams of the PEDOT:PSS pressure sensor after $N_2$ plasma surface modification for (a) parts A and B, (b) the final device, (c) the cross-sectional view of a pressure sensor, and (d) an image of the fabricated device. (Reprinted with permission from Ref. [58]. Copyright 2016, American Chemical Society.)

More specifically, the resistance temperature coefficient of the antimony tin oxide, which acts as a sensitive layer, was found to be enhanced when treated with $O_2$ plasma. Furthermore, a demonstrator for detecting the temperature of the human body was developed based on this temperature sensor. A plasma polymer-based relative humidity (RH) sensor was prepared in a dc plasma reactor by exposing polymethyl methacrylate to Ar plasma at 5 W for 5 min [60]. In a relative humidity range of very low to nearly 90%, the sensitivity of the developed sensor was not linear but, in the range of 0–40%, the sensor was found to be usable. This was the first time that an RH sensor based on plasma polymer was reported in a study. Another humidity sensor was synthesized by an inductively coupled $CF_4/O_2$ plasma dry etching process at 900 W for 3 min [61]. After the plasma treatment, residue removal of $TiO_2$ from the surface of a functional polymer mixed with $TiO_2$ microparticles was carried out using ultrasonic-treated acetone. The etching was done in a controlled manner which modified the roughness at the polymer surface. More recently, a capacitive humidity sensor has been developed by the combination of two types of plasma treatments, namely inductively coupled $O_2$ plasma and $O_2$ reactive-ion etching on a functional polymer mixed with $TiO_2$ microparticles [62]. These two plasma treatments are carried out to etch the polymer surface and increase the roughness at the polymer surface, which is required to achieve high sensitivity. Consequently, the synthesized sensor exhibited high performance with respect to hysteresis, sensitivity, temperature dependability, response time, and stable capacitance.

A flexible temperature–humidity-based sensor was synthesized from graphene woven fabrics, flexible PDMS (substrate), and cellulose acetate butyrate, which acts as a dielectric layer by the process of chemical vapour deposition [63]. This sensor was composed of two parts, temperature-sensing and humidity-sensing, which were stacked in layer forms. The sensor exhibited increased temperature sensitivity upon $O_2$ plasma treatment for 1.5 min, while for sensing humidity plasma treatment was unnecessary. In a temperature range of 20–55°C, the temperature sensor was stable and sensitive, showing its applicability in biomedical devices to monitor health and for environmental sensing. Thus, the need for temperature and humidity sensors possessing high accuracy and stability can be realized using plasma polymers.

Plasma polymers have been reported to sense the biomolecules and are used as sensors in living systems. For example, glucose, DNA, dopamine, and pathogens can be detected via plasma-treated polymers and/or composites [5]. The use of plasma polymer pyrrole for detecting DNA and also cell adhesion has been reported owing to its chemical composition, electrochemical property, and capability of adsorbing protein [64]. The changes in electrochemical properties of the deposited polymer film were reported using electrochemical impedance spectroscopy under three different plasma conditions at 5, 50, and 100 W. More specifically, the surface charge transfer resistance of the polymer film after immobilizing the DNA at different concentrations was measured. Furthermore, some other parameters such as equilibrium concentration, association, and dissociation constant were also deduced from the experimental data to further quantify the sensing properties of synthesized plasma-polymerized pyrrole. This plasma polymer finds its application as a DNA sensor and offers a choice for the sensitive layer in biosensors under the category of biomaterials. In another study, composites of polyfuran/chitosan deposited onto ITO electrodes by plasma glow discharge at atmospheric pressure were reported to act as potential glucose sensors [65]. The properties of developed films of polyfuran/chitosan composites were studied electrochemically. The composite films prepared by plasma treatment were found to be stable and exhibited better immobilization properties as enzyme sensors compared to pure polyfuran films. Furthermore, a comparative study of polyfuran/glucose oxidase enzyme electrodes and polyfuran/chitosan/glucose oxidase was also carried out for sensing glucose via the amperometric method, and the former displayed better sensing properties. A linear relation between current response and glucose concentration was observed for the polyfuran/chitosan/glucose oxidase electrode, which further suggests the applicability of this plasma-polymerized biomaterial composite for the development of a potential glucose sensor. In two other recent studies [66,67], low-density polyethylene (LDPE) upon treatment with cold plasma in an ambient atmosphere for 1–2 min was transformed into an electroactive substance that can sense dopamine and glucose, selectively. Plasma-polymerized LDPE can detect these biomolecules selectively because of different reactive species, such as N, O, $O^+$, etc. on the surface of the polymer. Besides LDPE, some other electrochemically inert polymers such as polypropylene, polyvinylpyrrolidone, polycaprolactone, poly(4-vinylphenol), and polystyrene were converted into electrochemical sensors with similar plasma treatment. The transformed electrochemically active plasma polymer LDPE could oxidize dopamine to dopamine-o-quinone. The importance of this conversion can be understood by the fact that dopamine is an important neurotransmitter linked to some neurological diseases such as Parkinson's and schizophrenia. Furthermore, some selected conducting polymers such as PEDOT and poly(N-cyanoethylpyrrole) upon cold plasma treatment showed better sensitivity and selectivity with a higher resolution and, especially the latter, could discriminate dopamine from L-ascorbic acid and uric acid (acts as interferents) on plasma treatment. The sensors based on plasma polymers, synthesized by a simple physical modification, can replace sensors prepared from sophisticated and costly chemical transformations. Thus, the fabricated plasma polymers offer a choice of making electrochemical sensors from cheap materials, which can play a significant role in poor countries where cost is a major concern for manufacturing and application. Interestingly, an $O_2$ plasma-treated biosensor to detect bacterial and viral pathogens was synthesized using an electrospun nitrocellulose nanofibrous membrane based on capillary separation and conductometric immunoassay [68]. After an $O_2$ plasma treatment (120 W at 13.6 MHz) the electrospun nitrocellulose nanofibres showed enhanced capillary performance because of the removal of nitrate groups present at the surface and, also, the water contact angle decreased from 135° to 56°. This plasma-treated polymer was then functionalized with antibodies to sense bacterial and viral pathogens. Owing to the unique composition and biocompatibility, the prepared electrospun biosensor displayed a linear response for both the considered microbial samples, *Escherichia coli* O157:H7 and bovine viral diarrhoea virus. After changing the antibodies, this pathogen biosensor can detect other antimicrobial and viral organisms.

## 5.6 SENSORS IN BIOMEDICAL SYSTEMS

The need for better sensing devices in biomedical systems is rapidly increasing to timely diagnose various diseases, so that patients can be treated effectively with proper medication. Also, a better healthcare system requires good-quality devices embedded with highly sensitive and selective sensors. Nowadays, monitoring personal health is becoming a trend, so that the individual's lifestyle is improved. The advances in technology and synthesis processes lead to the formation of desirable composites that are required for making good-quality sensors using different materials. However, the sensors in the biomedical systems should be biocompatible with humans, degrade controllably with no toxic residues, and possess high performance. The answer to all these requirements is the polymeric biomaterials or biopolymers, the next-generation materials for the sensor synthesis. A biosensor can be mounted on skin, implanted surgically into the body, or can be a part of a biomedical device. There are different types of biosensors that are utilized in biomedical systems such as cardiac, respiration, implantable, blood pressure, wearable, etc. [12].

In the area of respiration monitoring biosensors, a chitosan/PPy composite film-based, quartz crystal microbalance humidity sensor was fabricated via a facile physical modification process [69]. The prepared humidity sensor exhibited high sensitivity and selectivity, low hysteresis, short response time, good repeatability, and high stability. The improved sensing ability was due to the hydrophilic groups present at the surface and the imperfect coating structure. This humidity sensor could sense a change of pattern of breathing with respect to rate and depth, quickly and accurately. In

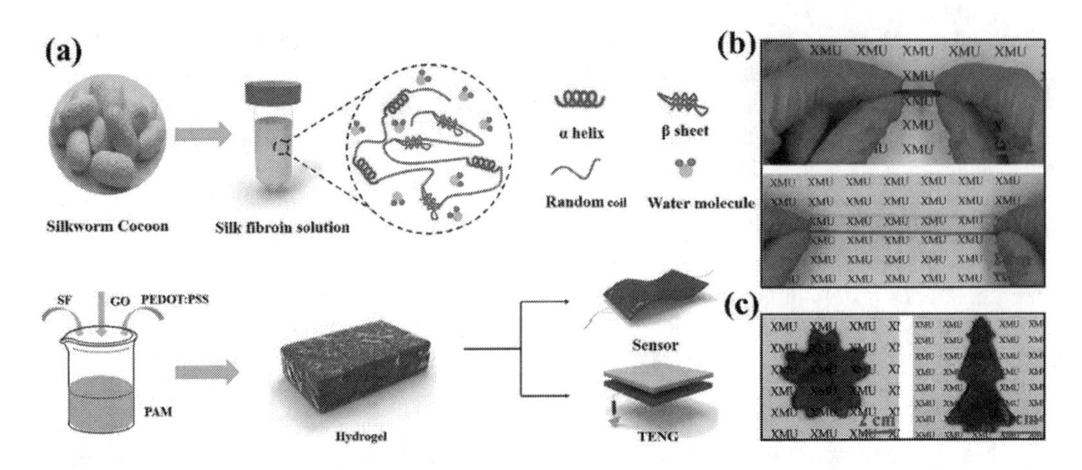


**FIGURE 5.7** (a) Schematic representation of SF-based conductive hydrogel, sensor, and triboelectric nanogenerator (TENG) from the starting materials. Digital photographs of the hydrogel (b) under stretching and (c) fabricated into different shapes. (Reprinted with permission from Ref. [70]. Copyright 2020, American Chemical Society.)

addition, an app was also designed which can accurately determine the rhythm of breathing and find its potential application in biomedical systems to monitor the respiratory pattern of patients. Another biosensor capable of monitoring various physical signals along with breathing was fabricated using SF-based conductive hydrogel [70]. The sensor and triboelectric nanogenerator was fabricated by mixing SF, polyacrylamide, graphene oxide, and PEDOT:PSS in a specific proportion, as shown in Figure 5.7.

The resultant wearable strain/pressure sensor was highly stretchable, compressible, multifunctional, sensitive, stable, and biocompatible, with a sensing range for strain and pressure of between 2–600% and 0.5–119.4 kPa, respectively. The sensor can sense facial gestures, joint movements, pulse, along with many other body signals. The synthesized SF-based conductive hydrogel can be used in soft robots, wearable electronics to monitor health and exercise, and in power sources due to its positive response to triboelectric nanogenerators. Similarly, based on SF conductive hydrogel, a degradable mechanical biosensor that can be integrated into the body and triggered by an enzyme was synthesized to monitor health and *in situ* treatment of chronic disorders [71]. The proposed wearable sensor possesses high flexibility, stretchability, and can be implanted into the human body due to its excellent and outstanding mechanical strength. In addition, when the AuNP-doped conductive hydrogel was combined with a temperature-activated papain enzyme, the sensor degradation can be activated by light. The sensor could detect multiple signals that are mechanical in nature such as strain, bending angles, and pressure. Furthermore, after integrating with a drug-loaded SF-based microneedle array, the sensor device was highly effective for real-time supervision and *in situ* treatment of a chronic disorder, epilepsy, when tested on a rodent model. This uniquely designed sensor can be potentially applied in the medication of chronic diseases, soft robotics, and patient therapy. Another wearable sensor based on silk film was fabricated to monitor the oral cavity and for *in situ* detection of food consumption [72]. The developed radiofrequency trilayer sensor could be mounted on a tooth, as shown in Figure 5.8.

For the synthesis of this dielectric sensor, a conformal radiofrequency construct was developed, composed of an active silk film or a responsive hydrogel [poly(N-isopropylacrylamide)] as a layer condensed between two reverse-facing split-ring resonators. The interlayer present in the construct swells on absorbing surrounding solvents, which result in changes in thickness and dielectric constant and, thereby, a change in the resonant frequency was observed in the sensor. The interlayer

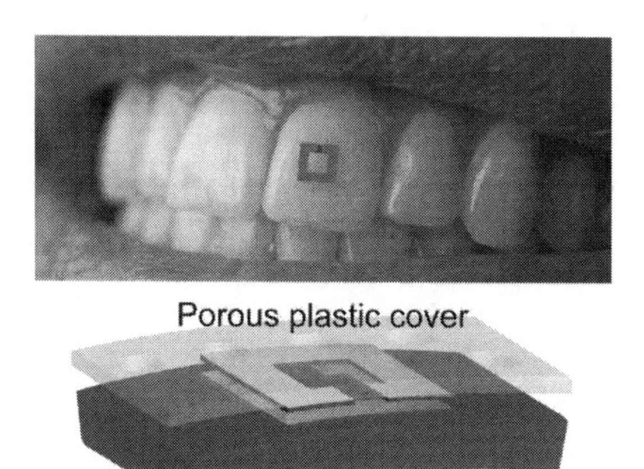


**FIGURE 5.8** SF-based radiofrequency trilayer dielectric sensor for tooth-mounted, wireless monitoring of the oral cavity and food consumption. (Reprinted with permission from Ref. [72]. Copyright 2018 WILEY-VCH Verlag GmbH & Co. KGaA, Weinheim.)

materials were responsive to pH, temperature, or analytes. The size of the dielectric sensing device limits their practical use, but in this study the size of the sensor was at a millimetre-scale (2 mm × 2 mm), which enhances the practical applicability of the sensor. The results of the *in vivo* demonstration of the device showed that the sensor possesses high sensitivity, excellent reproducibility, and good stability. Furthermore, the sensitivity of the sensor can be further increased by functionalization of the interlayer materials. The prepared radiofrequency wearable sensor offers a new direction to the application of such dielectric biosensors in different environments and permits multiplexed and distributed sensing by the trilayer structure. Moreover, the format of this sensor could be utilized for other interlayer materials that are normally employed in different sensors for a specific application.

Recently, the demand for artificial electronic skin (E-skin) has increased explosively owing to its significant potential in the field of wearable electronics to monitor health/fitness and disease diagnostics. The prime feature of E-skin is its highly sensitive pressure detection ability. However, to date, most E-skin is based on materials that have unknown biotoxicity, which thus limits their practical use. A sensitive and transparent E-skin was synthesized using active materials derived from SF due to its high flexibility and biocompatibility, as shown in Figure 5.9 [73].

In particular, carbonized silk nanofibre membranes (CSilkNM) were utilized in combination with unstructured PDMS films. The developed pressure sensor displayed excellent performance with a high sensitivity of 34.47 $kPa^{-1}$, a fast response time of less than 16.7 ms, a low detection limit of 0.8 Pa, and high durability of over 10,000 cycles. Owing to the remarkable performance of the proposed sensor, its demonstration to monitor human physiological signals such as pulse rate, respiration, etc., to sense delicate touch, and to detect the spatial distribution of pressure was shown in the study.

## 5.7 CONCLUSION

The polymeric biomaterials/biopolymers, whether they are natural or synthetic, are abundant in nature and can be easily functionalized to be utilized in specific applications. These materials are highly flexible, stretchable, stable, biocompatible, low toxic, low cost, easy to synthesize, degradable, sensitive, and selective. Moreover, the properties of the polymeric biomaterials can be tuned and modified depending on the need of an application. The polymeric biomaterials such as conducting biopolymers, LDHs, plasma polymers, or composites of these materials find their application in making various types of sensors. These materials can sense a change in pressure, strain,

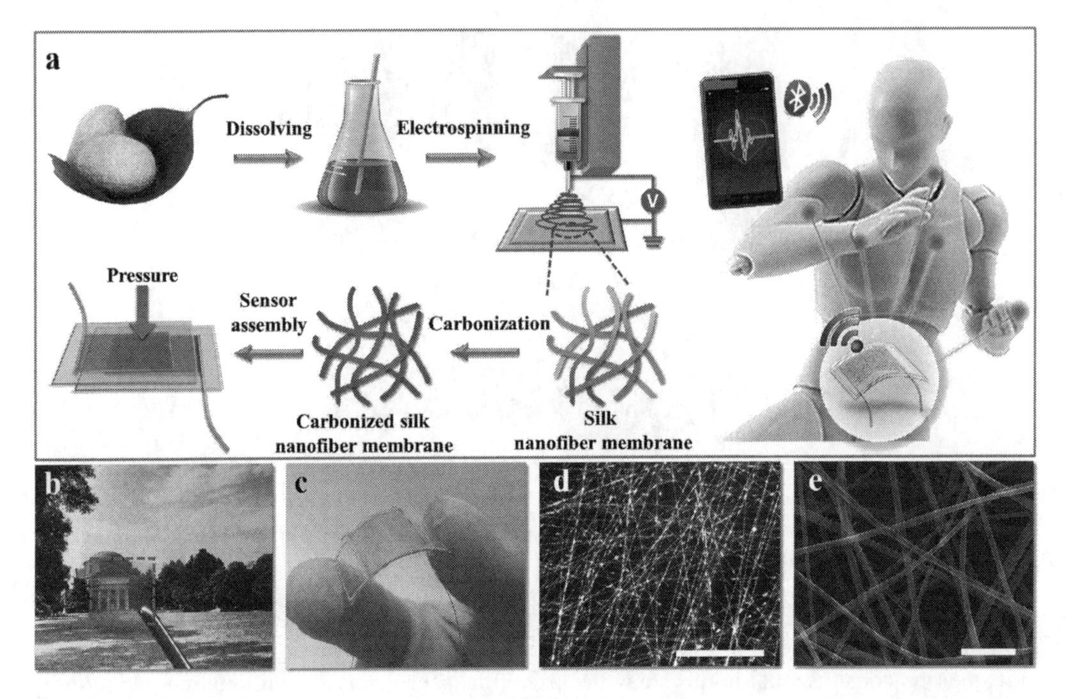


**FIGURE 5.9** Synthesis process and structure of the CSilkNM pressure sensor. (a) Schematic representation showing the fabrication process of CSilkNM pressure sensors. Photographs showing the (b) transparency, and (c) flexibility of the sensor. (d) Optical image of a silk nanofibre membrane. Scale bar: 100 μm. (e) SEM image of CSilkNM. Scale bar: 1 μm. (Reprinted with permission from Ref. [73]. Copyright 2017 WILEY-VCH Verlag GmbH & Co. KGaA, Weinheim.)

pH, temperature, concentration, light, sound, humidity, pathogen, DNA, glucose, lactate, etc. with high sensitivity, and high selectivity. Various types of sensors or biosensors that are fabricated using these biopolymers exhibited good performance when tested in real time, signifying the practical applications of these sensors. Most of the reported sensors fall into the category of optical and electrochemical due to their excellent detection ability for analytes with a low detection limit and short response time, which are the most desirable characteristics of a sensor. The sensors based on polymeric biomaterials are also employed in biomedical systems owing to their controllable degradability, low toxicity, and outstanding biocompatibility. However, the sensors in the biomedical systems are in their infancy due to some limitations, such as size, unknown biotoxicity that may occur, limited real-time experimentation, and in vivo study considering limited variables. However, these polymeric biomaterials could be a road to future sensors with more research in designing and fabrication.

## REFERENCES

1. Cui, C., Fu, Q., Meng, L., Hao, S., Dai, R., & Yang, J. (2021). Recent progress in natural biopolymers conductive hydrogels for flexible wearable sensors and energy devices: materials, structures, and performance. *ACS Appl. Bio Mater.*, 4, 85–121. doi: 10.1021/acsabm.0c00807
2. Shan, D., Gerhard, E., Zhang, C., Tierney, J.W., Xie, D., Liu, Z., & Yang, J. (2018). Polymeric biomaterials for biophotonic applications. *Bioact. Mater.*, 3, 434–445. doi: 10.1016/j.bioactmat.2018.07.001
3. Sun, Q., Qian, B., Uto, K., Chen, J., Liu, X., & Minari, T. (2018). Functional biomaterials towards flexible electronics and sensors. *Biosens. Bioelectron.*, 119, 237–251. doi: 10.1016/j.bios.2018.08.018

4. Annu, & Raja, A.N. (2020). Recent development in chitosan-based electrochemical sensors and its sensing application. *Int. J. Biol. Macromol.*, 164, 4231–4244. doi: 10.1016/j.ijbiomac.2020.09.012
5. Alemán, C., Fabregat, G., Armelin, E., Buendía, J.J., & Llorca, J. (2018). Plasma surface modification of polymers for sensor applications. *J. Mater. Chem. B*, 6, 6515–6533. doi: 10.1039/C8TB01553H
6. Ramanavicius, S., & Ramanavicius, A. (2021). Conducting polymers in the design of biosensors and biofuel cells. *Polymers*, 13, 49. doi: 10.3390/polym13010049
7. Huang, Y., Kormakov, S., He, X., Gao, X., Zheng, X., Liu, Y., et al. (2019a). Conductive polymer composites from renewable resources: an overview of preparation, properties, and applications. *Polymers*, 11, 187. doi: 10.3390/polym11020187
8. Ramdzan, N.S.M., Fen, Y.W., Anas, N.A.A., Omar, N.A.S., & Saleviter, S. (2020). Development of biopolymer and conducting polymer-based optical sensors for heavy metal ion detection. *Molecules*, 25, 2548. doi: 10.3390/molecules25112548
9. Terán-Alcocer, Á., Bravo-Plascencia, F., Cevallos-Morillo, C., & Palma-Cando, A. (2021). Electrochemical sensors based on conducting polymers for the aqueous detection of biologically relevant molecules. *Nanomaterials*, 11, 252. doi:10.3390/nano11010252
10. Arrabito, G., Pezzilli, R., Prestopino, G., & Medaglia, P.G. (2020). Layered double hydroxides in bioinspired nanotechnology. *Crystals*, 10(7), 602. doi:10.3390/cryst10070602
11. Grosu, E.F., Simiuc, D., & Froidevaux, R. (2018). Layered double hydroxides nanomaterials in biomedicine and (bio) sensing design. *Biomed. J. Sci. & Tech. Res.*, 2, 2747–2752. doi: 10.26717/BJSTR.2018.2.000786
12. Biswas, M.C., Jony, B., Nandy, P.K., Chowdhury, R.A., Halder, S., Kumar, D., et al. (2021). Recent advancement of biopolymers and their potential biomedical applications. *J. Polym. Environ.* 30, 51–74. doi: 10.1007/s10924-021-02199-y
13. Jadoun, S., Riaz, U., & Budhiraja, V. (2021). Biodegradable conducting polymeric materials for biomedical applications: a review. *Med. Devices Sens.*, 4, e10141. doi: 10.1002/mds3.10141
14. da Silva, A.C., & de Torresi, S.I.C. (2019). Advances in conducting, biodegradable and biocompatible copolymers for biomedical applications. *Front. Mater.*, 6, 98. doi:10.3389/fmats.2019.00098
15. Pal, R.K., Farghaly, A.A., Wang, C., Collinson, M.M., Kundu, S.C., & Yadavalli, V.K. (2016). Conducting polymer-silk biocomposites for flexible and biodegradable electrochemical sensors. *Biosens. Bioelectron.*, 81, 294–302. doi: 10.1016/j.bios.2016.03.010
16. Wang, C., Li, X., Gao, E., Jian, M., Xia, K., Wang, Q., et al. (2016a). Carbonized silk fabric for ultrastretchable, highly sensitive, and wearable strain sensors. *Adv. Mater.*, 28, 6640–6648. doi: 10.1002/adma.201601572
17. Farjana, S., Toomadj, F., Lundgren, P., Sanz-Velasco, A., Naboka, O., & Enoksson, P. (2013). Conductivity-dependent strain response of carbon nanotube treated bacterial nanocellulose. *J. Sens.*, 2013, 741248. doi: 10.1155/2013/741248
18. Hosseini, H., Kokabi, M., & Mousavi, S.M. (2018). Conductive bacterial cellulose/multiwall carbon nanotubes nanocomposite aerogel as a potentially flexible lightweight strain sensor. *Carbohydr. Polym.*, 201, 228–235. doi: 10.1016/j.carbpol.2018.08.054
19. Tsai, Y.C., Chen, S.Y., & Liaw, H.W. (2007). Immobilization of lactate dehydrogenase within multiwalled carbon nanotube-chitosan nanocomposite for application to lactate biosensors. *Sens. Actuators B Chem.*, 125, 474–481. doi: 10.1016/j.snb.2007.02.052
20. Shan, C., Yang, H., Han, D., Zhang, Q., Ivaska, A., & Niu, L. (2010). Graphene/AuNPs/chitosan nanocomposites film for glucose biosensing. *Biosens. Bioelectron.*, 25(5), 1070–1074. doi: 10.1016/j.bios.2009.09.024
21. Suginta, W., Khunkaewla, P., & Schulte, A. (2013). Electrochemical biosensor applications of polysaccharides chitin and chitosan. *Chem. Rev.*, 113, 5458–5479. doi: 10.1021/cr300325r
22. Mironenko, A., Modin, E., Sergeev, A., Voznesenskiy, S., & Bratskaya, S. (2014). Fabrication and optical properties of chitosan/Ag nanoparticles thin film composites. *Chem. Eng. J.*, 244, 457–463. doi: 10.1016/j.cej.2014.01.094
23. Chen, L.H., Li, T., Chan, C.C., Menon, R., Balamurali, P. Shaillender, M., Neu, B., et al. (2012). Chitosan based fiber-optic Fabry-Perot humidity sensor. *Sensor. Actuator. B Chem.*, 169, 167–172. doi: 10.1016/j.snb.2012.04.052

24. Voznesenskiy, S.S., Sergeev, A.A., Mironenko, A.Yu., Bratskaya, S.Yu., & Kulchin, Yu.N. (2013). Integrated-optical sensors based on chitosan waveguide films for relative humidity measurements. *Sensor. Actuator. B Chem.*, 188, 482–487. doi: 10.1016/j.snb.2013.07.043
25. Pasche, S., Angeloni, S., Ischer, R., Liley, M., Luprano, J., & Voirin, G. (2008). Wearable biosensors for monitoring wound healing. *Adv. Sci. Technol.*, 57, 80–87. doi: 10.4028/www.scientific.net/ast.57.80
26. Lee, M., Jeon, H., & Kim, S. (2015). A highly tunable and fully biocompatible silk nanoplasmonic optical sensor. *Nano Lett.*, 15, 3358–3363. doi: 10.1021/acs.nanolett.5b00680
27. Li, C., Wu, Z., Yang, H., Deng, L., & Chen, X. (2017). Reduced graphene oxide-cyclodextrin-chitosan electrochemical sensor: effective and simultaneous determination of o- and p-nitrophenols. *Sens. Act. B: Chemical*, 251, 446–454. doi: 10.1016/j.snb.2017.05.059
28. Yu, L., Yue, X., Yang, R., Jing, S., & Qu, L. (2016). A sensitive and low toxicity electrochemical sensor for 2, 4-dichlorophenol based on the nanocomposite of carbon dots, hexadecyltrimethyl ammonium bromide and chitosan. *Sens. Act. B: Chemical*, 224, 241–247. doi: 10.1016/j.snb.2015.10.035
29. Bakhsh, E.M., Ali, F., Khan, S.B., Marwani, H.M., Danish, E.Y., & Asiri, A.M. (2019). Copper nanoparticles embedded chitosan for efficient detection and reduction of nitroaniline. *Int. J. Biol. Macromol.*, 131, 666–675. doi: 10.1016/j.ijbiomac.2019.03.095
30. Pavinatto, A., Mercante, L.A., Leandro, C.S., Mattoso, L.H.C., & Correa, D.S. (2015). Layer-by-layer assembled films of chitosan and multi-walled carbon nanotubes for the electrochemical detection of 17α – ethinylestradiol. *J. Electroanal. Chem.*, 755, 215–220. doi: 10.1016/j.jelechem.2015.08.002
31. Xu, Z., Fan, X., Ma, Q., Tang, B., Lu, Z., Zhang, J., et al. (2019). A sensitive electrochemical sensor for simultaneous voltammetric sensing of cadmium and lead based on $Fe_3O_4$/multiwalled carbon nanotube/laser scribed graphene composites functionalized with chitosan modified electrode. *Mater. Chem. Phys.*, 238, 121877. doi: 10.1016/j.matchemphys.2019.121877
32. Saha, S., & Sarkar, P. (2016). Differential pulse anodic stripping voltammetry for detection of As (III) by chitosan-$Fe(OH)_3$ modified glassy carbon electrode: a new approach towards speciation of arsenic. *Talanta*, 158, 235–245. doi: 10.1016/j.talanta.2016.05.053
33. Hu, L., Zhang, L., Zhou, Y., Meng, G., Yu, Y., Yao, W., & Yan, Z. (2018). Chitosan-stabilized gold nano composite modified glassy carbon electrode for electrochemical sensing trace $Hg^{2+}$ in practice. *J. Electrochem. Soc.*, 165, B900–B905. doi: 10.1149/2.1101816jes
34. Salimi, A., Pourbahram, B., Majd, S.M., & Hallaj, R. (2015). Manganese oxide nanoflakes/multiwalled carbon nanotubes/chitosan nanocomposite modified glassy carbon electrode as a novel electrochemical sensor for chromium (III) detection. *Electrochim. Acta*, 156, 207–215. doi: 10.1016/j.electacta.2014.12.146
35. Bano, S., Ganie, A.S., Sultana, S., Khan, M.Z., & Sabir, S. (2021). The non-enzymatic electrochemical detection of glucose and ammonia using ternary biopolymer based-nanocomposites. *New J. Chem.*, 45, 8008–8021. doi: 10.1039/D1NJ00474C
36. Sharma, A., & Kumar, A. (2016). Study of structural and electro-catalytic behaviour of amperometric biosensor based on chitosan/polypyrrole nanotubes-gold nanoparticles nanocomposites. *Syn. Metals.*, 220, 551–559. doi: 10.1016/j.synthmet.2016.07.012
37. Choi, Y.B., Kim, H.S., Jeon, W.Y., Lee, B.H., Shin, U.S., & Kim, H.H. (2019). The electrochemical glucose sensing based on the chitosan-carbon nanotube hybrid. *Biochem. Eng. J.*, 144, 227–234. doi: 10.1016/j.bej.2018.10.021
38. Asrami, P.N., Mozaffari, S.A., Tehrani, M.S., & Azar, P.A. (2018). A novel impedimetric glucose biosensor based on immobilized glucose oxidase on a CuO-Chitosan nanobiocomposite modified FTO electrode. *Int J Biol Macromol.*, 118(Pt A), 649–660. doi: 10.1016/j.ijbiomac.2018.05.228
39. Chaichi, M.J., & Ehsani, M. (2016). A novel glucose sensor based on immobilization of glucose oxidase on the chitosan-coated $Fe_3O_4$ nanoparticles and the luminol-$H_2O_2$-gold nanoparticle chemiluminescence detection system. *Sens. Act. B: Chemical*, 223, 713–722. doi: 10.1016/j.snb.2015.09.125
40. Khalaf, N., Ahamad, T., Naushad, M., Al-Hokbany, N., Al-Saeedi, S.I., Almotairi, S., & Alshehri, S.M. (2020). Chitosan polymer complex derived nanocomposite (AgNPs/NSC) for electrochemical non-enzymatic glucose sensor. *Int J Biol Macromol.*, 146, 763–772. doi: 10.1016/j.ijbiomac.2019.11.193
41. Kangkamano, T., Numnuam, A., Limbut, W., Kanatharana, P., & Thavarungku, P. (2017). Chitosan cryogel with embedded gold nanoparticles decorated multiwalled carbon nanotubes modified electrode for highly sensitive flow based non-enzymatic glucose sensor. *Sens. Act. B: Chemical*, 246, 854–863. doi: 10.1016/j.snb.2017.02.105

42. Gomes, N.O., Carrilho, E., Machado, S.A.S., & Sgobbi, L.F. (2020). Bacterial cellulose-based electrochemical sensing platform: a smart material for miniaturized biosensors. *Electrochimica Acta*, 349, 136341. doi: 10.1016/j.electacta.2020.136341
43. Pontié, M., Mbokou, S.F., Bouchara, J., Razafimandimby, B., Egloff, S., Dzilingomo, O. et al. (2018). Paracetamol sensitive cellulose-based electrochemical sensors. *J. Renew. Mater.*, 6, 242–250. doi:10.7569/JRM.2017.634169
44. Chatterjee, A., Bharadiya, P., & Hansora, D. (2019). Layered double hydroxide based bionanocomposites. *Appl. Clay Sci.*, 177, 19–36. doi: 10.1016/j.clay.2019.04.022
45. Baig, N., & Sajid, M. (2017). Applications of layered double hydroxides based electrochemical sensors for determination of environmental pollutants: a review. *Trends Environ. Anal. Chem.*, 16, 1–15. doi: 10.1016/j.teac.2017.10.003
46. Munyemana, J.C., Chen, J., Han, Y., Zhang, S., & Qiu. H. (2021). A review on optical sensors based on layered double hydroxides nanoplatforms. *Microchim Acta*, 188, 80. doi: 10.1007/s00604-021-04739-8
47. Wang, F., Zhang, Y., Liang, W., Chen, L., Li, Y., & He, X. (2018). Non-enzymatic glucose sensor with high sensitivity based on Cu-Al layered double hydroxides. *Sens. Actuators B Chem.*, 273, 41–47. doi: 10.1016/J.SNB.2018.06.038
48. Cui, J., Li, Z., Liu, K., Li, J., & Shao, M. (2019). A bifunctional nonenzymatic flexible glucose microsensor based on CoFe-Layered double hydroxide. *Nanoscale Adv.*, 1, 948–952. doi: 10.1039/C8NA00231B
49. Samuei, S., Fakkar, J., Rezvani, Z., Shomali, A., & Habibi, B. (2017). Synthesis and characterization of graphene quantum dots/CoNiAl-layered double-hydroxide nanocomposite: application as a glucose sensor. *Anal Biochem.*, 521, 31–39. doi: 10.1016/j.ab.2017.01.005
50. Wang, X., Zheng, Y., Yuan, J., Shen, J., Hu, J., Wang, A.J., et al. (2017a). Three-dimensional NiCo layered double hydroxide nanosheets array on carbon cloth, facile preparation and its application in highly sensitive enzymeless glucose detection. *Electrochim Acta*, 224, 628–635. doi: 10.1016/j.electacta.2016.12.104
51. Fu, S., Fan, G., Yang, L., & Li, F. (2015). Non-enzymatic glucose sensor based on Au nanoparticles decorated ternary Ni-Al layered double hydroxide/single-walled carbon nanotubes/graphene nanocomposite. *Electrochim Acta*, 152, 146–154. doi: 10.1016/j.electacta.2014.11.115
52. Wu, Y.T., Tsao, P.K., Chen, K.J., Lin, Y.C., Aulia, S., Chang, L.Y., et al. (2021). Designing bimetallic Ni-based layered double hydroxides for enzyme-free electrochemical lactate biosensors. *Sens. Actuators B Chem.*, 346, 130505. doi: 10.1016/j.snb.2021.130505
53. Amini, R., & Zeynali, K.A. (2019). Layered double hydroxide nanoparticles embedded in a biopolymer: a novel platform for electroanalytical determination of diazepam. *New J. Chem.*, 43, 7463–7470. doi: 10.1039/C8NJ06325G
54. Vigna, L., Nigro, A., Verna, A., Ferrari, I.V., Marasso, S.L., Bocchini, S., et al. (2021). Layered double hydroxide-based gas sensors for VOC detection at room temperature. *ACS Omega*, 6(31), 20205–20217. doi: 10.1021/acsomega.1c02038
55. Tcheumi, H.L., Kameni, W., Aude, P., Tonle, I.K., & Ngameni, E. (2020). A low-cost Layered Double Hydroxide (LDH) based amperometric sensor for the detection of isoproturon in water using carbon paste modified electrode. *J Anal Methods Chem*, 2020, 8068137. doi: 10.1155/2020/8068137
56. Huang, S.H., Liu, S.J., & Uan, J.Y. (2019b). Controllable luminescence of a Li–Al layered double hydroxide used as a sensor for reversible sensing of carbonate. *J. Mater. Chem. C*, 7, 11191–11206. doi: 10.1039/C9TC00870E
57. Choi, T.Y., Hwang, B.U., Kim, B.Y., Trung, T.Q., Nam, Y.H., Kim, D.N., et al. (2017). Stretchable, transparent, and stretch-unresponsive capacitive touch sensor array with selectively patterned silver nanowires/reduced graphene oxide electrodes. *ACS Appl. Mater. Interfaces*, 9, 18022–18030. doi: 10.1021/acsami.6b16716
58. Wang, J.C., Karmakar, R.S., Lu, Y.J., Wu, M.C., & Wei, K.C. (2016b). Nitrogen plasma surface modification of poly(3,4-ethylenedioxythiophene): poly(styrenesulfonate) films to enhance the piezoresistive pressure-sensing properties. *J. Phys. Chem. C*, 120, 25977. doi: 10.1021/acs.jpcc.6b09642
59. Aliane, A., Fischer, V., Galliari, M., Tournon, L., Gwoziecki, R., Serbutoviez, C., et al. (2014). Enhanced printed temperature sensors on flexible substrate. *Microelectronics J*, 45, 1621–1626. doi: 10.1016/j.mejo.2014.08.011

60. Dabhade, R.V., Bodas, D.S., & Gangal, S.A. (2004). Plasma-treated polymer as humidity sensing material-a feasibility study. *Sens. Actuators B*, 98, 37–40. doi: 10.1016/j.snb.2003.08.020
61. Liu, M.Q., Wang, C., Yao, Z., & Kim, N.Y. (2016). Dry etching and residue removal of functional polymer mixed with $TiO_2$ microparticles via inductively coupled $CF_4/O_2$ plasma and ultrasonic-treated acetone for humidity sensor application. *RSC Adv.*, 6, 41580–41586. doi: 10.1039/C6RA07688B
62. Qiang, T., Wang, C., Liu, M.Q., Adhikari, K.K., Liang, J.G., Wang, L., et al. (2018). High-Performance porous MIM-type capacitive humidity sensor realized via inductive coupled plasma and reactive-ion etching. *Sens. Actuators B*, 258, 704–714. doi: 10.1016/j.snb.2017.11.060
63. Zhao, X., Long, Y., Yang, T., Li, J., & Zhu, H. (2017). Simultaneous high sensitivity sensing of temperature and humidity with graphene woven fabrics. *ACS Appl. Mater. Interfaces*, 9, 30171–30176. doi: 10.1021/acsami.7b09184
64. Zhang, Z., Liu, S., Shi, Y., Dou, J., & Fang, S. (2014). DNA detection and cell adhesion on plasma-polymerized pyrrole. *Biopolymers*, 101, 496–503. doi: 10.1002/bip.22408
65. Turkaslan, B.E., Aktan, T., Oksuz, L., & Oksuz, A.U. (2016). Use of polyfuran/chitosan composite films deposited by atmospheric pressure plasma glow discharge as glucose sensors. *Asian J. Chem.*, 28, 941–946. doi: 10.14233/ajchem.2016.19093
66. Buendía, J.J., Fabregat, G., Castedo, A., Llorca, J., & Alemán, C. (2017). Plasma-treated polyethylene as electrochemical mediator for enzymatic glucose sensors: toward bifunctional glucose and dopamine sensors. *Plasma Process Polym.*, e1700133. doi: 10.1002/ppap.201700133
67. Fabregat, G., Osorio, J., Castedo, A., Armelin, E., Buendía, J.J., Llorca, J., & Alemán, C. (2017). Plasma functionalized surface of commodity polymers for dopamine detection. *Appl. Surf. Sci.*, 399, 638–647. doi: 10.1016/j.apsusc.2016.12.137
68. Luo, Y., Nartker, S., Miller, H., Hochhalter, D., Wiederoder, M., Wiederoder, S., et al. (2010). Surface functionalization of electrospun nanofibers for detecting E. coli O157:H7 and BVDV cells in a direct-charge transfer biosensor. *Biosens. Bioelectron.*, 26, 1612–1617. doi: 10.1016/j.bios.2010.08.028
69. Liu, X., Zhang, D., Wang, D., Li, T., Song, X., & Kang, Z. (2021). A humidity sensing and respiratory monitoring system constructed from quartz crystal microbalance sensors based on a chitosan/polypyrrole composite film. *J. Mater. Chem. A*, 9, 14524–14533. doi: 10.1039/D1TA02828F
70. He, F., You, X., Gong, H., Yang, Y., Bai, T., Wang, W., et al. (2020). Stretchable, biocompatible, and multifunctional silk fibroin-based hydrogels toward wearable strain/pressure sensors and triboelectric nanogenerators. *ACS Appl. Mater. Interfaces*, 12, 6442–6450. doi: 10.1021/acsami.9b19721
71. Zhang, S., Zhou, Z., Zhong, J., Shi, Z., Mao, Y., & Tao, T.H. (2020). Body-integrated, enzyme-triggered degradable, silk-based mechanical sensors for customized health/fitness monitoring and in situ treatment. *Adv. Sci.*, 7, 1903802. doi: 10.1002/advs.201903802
72. Tseng, P., Napier, B., Garbarini, L., Kaplan, D.L., & Omenetto, F.G. (2018). Functional, RF-trilayer sensors for tooth-mounted, wireless monitoring of the oral cavity and food consumption. *Adv. Mater.*, 30, e1703257. doi: 10.1002/adma.201703257
73. Wang, Q., Jian, M.Q., Wang, C.Y., & Zhang, Y.Y. (2017b). Carbonized silk nanofiber membrane for transparent and sensitive electronic skin. *Adv. Funct. Mater.*, 27, 1605657. doi: 10.1002/adfm.201605657

# 6 Biocomposites in Food Packaging

*Garima Gupta*

## CONTENTS



## 6.1 INTRODUCTION

Packaging is considered as a most important unit operation in the production of industrial products as it preserves the quality of food during storage, transportation, redistribution and consumption [1]. The most common raw materials for manufacturing plastic packaging are petroleum-based polymers such as polypropylene (PP), polyester (PET), polyethylene (PE) and polystyrene (PS). Most of these traditional plastics are recyclable, but many countries are experiencing technical and economic challenges in reusing these types of plastic packaging [2,3]. The nonbiodegradable nature of petroleum-based plastics has generated major waste accumulation and pollution

DOI: 10.1201/9781003240884-6

problems in clean water systems and residential areas around the world [4]. The majority of single-use plastic packaging from the food industry ends up being dumped in water streams or landfills and, finally, reaches the sea [2,5]. Around 6300 million metric tons (Mt) of nondegradable petroleum-based plastic waste was produced in 2015 worldwide, with only 9% recycled, 12% combusted, and most of it (79%) either being accumulated in waste dumps or distributed in the ecosystem [(4]. By 2050, it is expected that the rate of plastic waste recycling will increase by 40%, but the accumulation of plastic waste in the ecosystem and landfills will also rise to approximately 12,000 Mt per year. It is predicted that, by 2025, for every 3 tons of marine fish there will be one ton of plastic waste, and by 2050 the amount of plastic waste will exceed the amount of fish [6]. At the same time, continued dependence on fossil fuels will cause a depletion of raw materials and thus increase the cost of products [7]. Manufacturers, retailers and consumers, in line with the current regulatory guidelines of governments on sustainable environment, are working to design an alternative to petroleum-based plastic which will be considered as green packaging. In order to completely develop the packaging sector and experience sustainable growth, strong renewable resources are required [7]. Biodegradable materials are now being given greater prominence for their carbon-reduction benefits compared with traditional plastics. Biodegradable plastics aims to recycle the biomass life cycle that includes the conservation of fossil fuels, water and $CO_2$ production [6,7].

Economies based upon biomaterials are growing exponentially around the world and with increasing support from global collaborations and government policies that have shown potential for a sustainable future. Plans have been developed within North America, Asia Pacific and Europe to maximize the utilization of biomass for the development of sustainable commercial systems [8]. At present, the consumption of biomass worldwide accounts for only 3 percent, which opens up the opportunity of utilizing lignocellulosic biomass as a starting material for organic polymer, biofuel and bioenergy production. Applications of biomass in revolutionizing industries with reference to recent global issues of fuel, energy, sustainability and food has always been on the backfoot because of the easy availability of fossil fuels. In order to enhance its availability, we require improving distribution chains, well-designed new business models with the support of the government and the consumers. Application of biomass in the food packaging industry involves the construction of polymers and fillers of existing plastics in order to make them environment friendly. The major factors governing innovation in food safety, quality and freshness are the increased demand for packaging material which can resist microbial contamination against foodborne pathogens like *Salmonella* spp., *E. coli* and *Listeria* [9–11]. Antibacterial packaging that increases the shelf life of minimally processed foods is preferred over traditional ingredients and manufacturing practices. Current innovations in the field of active packaging which utilize antimicrobial agents to provide resistance towards microbial contamination of food products [11] are revolutionizing the market of packaging materials. The main objective of food packaging is to provide affordable food by satisfying the needs of industry and consumers while at the same time maintaining food security and minimizing the environmental impact [12]. Packaged food products contaminate the plastic packaging, which requires extra extraction steps before recycling, thus increasing the cost of recycling [13]. The annual filling of plastic packaging has raised environmental concerns, leading to the evolution of biodegradability which proves to be a sustainable solution for synthetic packaging.

Biocomposites are compounds made up of two or more polymeric materials such as matrix and fibres. These components, when combined together, show enhanced mechanical strength. Organic materials have received huge attention in the past few years and research has shown that they are extremely promising solutions to the needs of the product in the design process [14,15]. Biocomposites not only have the advantages of their remarkable mechanical properties, but also provide a number of processing benefits. Other benefits are lower cost, easy availability, less weight, more natural flexibility for regeneration and reduced weight loss features [16].

## 6.2 SIGNIFICANCE OF BIOCOMPOSITES IN THE PACKAGING INDUSTRY

The food packaging global market was estimated at USD 303.26 billion in 2019, showing a compound annual growth rate (CAGR) of 5.2% for 2027. The fast pace of life and revolution in eating habits have increased the demand for packaged food and impacted this market growth. The various benefits of packaged foods including increased barrier properties, enhanced shelf life and food safety have played a major role in this market growth. This can be further benefitted by designing active and intelligent materials for packaging to reduce contamination by microbes and improve the quality of food. Social factors like increased disposable income, nuclear families and increasing populations have also contributed significantly to the market. Flexible packaging materials which can retain high levels of moisture are mostly required in the bakery and confectionary industry as they are printable, light weight and cost effective. With the increase in a nuclear family culture, the demand for small pack sizes in the dairy industry is believed to augment the market demand. Increased demand for dairy products like ice cream and yoghurt, along with the strategy for attractive packaging to attract consumers, has boosted the demand for packaging material. A lot of attention has been drawn on healthy lifestyles, leading to increased interest in exotic fruits and vegetables that has driven the demand for active packaging material to help retain freshness and shelf life during storage and transportation (Figure 6.1).

The biggest share of this market is dominated by Asia Pacific, followed by Europe and North America. This market trend is assumed to be due to the increase in demand from the emerging economies like India, China and Brazil. The large population and growth in the economy make China the largest consumer. The retail chain product demand is continuously increasing in India leading to the growth of the market [18].

To increase their revenue, companies are providing a lot of customization which ultimately has increased their share in the market. To meet the government standard, industries are replacing petroleum-based products with biodegradable ones. The global market size of biocomposites in 2016 was around USD15.99 billion and it is expected to grow at a CAGR of 12.5% by 2025. The major factors responsible for this growth are the increasing efforts by government policies to introduce alternatives to petroleum-based synthetic plastics and the awareness of the community regarding eco-friendly solutions and sustainability [19].

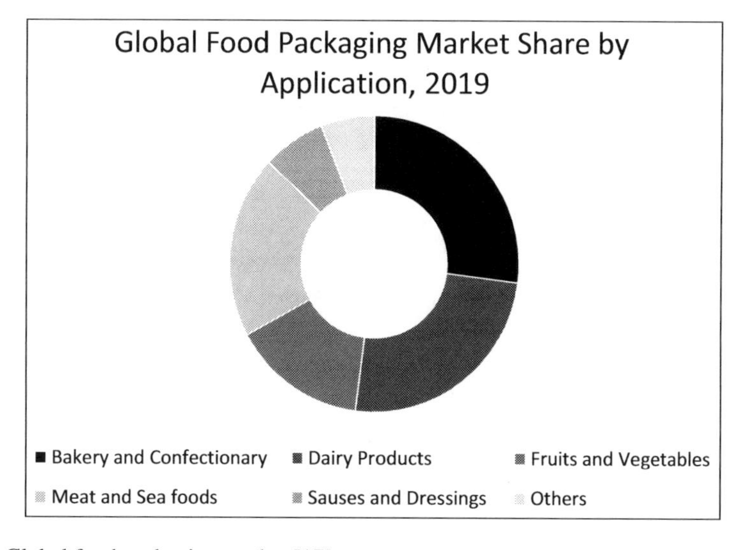


**FIGURE 6.1** Global food packaging market [17].

Biocomposites have several advantages over traditional plastics, such as being biodegradable and light weight, which makes them more acceptable to both consumers and producers. The accumulation of plastic waste is a critical situation globally as it is non-degradable and takes decades to decompose. To overcome this governments are imposing strict penalties on plastic consumption around the globe and are promoting the use of biodegradable alternatives. These drives will revolutionize the packaging industry in the coming decades. The major drawback to this growing market could be the high cost of biocomposites over petroleum-based plastic as they are made of natural polymers [20].

Preservation, containment and promotion are the main functions of food packaging which are achieved by reducing microbial contamination and protection from environmental conditions. This helps in preserving the quality of food and enhancing the shelf life. These packaging materials also provide details about the food products through labelling [21,22]. The art of designing the packaging of the material in such a way as to provide barrier properties, reduce transportation, material compatibility and shelf-life extension properties is demanded by the market and consumers. The 12th goal of sustainable development as outlined by the United Nation includes biodegradable packaging material as a major part, as it aims to reduce plastic pollution in the oceans and reduce food losses. As the awareness on issues related to sustainability is increasing, the demand to replace petroleum-based polymers with biocomposites is increasing globally among the packaging industries and consumers. There is a constant demand for the development of innovative packaging materials to benefit consumers in terms of ready-to-eat food, retention of food quality, convenience and shelf stability [23]. This has led to innovations in food packaging from remarkable materials to compostable and ultimately biodegradable products. Further developments are found in the area of designing active and intelligent packaging materials [24].

## 6.3 TYPES OF BIOCOMPOSITES

Biodegradable film-based packaging material can be divided into three groups based on their sources of origin and production procedure (Figure 6.2) (25).

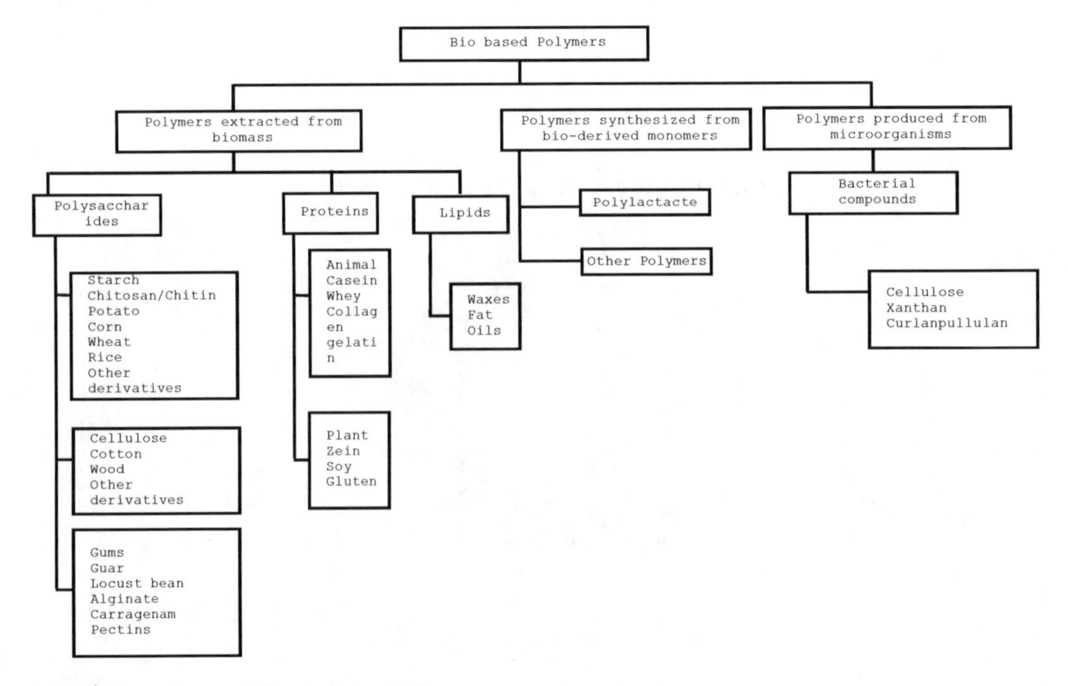


**FIGURE 6.2** Types of biocomposites [26].

### 6.3.1 Group 1

The polymers which are extracted or isolated directly from biomass are categorized into this group, which includes starch, polysaccharides like cellulose and proteins like gluten, casein, etc. [27]. All of these are hydrophilic and somewhat crystalline in nature. Their use is not suggested for packaging of moist food products or items. They present excellent gas barrier properties, and for this reason they can be used in some food packaging industries [27].

Agriculture polymers can fall into this category. They are used for packaging and their source of origin is agriculture, such as starch protein and grains. Because of the easy availability of raw materials, they are quite economical. However, they have one limitation which is with the increase in the level of biodegradability, the product will exhibit lower barrier performance [25].

### 6.3.2 Group 2

Polymers which are produced by a classical polymerization process fall into this category, such as aliphatic aromatic copolymers, aliphatic polyesters, polylactide, aliphatic copolymers (CPLA), and renewal biomass monomers like oil and oil-based monomers, for example, polycaprolactones. Poly-lactic acid is also a very good example of classical chemical synthesis using renewal biobased monomers. The first commercial biobased product was PLA, which is used in the production of blown films [27]. These polymers are a combination of synthetic and natural biopolymers and are available in regular starch (5–20%) along with synthetic polymer additives which are autoxidative and prooxidative in nature [25].

### 6.3.3 Group 3

Polymers which are produced by genetically modified bacteria or other microorganisms fall under this group. The most prominent member of this biobased polymer group, till date, is polyhydroxyalkenoates. Recently, the production of bacterial cellulose and polysaccharides has seen a lot of developments [28]. A fermentation process can be used to produce these microbial polymers by using agricultural components as raw materials. Despite being biodegradable in nature these polymers are not widely used in the packaging industry because of their high cost [29].

## 6.4 ANTIMICROBIAL BIOCOMPOSITES

Microbial contamination is the greatest challenge in the food industry and antimicrobial active packaging is one of the most promising strategies to overcome this and enhance the quality and shelf life of packaged food products [30]. Active antimicrobial food packaging provides a support matrix for the incorporation of active substances like antioxidants and antimicrobials at the time of production [31,32]. As per the guidelines of regulatory authorities, incorporation of antimicrobials directly into food products is prohibited, thus there is current interest in the food industry to introduce these antimicrobials in the packaging material. The mode of action of these packaging materials follows two strategies, either the antimicrobial compound is coated on the surface of the film or it is added into the film. The partial or complete migration of antimicrobial substances occurs by slow diffusion into the food or headspace in order to exert its inhibitory action. If the active substance does not migrate then it is coated on the film surface and acts when food and microorganisms are in contact [33,34].

Figure 6.3 illustrates the major strategies to formulate active antimicrobial packaging. The antimicrobial compounds are: (A) added in packaging film; (B) direct contact by pads; (C) diffused from a sachet; or (D) coated on the packaging film.

The major sources of antimicrobial agents involved in active packaging are chemical in nature or biological, such as from animals, microbes and plants. Traditional synthetic preservatives include

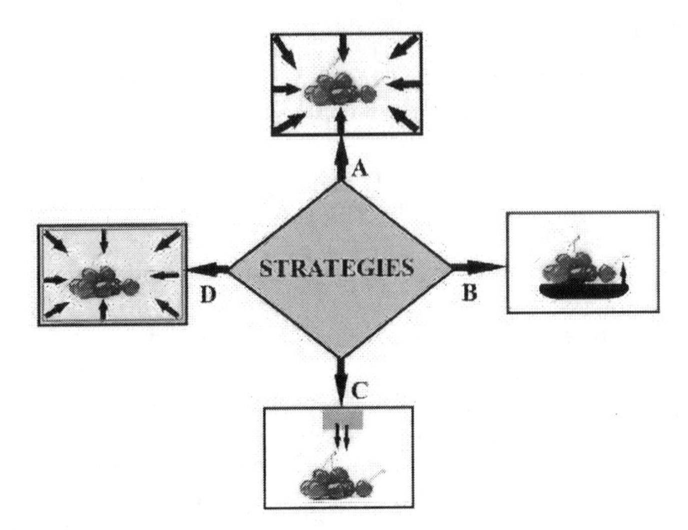


**FIGURE 6.3** Main strategies employed to obtain antimicrobial packaging [35].

organic acids, ethanol and salts of organic acids like sorbates, benzoates, propionates, etc. These are the preferred choice of industry owing to their cost effectiveness and ease of handling. Current research aims at replacing these synthetic preservatives with natural antimicrobial compounds which includes chitin and its derivatives, bacteriocins, chitosan, essential oils, enzymes, etc. [36–39].

Potential antioxidant and antimicrobial activity is shown by natural extracts like plant extracts or essential oils from different spices, fruits and plants. The extrusion method is considered to be the most effective method to incorporate these natural extracts in the packaging film [40]. Incorporating these bioactive compounds before extrusion will lead to effective and homogeneous distribution due to the high temperatures involved in extrusion. These high temperatures can cause thermal degradation of the bioactive compounds leading to reduced activity. Therefore, heat-labile compounds are incorporated by nonthermal methods like surface coating and electrospinning. The surface coating method is simple to perform at low temperatures but when applied to plastics as active packaging, shows poor adhesion properties. Such coatings are applied directly on the food samples for direct contact, such as PP/LDPE films coated with rosemary oil, garlic oil and cinnamaldehyde [41], PP films which are coated with essential oils and chitosan [42], chitosan-coated film, oregano volatile oil and citral-coated PP/EVOH film and thyme- and oregano-coated LDPE [41].

### 6.4.1 Antimicrobial Agents in Food Packaging

There is a continuous interaction between active antimicrobial packaging and food products till the defined shelf life which will actively alter the internal environment [43]. The antimicrobial agents present in these active biocomposite films will restrict the microbial growth rate by prolonging the lag period [44]. Recent research in this area involves various organic and inorganic sources of antimicrobial agents from essential oils (extracted from oregano, clove and thyme), natural extracts (green tea), polymer (chitosan), bacteriocins (nisin), enzyme (lysozyme), organic acid (includes lactic acid, acetic acid and benzoic acid) and metal ions (like silver nanoparticles and zinc oxide).

#### 6.4.1.1 Antimicrobial Agents from Organic Sources

The main organic sources for antimicrobials involve plant, animal and microbial metabolites along with organic acids. They can inhibit microbial growth and inactivate enzymes without altering the nutritional and organoleptic properties of the packaged food products [45].

#### 6.4.1.1.1 *Antimicrobial Agents from Plant Sources*

Phenolic compounds are the major components of the antimicrobial agents extracted from plants which interfere with functions of cell membrane by altering its permeability. The membrane functions affected by this are protein synthesis, electron transport, nutrient uptake, enzyme activity and loss of biomolecules like sodium glutamate and ribose from the cell [46]. Essential oils are the most researched compound for their antimicrobial property, with which they interact with cell membrane lipids owing to their molecular hydrophobicity. The interaction with membrane lipids will increase the membrane permeability leading to loss of its integrity and function leading to leakage of cellular contents and ions [47]. From recent research it is assumed that most essential oils contain compounds which show good antimicrobial properties which can enhance the stability, quality and shelf life of food products.

#### 6.4.1.1.2 *Antimicrobial Agents from Animal Sources*

Animal-derived antimicrobial agents display better resistance and growth inhibition of microorganisms, and thus are preferred over other sources. In an antimicrobial system these agents are developed as a defence mechanism and mostly emerge as antimicrobial peptides like lactoferrin, pleurocidin, protamine and defensins [48]. The antibiotic-resistant nature of these peptides is explained by their ability to inhibit the lipid bilayer membrane of cells which can interfere with the phenomenon of mutation in fast-growing microorganisms. These antimicrobial peptides have an antiviral, antifungal and antibacterial nature against both Gram-negative and Gram-positive bacteria [49]. Lysozymes are another type of enzyme showing effective antimicrobial activity, and can be extracted from milk, egg white and blood.

### 6.4.1.2 Antimicrobial Agents from Metallic Sources

The antimicrobial nature of metals dates back a long time due to their capacity to destroy microorganisms by causing membrane damage, protein dysfunction and oxidative stress [50]. These antimicrobial properties are shown by many metal ions like palladium, titanium, copper, zinc and silver against a wide range of fungi, bacteria and yeasts [51]. As organic sources are not stable at higher temperatures, they have limited applications in food packaging. Metallic sources being stable at higher temperatures are preferred antimicrobial agents in food packaging [45].

#### 6.4.1.2.1 *Antimicrobial Agents from Zinc Particles*

Zinc oxide is the main zinc particle commercially employed as an antimicrobial agent and in various other applications because of its ability to survive under extreme adverse conditions. Emission of zinc ions ($Zn^{2+}$) from zinc oxide is the main cause of antimicrobial activity as they penetrate bacterial cell membranes and alter the cytoplasmic content leading to cell death. Zinc oxide nanoparticles incorporated in a gelatine film have shown better inhibition against *Pseudomonas aeruginosa* (Gram-negative bacteria) as compared to *Enterococcus faecalis* (Gram-positive bacteria) [52]. The mechanism of action related to this phenomenon states that the photocatalytic mechanism induced by zinc oxide is associated with semi-conductive properties shown by zinc oxide leading to the formation of $H_2O_2$ and reactive oxygen species which will ultimately damage the bacterial cell membrane [53,54]. Other studies claim that reactive oxygen species are more effective towards the lipid bilayer membrane of Gram-negative bacteria as compared to the peptidoglycan protective layer of Gram-positive bacteria [53].

#### 6.4.1.2.2 *Antimicrobial Agents from Silver Particles*

The antimicrobial properties of silver particles have been recognized globally, leading to their use in active antimicrobial packaging. Silver nanoparticles have shown enhanced effects as antimicrobial agents against foodborne pathogens due to their higher surface area to volume ratio leading to increased catalytic activity [54]. When silver nanoparticles (AgNPs) were incorporated in gelatine films they showed enhanced inhibitory effects against *Escherichia coli* (Gram-negative bacteria) and *Listeria monocytogenes* (Gram-positive bacteria) [(55]. The mechanism behind this is believed to be the interaction of phosphorus and sulphur groups of DNA and protein, respectively, with AgNPs, resulting in cell death due to DNA replication inhibition. Charged bacterial cell membranes can be the target for binding of charged AgNPs causing cell wall disruption by cytoplasm shrinkage and detachment of the membrane leading to necrosis [56]. Another mechanism could be inactivation of enzymes by penetrating AgNPs which will induce $H_2O_2$ assembly and cause cell death [55].

## 6.4.2 Active Antimicrobial Packaging

The interaction between packaging, product and environment forms the basis of active packaging. These systems are designed to extend the shelf life and sustainability of packaged food products by involving physical, chemical and biological activities. Active packaging also helps in retaining organoleptic properties and enhancing microbiological safety along with the quality of packaged food products [55]. Active packaging involves substances which can either release (emitters) or absorb (scavengers) gases or substances to continuously change the inside environment of packaging. Unwanted substances like moisture, oxygen, odours, ethylene and carbon dioxide are absorbed by the scavengers from the internal environment of the package, whereas the substances having a positive impact on the packaged food like antioxidants, flavours and antimicrobials are released by the emitters [55]. Antimicrobial packaging has the widest application as it involves incorporating antimicrobial agents in the packaging films which will control the microbial contamination and help in retaining the quality of food products and enhance the shelf life [57].

### 6.4.2.1 Active Films with Protein

The production of protein-based activity can be achieved by either dispersion when the solvent/carrier evaporates or from protein solutions. The carriers/solvents are mostly ethanol, water or water–ethanol mixtures [58]. The water-resisting property of these films is much less than that of polysaccharides, however they can form films with better mechanical and barrier properties [59]. Protein-based films can also act as active packaging by incorporating antimicrobial compounds and diffusing them on the surface of the product to enhance its shelf life. Several factors affect the rate of diffusion and antimicrobial activity of active compounds on the food surface including chemical compatibility with polymer matrix, physicochemical properties of the chemical, headspace humidity, release temperature and solubility of antimicrobial compounds in food products. The various sources of protein-based edible films acting as active antimicrobial packaging are gelatine, casein, whey and corn zein [60].

#### 6.4.2.1.1 Active Films with Gelatine

Gelatine protein is extracted by hydrolysis of collagen, which is found in the skin and bones of animals. The major factors influencing the physical and chemical properties of gelatine are the source, collagen type, age of the animal and method of extraction [61]. The film-forming properties of gelatine are utilized in food packaging to retain the quality of food and enhance its shelf life, and thus they have encouraged researchers to extensively study this protein. Gelatine-based films can be incorporated with various antimicrobial compounds like essential oils, polymers, metal ions, bacteriocins, natural extracts and organic acids to restrict microbial spoilage by inhibiting the growth of microorganisms. Silver nanoparticle (AgNP)-incorporated gelatine-based active nanocomposite films show good antimicrobial activity against both Gram-positive (*Listeria monocytogenes*) and Gram-negative (*E. coli*) bacteria [62].

#### 6.4.2.1.2 Active Films with Casein

Active edible films with casein have been studied extensively as they demonstrate enhanced nutritional quality, along with good organoleptic properties. The most common sources of casein are dairy products or mammalian milk. Casein proteins can be precipitated from skim milk by either treating it with rennet to form rennet casein or by acid treatment to precipitate casein at its isoelectric point (4.6) [58]. Edible films can be cast with casein and caseinates in aqueous solutions. Good barrier properties of casein-based films against oxygen along with other nonpolar molecules are observed as they provide a lot of polar functional groups towards the film matrix like amino and hydroxyl groups. Therefore, casein-based films can act as active packaging and can protect food products prone to moisture and oxygen by combining them with other packaging materials [56].

##### 6.4.2.1.3 *Active Films with Whey Protein*

In the cheese-making process, after casein coagulation at pH 4.6 and 20°C temperature, the residual left in the milk serum is called whey, which is a by-product of the cheese industry. The proteins which together comprise to form whey protein are alpha-lactalbumin, immunoglobulins, beta-lactoglobulin and bovine serum albumin [63]. Recent developments in the food industry involve the utilization of whey proteins for the formulation of edible coatings and films on the surface of packaged food products. The various beneficial properties exhibited by active films with whey protein include good barrier characteristics with respect to oxygen and lipids, transparency and odourlessness. These films can also act as active antimicrobial films by providing a good matrix for incorporating antimicrobial or antioxidant agents [64,65].

##### 6.4.2.1.4 *Active Films with Zein Protein*

Zein protein is classified as a prolamin protein which is the major protein found in corn and is hydrophobic and thermoplastic in nature. The hydrophobic nature is associated with the high amount of nonpolar amino acids present in this protein. The hydrophobic grease-proof film formation helps in developing resistance against microbial contamination and thus makes it a good coating material for use in pharmaceutical and food products. Recent studies have shown that zein proteins could be utilized for the manufacture of biobased degradable packaging films which have hydrophobic characteristics and thus can act as a barrier against water vapour permeability [58]. These films can also act as carriers for antimicrobial agents like bacteriocins, lysozyme, phenolics, lactoperoxidase, essential oils and glucose oxidase [59].

## 6.5 USE OF BIOMOLECULES IN FOOD PACKAGING

Packaging films are designed to enhance the shelf life of food products by retaining mechanical strength and barrier properties towards water vapour and gases. These barrier properties help to retain the modified atmosphere present in the package and inhibit water vapour permeability to further reduce the organoleptic degradation. Apart from mechanical strength, packaging material should also possess the property of plasticity to attain the deformation caused by the filling (Figure 6.4).

Composites are made up of more than one type of polymer to attain various characteristic features such as barrier property, mechanical strength and plasticity. These can be made by combining proteins, polysaccharides or lipids in various combinations to achieve the desired packaging film. Enhanced mechanical strength is provided by using proteins or polysaccharides, while lipids show increased barrier property. Based on the biochemical composition, these polymers are classified as polysaccharides, lipids, proteins and complex polymers. The poor vapour barrier characteristics of polysaccharides and proteins are the reason for lipids being the choice of candidate for film formation even though they have comparatively less optical and mechanical strength.

### 6.5.1 Lipids

Lipids are widely used for generating a vapour barrier leading to the prevention of bacterial and physicochemical degradation. The various factors affecting the use of lipids as an effective moisture barrier are their hydrophobic nature, composition, atmospheric conditions, etc. (Figure 6.5) [67].

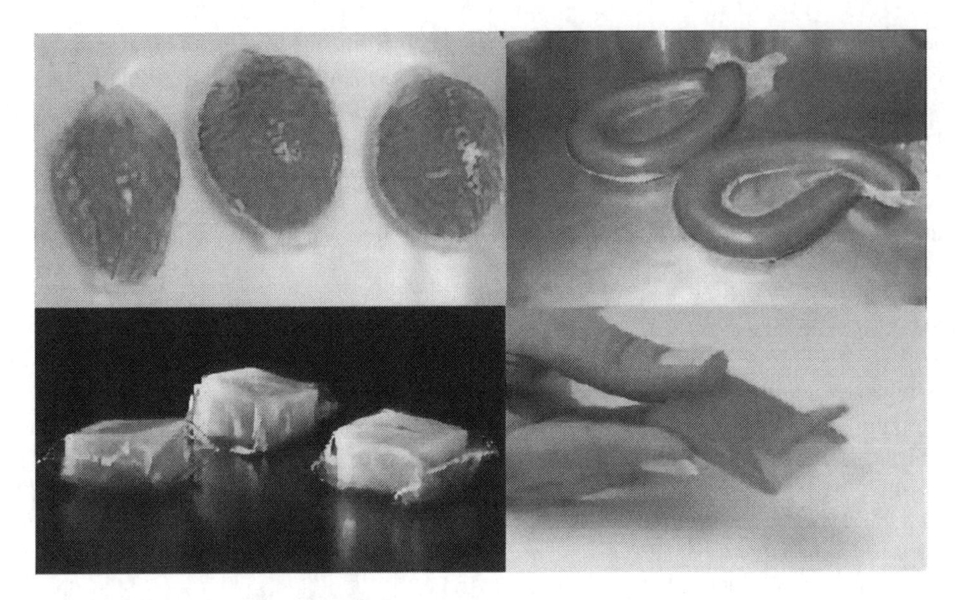

**FIGURE 6.4** Edible biocomposite films [66].

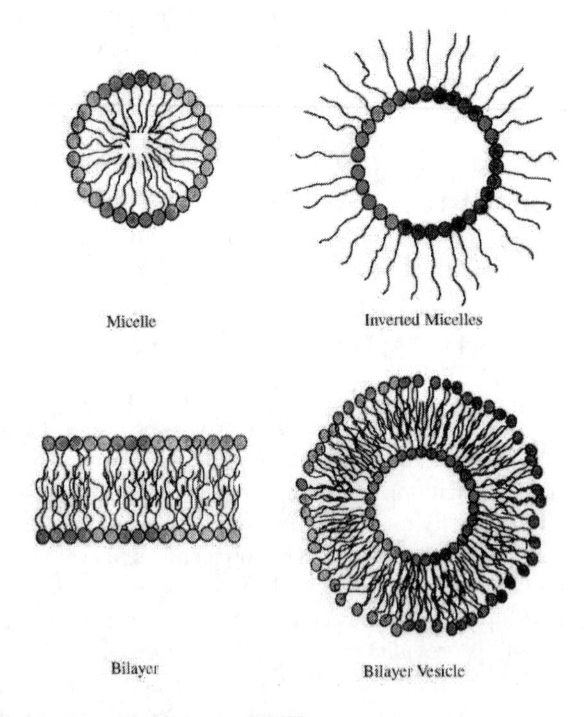


**FIGURE 6.5** Different micelles of lipids in water [67].

Lipid interactions with any food substance should be well understood. Lipids in water form various types of micelles due to the amphipathic nature of lipids [68]. Lipids can be classified as polar or nonpolar. Waxes do not have any polar groups such as hydrocarbons, which is why they have no contact with water and fall under the nonpolar category. They do not completely dissolve in water and do not form a single monolayer when spread on a surface. Thus, waxes are a perfect lipid barrier that works by restricting the transfer of water vapour through edible films. Due to the hydrophobic nature of waxes, they reduce the water vapour permeability to a great extent and hence are considered as the best candidate for packaging film [69].

A visibly stable monolayer is formed by spreading triglycerides, even though they are mostly water-insoluble. Based on this characteristic they are classified as class I polar lipids. The long-chain triglycerides do not dissolve in water, while the triglycerides of short chains are partially dissolved [68]. To improve the vapour barrier properties of edible films, lauric, palmitic, stearic acid and stearyl alcohol can be added [69].

Based on the length of the carbon chain, monoglycerides can be classified as class II or class III. Lipids of class II are not water-soluble, but water-associated with their hydrophilic regions leading to inflammation. In a water-deficit state, it dissolves slightly in a normal organic solvent such as hexane [68]. Monoglycerides act as an emulsifier in edible films and provide stability to the emulsified film. As an emulsifier, monoglycerides work by strengthening the adhesion of two hydrophobically different components. Adhesions between food components and edible film or between the hydrocolloid frame and lipid edible film are some of the examples of monoglycerides acting as an emulsifier [69]. Monoglycerides with attached acetic acid are described as acetylated monoglycerides, which also function as emulsifiers in edible film coatings and exhibit plastic properties by coagulating cold foods (Table 6.1) [70].

#### 6.5.1.1 Lipids in Edible Film

The packaging film made entirely from lipids shows good barrier properties but is brittle in nature [28]. The fatty acids with carbon numbers 14–18 are mostly used in packaging films which include

**TABLE 6.1**
**Classification of Biological Lipids**

| Category | Bulk Features | Surface Characteristics | Examples |
|---|---|---|---|
| **Non-Polar** | Not soluble | Monolayer is not formed when spread on surface | Long-chain fatty acids, waxes and paraffin oil |
| **Polar** | | | |
| Class I: These compounds neither solubilize nor swell and are amphiphilic in nature | Not soluble or less soluble | Monolayer is formed when spread on surface | Vitamins A, D, E, K, cholesterol, tri-diglycerides |
| Class II: These compounds neither solubilize nor swell and are amphiphilic in nature | They do not solubilize in water but swell to produce lyotropic liquid crystals | Stable monolayer is formed when spread on surface | Phospholipids, monoglycerides |
| Class IIIA: These compounds are soluble and amphiphilic in nature and have lyotropic mesomorphism properties | They solubilize beyond the critical micellar concentration to produce micelles, and at low water concentration produce liquid crystals | A stable monolayer is not formed when spread on the surface because they are soluble in water | Salts of long-chain fatty acids, detergents, lysolecithin |
| Class IIIB: These compounds are soluble and amphiphilic in nature but do not have lyotropic mesomorphism properties | These compounds form micelles but do not produce liquid crystals | A stable monolayer is not formed when spread on the surface because they are soluble in water | Saponins, bile salts, rosin soaps |

fatty alcohols, waxes and vegetable oil (non-hydrogenated). Lipids are generally used in formulating emulsion-based coatings and edible films. The major food groups utilizing lipid-based biocomposites include vegetables and fruits, dairy products, meat and meat products and cereal products. (71)

##### 6.5.1.1.1 Vegetables and Fruits

There has been a several-fold increase in global demand for imported vegetables and fruits in the recent past. The major concern lies in the transport and preservation of these fresh products. To overcome this problem researchers have developed packaging material that can retain the freshness and quality of these products and enhance their shelf life. The water content of these fruits and vegetables ranges between 80–90%, thus the biocomposite design should exhibit good barrier properties against water vapour [72]. Natural waxes or a paraffin coating on fruits increases the water retention in storage conditions [73].

These coatings can also protect fruits and vegetables from microbial spoilage as fungicides and other antimicrobial compounds can be mixed in the coatings. Apart from water content, respiration is another factor that alters the organoleptic characteristics of stored fruits and vegetables. The heat and $CO_2$ generated during respiration can cause a change in colour, odour and taste [74], The stability of lipid-based biocomposite films makes them a good candidate of choice but the change in organoleptic properties still makes it an area requiring extensive research [75].

##### 6.5.1.1.2 Dairy Products

The most widely used application of edible biocomposite film is in the dairy industry, which is mostly dominated by the cheese industry. The shelf life of cheese is defined by the type of microorganism and processing steps involved. As cheeses are prone to surface contamination by bacteria and fungi, their shelf life is quite limited. The edible biocomposite film incorporating lipids, proteins, plasticizers and additives seems to be a promising solution to this problem. Several researches have shown the enhanced water-retention properties and reduced growth of microorganisms on the surface of cheese when using these biocomposite films. The incorporation of compounds with antimicrobial activity in these films will increase their application by reducing microbial growth on the surface and retaining the organoleptic property [76].

##### 6.5.1.1.3 Meat and Meat Products

Meat and meat products are highly susceptible to microbial spoilage due to the high-water content and favourable environment in the tissue system. Cellulose derivatives-based hydrocolloid coatings with additives and plasticizers can retain the water content present in the tissue. Incorporating antimicrobial agents will reduce the microbial spoilage and thus enhance the shelf life of these products. Biocomposite films based on chitosan show a reduction in metmyoglobin content [77]. These films can also be used in the packaging of sausages in order to retain the water content and enhance the shelf life [78].

##### 6.5.1.1.4 Bakery Products

The major concern in the bakery industry is to retain the texture of cereals and biscuits by restricting water permeability. The transfer of water can occur not only from the atmosphere but also from the dried fruits added to cereals and can destroy the crisp texture of cereals. To limit this, dried fruits can be coated with edible biocomposite films made by lipids like chitosan and caseinates, which will help in retaining the texture of cereals and thereby enhance their shelf life [79].

### 6.5.2 Protein

Protein-based biocomposite films are foreseen as a good alternative to petroleum-based polymers for a sustainable environment. These biocomposites, being eco-friendly, pose no threat of environmental damage and still have similar physicochemical properties to synthetic polymers [80]. The protein-based edible biocomposite films act as a barrier for moisture and atmospheric gases. These films are made of polymer matrix which provides mechanical strength and structural integrity and helps in improving the rheological characteristics of the product [81]. Further addition of antioxidants and antimicrobial compounds into these edible films will enhance the quality and shelf life of the coated products. For the ability to exhibit extensive intermolecular binding between molecules, the protein-based biocomposite films possess better mechanical and barrier properties as compared to lipid- or polysaccharide-based biocomposites. Apart from good binding properties, these protein-based films also have nutritional value and are transparent in nature [82]. These protein-based biocomposites are broadly categorized under plant or animal origin. Gelatine, collagen, whey protein and fish myofibrillar coat proteins are the most commonly employed animal-based proteins to formulate edible biocomposite films. The polymers extracted from plants are mainly obtained from wheat gluten, soy, peanut, corn zein protein, Amaranth, pea, etc. [83]. The solvent casting method, which involves dispersion of protein in a desired solvent and later evaporation of solvent to form a firm layer, is the method of choice. The other method used for film production is the extrusion method, which works at high temperature but is economical in terms of energy inputs as compared to solvent casting methods [31].

### 6.5.3 Protein-based Edible Coatings

These edible film wraps are coated on food products directly and create a modified atmosphere around it which acts as a barrier for gases and moisture. These edible coatings can also incorporate active compounds such as antimicrobials and antioxidants to preserve the quality and increase the shelf life of packaged food products. They can also impart gloss and colour to the product to improve the aesthetic characteristics and consumer acceptance [64,81]. Wet and dry methods are used to process the protein coatings. In wet coating methods the polymer solution is sprayed on the food product and kept for drying under microwave, hot air, or infrared. The coating process defines the rheological properties of the film and the drying method influences the appearance, morphology, mechanical and barrier properties [84]. Extrusion is the most commonly used dry method, which involves melting of polymer at high temperature followed by casting in a die of the desired shape. Protein-based biocomposite edible films are sensitive to moisture due to their hydrophilic nature which leads to poor barrier properties for water vapour. Incorporation of plasticizers, cross-linkers, and additives helps in improving the characteristics of these films [85].

### 6.5.4 Polysaccharides

Polysaccharide-based biocomposites show good barrier properties towards water vapours and gases, leading to improved organoleptic characteristics and enhanced shelf life. Gum arabic used in the formulation of these films shows antimicrobial properties and helps in reducing microbial spoilage. These coatings form a modified atmosphere around fruits and vegetables to form a barrier against $CO_2$ and $O_2$, thus reducing the effect of respiration. Alginate gum incorporated in these films delays the ripening of fruits by inhibiting ethylene production. These biocomposite films reduce the microbial growth, retain organoleptic properties, enhance antioxidant activity, reduce weight loss and control ethanol concentration [86].

Polysaccharides can be obtained from some plant sources and algae, e.g., starch, alginates, cellulose, etc. or they can be extracted from animal sources like glycogen and chitin. Some polysaccharides are even produced by microorganisms, e.g., alginates, cellulose, and some exopolysaccharides [87].

#### 6.5.4.1 Polysaccharides of Plant Origin

##### *6.5.4.1.1 Starch*

Starch is the main storage polysaccharide present in plants. Based upon its water solubility characteristics it is used in the production of the textile, adhesive and paper industries [88]. The film-forming characteristics of starch are based on its amylose/amylopectin ratio and molecular weight [89]. The mechanical properties of the packaging film are determined by the amylose content. The elongation percentage and tensile strength of the film increases with an increase in amylose content, e.g., corn with a high amylose content is the choice of polymer for the formulation of biocomposites. The starch-based films are thermoplastic and transparent with good mechanical strength and slight gas barrier properties. Pertaining to its moisture sensitivity, starch-based packaging films have a defined life span [90].

##### *6.5.4.1.2 Cellulose*

Cellulose-based films are hygroscopic in nature and this property is regulated by the relative humidity. The hydrophilic property of the cellulose derivative-based biofilms depends on the degree of hydroxyl group substitution and the number of free hydroxyl groups adversely affecting the mechanical strength [91]. The film-forming properties obtained through these modifications are not comparable to polyolefin-based film and they do not possess antimicrobial and antioxidant properties [87]. Antimicrobial characteristics can be obtained by supplementing cellulose acetate films with lysozymes [92]. The rate of release of active compound from cellulose films can be regulated by modifying the composition of the polymer solution and the degree of asymmetry to obtain the desired pore size of the film [93].

##### *6.5.4.1.3 Hemicelluloses*

Hemicelluloses obtained from waste water and agricultural residues are considered to be the best alternatives to petroleum-based plastics [94]. Biocomposite film produced by blending carboxymethylcellulose or alginate with hemicellulose shows enhanced mechanical strength and barrier properties which are comparable with those of amylose or amylopectin films [87]. Films showing oxygen barrier properties can be produced from hemicellulose obtained from hardwood and softwood [95]. Xylan biocomposites such as arabinoxylan and glucuronoxylan are used for packaging oxygen-sensitive products due to their low oxygen permeability [96]. Cellulose esters can be incorporated in a xylan solution to enhance the mechanical properties of biocomposite films [97].

##### *6.5.4.1.4 Pectin*

Pectin is soluble in water and naturally obtained from the primary cell wall of cells present in the intracellular regions of plants. These hygroscopic polysaccharides are extracted from lemon peel. Pectins are not much reported on as a packaging biopolymer either alone or in combination with other biopolymers. Pectin molecules can form gel in the presence of sugar and acid, whereas methoxyl pectin needs calcium ion for jellification. The formation of ionic bonds between the carboxylic group of pectin and calcium ions at acidic pH results in gel formation in methoxyl pectin [90].

##### *6.5.4.1.5 Carrageenans and Alginates*

Carrageenans having the basic structure of 1-4-linked-α-D-galactopyranose are a complex mixture of various polysaccharides. Thermoreversible gel-based coatings of carrageenans are used extensively to enhance the shelf life of poultry products and fish [97]. Alginates are linear unbranched polymer comprising α-(1->4)-linked L-guluronic acid and β-(1->4)-linked D-mannuronic acid residue. They are extracted from seaweed and made of blocks of identical and alternating residues. The wide application of normal thermoreversible gel of alginates in the food packaging industry is because of their good film-forming property. The gelling agents used for film formation are divalent cations and acids [98].

#### 6.5.4.2 Polysaccharides of Animal Origin

##### 6.5.4.2.1 *Glycogen*

Glycogen is similar in structure to amylopectin but is more branched. It is the major storage polysaccharide found in animals [99]. The applications of glycogen as a base matrix for the formulation of packaging film are not much reported. It is mostly used as a collagen cross-linker in biomedical biomaterials [100].

##### 6.5.4.2.2 *Chitosan*

Chitosan is a natural polymer composed of monomers glucosamine and N acetylglucosamine which are linked with each other by β-1,4-glycosidic bonds. Chitin is deacetylated in an alkaline medium for the commercial production of chitosan [101]. The film-forming properties of chitosan are regulated by the heterogeneous distribution of an acetyl group in the main chain. Natural sources of chitosan include some fungi and silkworm. Microencapsulation and films based on chitosan can act as active carriers for antibiotics and antioxidants [102].

#### 6.5.4.3 Polysaccharides of Bacterial Origin

Cellulose obtained from microbial sources is structurally similar to plant cellulose but differs in chemical and physical properties. *Acetobacter xylinum* and acetic acid-forming bacteria are the major sources of bacterial cellulose. These bacterial celluloses can replace plant cellulose due to their rapid production rate [90].

### 6.5.5 Complex Biopolymers

#### 6.5.5.1 Polyhydroxyalkanoates (PHAs)

During adverse growth conditions, many bacterial species accumulate polyhydroxyalkanoates (PHAs) like bacterial polyesters. These can also be produced by fermentation of lipids and sugar by bacterial species. PHAs are polymers of 3-hydroxy fatty acid monomers which have similar chemical properties to polypropylene like synthetic polymers. The various properties which make PHAs molecules of interest in the packaging industry include biocompatibility, petroleum displacement, greenhouse gas minimization, biodegradable and the fact that they can be procured from renewable sources. The applications of PHAs for the formation of biodegradable packaging include production of films, bottles, sheets, laminates, fibres, containers and coatings. The major properties exhibited by PHAs in food packaging which enhance its applications are good tensile strength, barrier properties for flavour and odour, printability, oil and grease resistance, temperature stability, sealing ability and ease to dye.

#### 6.5.5.2 Polylactic Acid (PLA)

Recently, polylactic acid (PLA) has attained immense attention in the food packaging industry because of its commercial and economic viability while processing. PLA comes under the family of aliphatic polyesters and comprises alpha hydroxy acids which incorporate poly-mandelic acid or polyglycolic acid. PLA is produced by depolymerization of lactic acid monomer under controlled conditions using corn and sugar feedstock as raw material, and it is biodegradable in nature. The properties of PLA which make it a good alternative to synthetic plastic material include better processability, high molecular weight, resistance towards solubility, biodegradability, odour and flavour barrier properties, good tensile strength, printability, temperature resistance and grease resistance. In unmodified conditions PLA is very brittle and easily degrades with a rise in temperature.

## 6.6 COMMERCIAL PRODUCTS

The food packaging industry is revolutionizing with respect to the development of environment-friendly biocomposite materials in order to replace petroleum-based plastic. These biodegradable polymers are considered safe and ensure enhanced shelf life of packaged food products with the addition of active chemical compounds like antioxidants, antimicrobials, etc. (Figure 6.6).

The renowned brands which have introduced these biocomposite packaging include Coca Cola (Plant Bottle), Heinz, Danone (Actimel, Activia, Volvic), PepsiCo, L'Occitane, Tetra Pak, etc. [103]. A list of companies manufacturing biodegradable films based on various biocomposites is given in the Table 6.2. [104].

The biopolymers commercially involved in packaging can be categorized into rigid packing and flexible packing. The main objective of rigid packing is to provide structure and support to the food products. These biocomposites cannot be forced out of shape and are characterized by their strength. Flexible biocomposite films can easily bend without breaking. These biocomposites can be easily moulded, folded, shaped and bent. The commercially used rigid and flexible packaging polymers are listed in Table 6.3 [105].

The commercial applications of biocomposites in the food packaging sector are categorized in the following sections.

### 6.6.1 Biocomposite-based Trays and Films

The recent era has witnessed the development of biocomposite-based trays and films [106]. Monolayer films manufactured from wheat gluten by the extrusion method have shown enhanced tensile strength characteristics at high pH levels [107]. Several techniques like sheet extrusion, injection moulding and thermo-folding are used at high temperatures to synthesize bioplastics of plant starch origin like BIOME Bioplastic (Biome Bioplastics Limited, Southampton, United Kingdom) [101]. The biocomposite film developed by Novamont S.p.A. (Novara, Italy) named MATER-B1 is produced from corn starch and oil. This is used for the production of trays, film, cups, bags, etc. and as a packaging material [108].

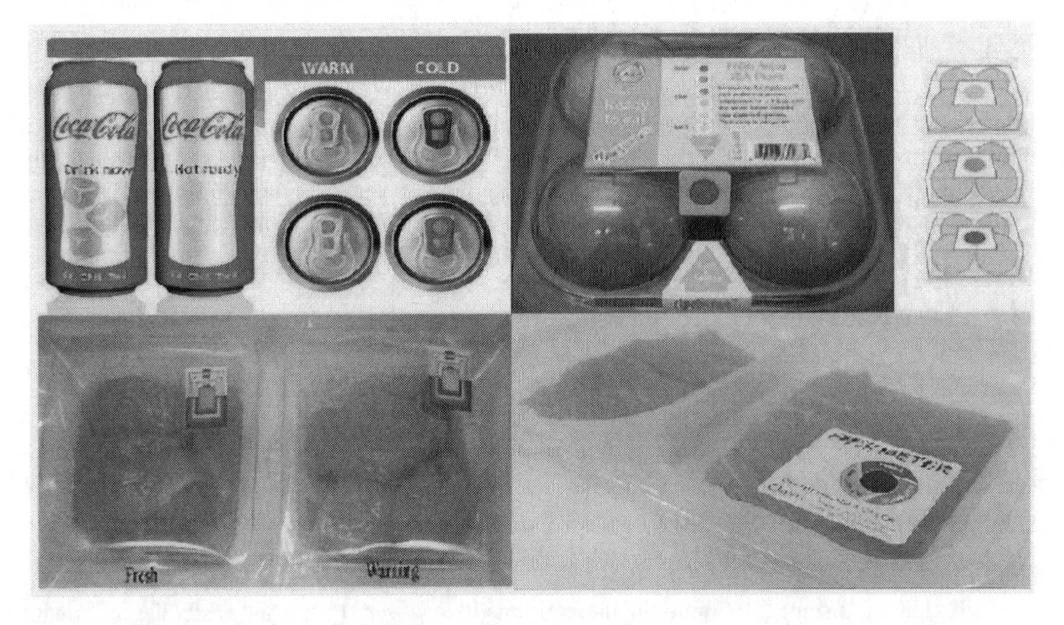


**FIGURE 6.6** Commercial biocomposites in food packaging (66).

**TABLE 6.2**
**List of Companies Producing Biodegradable Films**

| Starch-based Films | | | PLA-based Films | | | PLA/Starch-based Films | | |
|---|---|---|---|---|---|---|---|---|
| **Trade Name** | **Company** | **Country** | **Trade Name** | **Company** | **Country** | **Trade Name** | **Company** | **Country** |
| Bioplast | Biotec | USA | PURAC | PURAC Co. | Thailand | Bio-Flex® | BASF | Germany |
| Solanyl | Rodenburg Biopolymers | Japan | Ingeo | Nature Works | USA | Compole | Japan corn starch | Japan |
| Plantic | Plantic Technologies | USA | Pyramid | Tate & Lyle | Denmark | Plantic | Plantic Co. | USA |
| Biopar | Biop | Germany | HiSun | Revoda | Canada | Biolicem | Limagrain | France |
| Mater Bi | Novamont | USA | BIOFRONT | Teijin | Japan | Ecovio® | Nature Works | USA |

*Source:* (104).

**TABLE 6.3**
**Biopolymers Used in Commercial Packaging**

| Rigid Packaging | | | Flexible Packaging | | |
|---|---|---|---|---|---|
| **Natural Products** | **Synthetic Plastic** | **Applications** | **Natural Products** | **Synthetic Plastic** | **Applications** |
| Bio-PP | PP | Yoghurt cups | PBAT | PS | Shrink wraps and films |
| Bio-PET | PET | Carbonated beverages bottles | Blends of PLA/PHA/PBAT | PE, PP, PS | Garbage/shopping bags |
| Cellulose, starch blends | PET, PS | Plates, bowls, glasses | PLA/blends of PLA/bio-PET | PS/PP | Clingfilm wraps for vegetables and fruits |
| PLA blends | PE, PP | Caps | Bio-PE | PE | Stretchable wraps and films |
| Bio-PP, PLA | PET, PP, PS | Trays | PLA/PHB | PS/PP | Pouches and capsules of coffee |

### 6.6.2 Biocomposite-Based Bags and Pouches

High cost and extensive processing have limited the variety of biocomposite-based bags and pouches in the market. Multilayer pouches made of paper, starch and cellulose were produced by BioBag International AG (Askim, Norway) [109]. The first biocomposite-based spouted pouch made by GualapackGroup (Castellazzo, Italy) was 100% organic and targeted children as the major consumers. These "Premade multi-layered pouches" were made from fruit and vegetable pulp and composed of Bio-PE [110]. These PE pouches can be replaced by pouches made from mango kernel starch for packaging red chilli powder. The pungency and colour of the packaged chilli powder was better retained in biocomposite film-based pouches made from mango kernel starch as compared to PE pouches. Intensive studies are required to make these biocomposite-based pouches more environment friendly and cost effective [111].

## REFERENCES


1. Abeer, A., and Choudhary, R. (2013) Current practices in bread packaging and possibility of improving bread shelf life by nanotechnology. *International Journal of Food Sciences and Nutrition Engineering*. 3(4), 55–60.
2. Ayoub, Anjum, Hameed, Fozia, and Julie, D. (2018) Bandral food packaging technology and its emerging trends: a review. *International Journal of Current Microbiology and Applied Sciences*. 7(10), 3363–3378.
3. Su, Y., Yang, B., Liu, J., Sun, B., Cao, C., Zou, X., et al. (2018) Prospects for replacement of some plastics in packaging with lignocellulose materials: a brief review. *BioResources*. 13(2), 4550–4576.
4. Soroudi, A., and Jakubowicz, I. (2013) Recycling of bioplastics, their blends and biocomposites: A review. *European Polymer Journal*. 49(10), 2839–2858.
5. Geyer, R., Jambeck, J. R., and Law, K. L. (2017) Production, uses, and fate of all plastics ever made. *Science Advances*. 3(7), 5.
6. Majeed, K., Jawaid, M., Hassan, A., Abu Bakar, A., Abdul Khalil, H. P. S., Salema, A. A. et al. (2013) Potential materials for food packaging from nanoclay/natural fibres filled hybrid composites. *Materials and Design, Elsevier.* 46, 391–410.
7. Ellen MacArthur Foundation. (2016) The new plastics economy: rethinking the future of plastics. *World Economic Forum. New Plastics Economy*. 120.

8. Johansson, C. et al. (2012) Renewable fibers and bio-based materials for packaging applications – a review of recent developments. *Bioresources.* 7(2), 1–47.
9. Khan, B. A., Wang, J., Warner, P., and Wang, H. (2015) Antibacterial properties of hemp hurd powder against E. coli. *Journal of Applied Polymer Science.* 132, 41588.
10. Khan, B. A., Warner, P., and Wang, H. (2014) Antibacterial properties of hemp and other natural fibre plants: a review. *BioResources.* 9, 3642–3659.
11. Appendini, P., and Hotchkiss, J. H. (2002) Review of antimicrobial food packaging. *Innovative Food Science & Emerging Technologies.* 3, 113–126.
12. Su Cha, D., and Chinnan, M. S. (2004) Biopolymer-based antimicrobial packaging: a review. *Critical Reviews in Food Science and Nutrition.* 44, 223–237.
13. Marsh, K., and BUGUSU, B. (2007) Food packaging – roles, materials, and environmental issues. *Journal of Food Science.* 72, 39–55.
14. Siracusa, V., Rocculi, P., Romani, S., and Rosa, M. D. (2008) Biodegradable polymers for food packaging: a review. *Trends in Food Science & Technology.* 19, 634–643.
15. Mukherjee, T., and Kao, N. (2011) PLA based biopolymer reinforced with natural fibre: a review. *Journal of the Polymers and the Environment.* 19(3), 714–725.
16. Mansor, M. R., Salit, M. S., and Zainudin, E. S. (2015) *Life Cycle Assessment of Natural Fiber Polymer Composites.* Switzerland: Springer International. doi: 10.1007/978-3-319-13847-3_6
17. Grand View Research (2020) *Food Packaging Market Size, Share&Trends Analysis Report by Type (Rigid, Flexible), by Material (Paper, Plastic), by Application (Bakery and Confectionery, Dairy Products), by Region, and Segment Forecasts, 2020–2027.* Available online: www.grandviewresearch.com/industry-analysis/foodpackaging-market. Report ID: GVR-2-68038-365-2.
18. Al-oqla, F. M., and Omari, M. A. (2017) *Green Biocomposites. Sustainable Biocomposites: Challenges, Potential and Barriers for Development.* Switzerland: Springer International. doi: 10.1007/978-3-319-46610-1_2
19. *Food Packaging Market Size, Share&Trends Analysis Report by Type (Rigid, Flexible), by Material (Paper, Plastic), by Application (Bakery and Confectionery, Dairy Products), by Region, and Segment Forecasts, 2020–2027.* Available online: www.grandviewresearch.com/industry-analysis/foodpackaging-market (accessed on 26 September 2020).
20. Biocomposites Market by Fiber (Wood Fiber and Non-Wood Fiber), *Polymer (Synthetic and Natural), Product (Hybrid and Green), End-Use Industry (Transportation, Building & Construction and Consumer Goods), and Region – Global Forecast to 2022.*
21. Wohner, B., Gabriel, V. H., Krenn, B., Krauter, V., and Tacker, M. (2020) Environmental and economic assessment of food-packaging systems with a focus on food waste. Case study on tomato ketchup. *Science of the Total Environment.* 738. https://doi.org/10.1016/j.scitotenv.2020.139846
22. Mathlouthi, M. (2013) *Food Packaging and Preservation.* Germany: Springer Science & Business Media.
23. Han, J. W., Ruiz-Garcia, L., Qian, J. P., and Yang, X. T. (2018) Food packaging: a comprehensive review and future trends, *Comprehensive Reviews in Food Science and Food Safety.* 17(4), 860–877.
24. Majid, I., Ahmad Nayik, G., Mohammad Dar, S., and Nanda, V. (2018) Novel food packaging technologies: innovations and future prospective. *Journal of the Saudi Society of Agricultural Sciences.* 17(4), 454–462.
25. Ivankovic, A., Zeljko, K., Talic, S., Bevanda, A. M., and Lasic, M. (2017) Biodegradable packaging in the food industry. *Archiv für Lebensmittelhygiene.* 68, 26–38.
26. Weber, C. J., Haugaard, V., Festersen, R., and Bertelsen, G. (2002) Production and application of biobased packaging materials for the food. *Food Additives and Contaminants.* 19(Suppl):172–177. doi: 10.1080/02652030110087483
27. Milovanovic, M., and Picuric-Jovanovi, K. (2001) Lipids and biopackaging. Usage of lipids in edible films. *Journal of Agricultural Sciences.* 46(1), 79–87.
28. Averous, L., and Pollet, E. (2012) *Environmental Silicate Nano Biocomposites.* London Heidelberg, New York: Springer.
29. Gontard, N., Duchez. C., Cuq. J. L., and Guilbert, S. (1994) Edible composite films of wheat gluten and lipids: water vapour permeability and other physical properties. *International Journal of Food Science and Technology.* 29(1), 39–50.
30. Webb, H. K., Arnott, J., Crawford, R. J., and Ivanova, E. P. (2013) Plastic degradation and its environmental implications with special reference to poly (ethylene terephthalate). *Polymers.* 5(13), 1–18.

31. Said, Nurul Saadah, and Sarbon, Norizah Mhd (2019) Protein-based active film as antimicrobial food packaging: a review. *Active Antimicrobial Food Packaging*. 53–70, doi: http://dx.doi.org/10.5772/intechopen.80774
32. Kuorwel, K. K., Cran, M. J., Orbell, J. D., Buddhadasa, S., and Bigger, S. W. (2015) Review of mechanical properties, migration, and potential applications in active food packaging systems containing nanoclays and nanosilver. *Comprehensive Reviews in Food Science and Food Safety*. 14, 411–430. doi: 10.1111/1541-4337.12139
33. Bassani, A., Montes, S., Jubete, E., Palenzuela, J., Sanjuan, A. P., and Spigno, G. (2019) Incorporation of waste orange peels extracts into PLA films. *Chemical Engineering Transactions*. 74, 1063–1068. doi: 10.3303/CET1974178
34. Vermeiren, L., Devlieghere, F., and Debevere, J. (2002) Effectiveness of some recent antimicrobial packaging concepts. *Food Additives and Contaminants*. 19, 163–171. doi: 10.1080/02652030110104852
35. Motelica, Ludmila, Ficai, Denisa, Ficai, Anton, Oprea, Ovidiu Cristian, Kaya, Durmuş Alpaslan, and Andronescu, Ecaterina. (2020) Biodegradable antimicrobial food packaging: trends and perspectives. *Foods*. 9, 1438 doi:10.3390/foods9101438
36. Brockgreitens, J., and Abbas, A. (2016) Responsive food packaging: recent progress and technological prospects. *Comprehensive Reviews in Food Science and Food Safety*. 15, 3–15. doi: 10.1111/1541-4337.12174
37. Holley, R. A., and Patel, D. (2005) Improvement in shelf-life and safety of perishable foods by plant essential oils and smoke antimicrobials. *Food Microbiology*. 22, 273–292. doi: 10.1016/j.fm.2004.08.006
38. Aider, M. (2010) Chitosan application for active bio-based films production and potential in the food industry: review. *LWT – Food Science and Technology*. 43, 837–842. doi: 10.1016/j.lwt.2010.01.021
39. van den Broek, L. A. M., Knoop, R. J. I., Kappen, F. H. J., and Boeriu, C. G. (2015) Chitosan films and blends for packaging material. *Carbohydrate Polymers*. 116, 237–242. doi: 10.1016/j.carbpol.2014.07.039
40. Mlalila, N., Hilonga, A., Swai, H., Devlieghere, F., and Ragaert, P. (2018) Antimicrobial packaging based on starch, poly (3-hydroxybutyrate) and poly (lactic-co-glycolide) materials and application challenges. *Trends in Food Science and Technology*. 74, 1–11. doi: 10.1016/j.tifs.2018.01.015
41. Torlak, E., and Nizamlioğlu, M. (2011) Antimicrobial effectiveness of chitosan-essential oil coated plastic films against foodborne pathogens. *Journal of Plastic Film and Sheeting*. 27, 235–248. doi: 10.1177/8756087911407391
42. Gómez-Estaca, J., López-de-Dicastillo, C., Hernández-Muñoz, P., Catalá, R., and Gavara, R. (2014) Advances in antioxidant active food packaging. *Trends in Food Science and Technolgy*. 35, 42–51. doi: 10.1016/j.tifs.2013.10.008
43. Muriel-Galet, V., Cerisuelo, J. P., López-Carballo, G., Aucejo, S., Gavara, R., and Hernández-Muñoz, P. (2013) Evaluation of EVOH-coated PP films with oregano essential oil and citral to improve the shelf-life of packaged salad. *Food Control*. 30, 137–143. doi: 10.1016/j.foodcont.2012.06.032
44. Malhotra, B., Keshwani, A., and Kharkwal, H. (2015) Antimicrobial food packaging: Potential and pitfalls. *Frontiers in Microbiology*. 6(611), 1–9
45. Deng, X., Nikiforov, A. Y., and Leys, C. (2017) Antimicrobial nanocomposites for food packaging. *Nanotechnology in the Agri-food Industry, Food Preservation*. 6(1), 34.
46. Shankar, S, Jaiswal, L., and Rhim, J. W. (2016) Gelatin-based nanocomposite films: potential use in antimicrobial active packaging. In: Jorge Barros-Velazquez (ed.) *Antimicrobial Food Packaging*. Netherlands: Elsevier. 339–348. doi.org/10.1016/B978-0-12-800723-5.00027-9
47. Pisoschi, A. M., Pop, A., Georgescu, C., Turcuş, V., Olah, N. K., and Mathe, E. (2018) An overview of natural antimicrobials role in food. *European Journal of Medicinal Chemistry*. 143, 922–935.
48. Seow, Y. X., Yeo, C. R., Chung, H. L., and Yuk, H. G. (2014) Plant essential oils as active antimicrobial agents. *Critical Reviews in Food Science and Nutrition*. 54, 625–644.
49. Hayek, S. A., Gyawali, R., and Ibrahim, S. A. (2013) Antimicrobial natural products. *Microbial Pathogens and Strategies for Combating Them: Science, Technology and Education, Formatex Research Center*. 2, 910–921.
50. Lemire, J. A., Harrison, J. J., and Turner, R. J. (2013) Antimicrobial activity of metals: mechanisms, molecular targets and applications. *Nature Reviews Microbiology*. 11(6), 371–384.

51. Martucci, J. F., and Ruseckaite, R. A. (2017) Antibacterial activity of gelatin/copper (II)-exchanged montmorillonite films. *Food Hydrocolloids*. 64, 70–77.
52. Divya, M., Vaseeharan, B., Abinaya, M., Vijayakumar, S., Govindarajan, M., Alharbi, N. S., et al. (2018) Biopolymer gelatincoated zinc oxide nanoparticles showed high antibacterial, antibiofilm and anti-angiogenic activity. *Journal of Photochemistry and Photobiology, B: Biology*. 178, 211–218.
53. Pasquet, J., Chevalier, Y., Couval, E., Bouvier, D., Noizet, G., Morlière, C., et al. (2014) Antimicrobial activity of zinc oxide particles on five micro-organisms of the challenge tests related to their physico-chemical properties. *International Journal of Pharmaceutics*. 460, 92–100.
54. Sirelkhatim, A., Mahmud, S., Seeni, A., Kaus, A. H. M., Ann, L. C., Bakhori, S. K. M., et al. (2015) Review on zinc oxide nanoparticles: antibacterial activity and toxicity mechanism. *Nano-Micro Letters*. 7(3), 219–242.
55. Dakal, T. C., Kumar, A., Majumdar, R. S., and Yadav, V. (2016) Mechanistic basis of antimicrobial actions of silver nanoparticles. *Frontiers in Microbiology*. 7, 1831.
56. Kanmani, P., and Rhim, J. W. (2014) Physical, mechanical and antimicrobial properties of gelatin based active nanocomposite films containing AgNPs and nanoclay. *Food Hydrocolloids*. 35, 644–652.
57. Wyrwa, J., and Barska, A. (2017) Innovations in the food packaging market: active packaging. *European Food Research and Technology*. 243, 1681–1692.
58. Gonçalves, A. A., and Rocha, M. D. O. C. (2017) Safety and quality of antimicrobial packaging applied to seafood. *MOJ Food Processing and Technology*. 4(1), 00079.
59. Wittaya, T. (2012) Protein-based edible films: characteristics and improvement of properties. In Ayman Amer Eissa (ed.) *Structure and Function of Food Engineering*. London: InTech. 43–71.
60. Mellinas, C., Valdes, A., Ramos, M., Burgos, N., Garrigos, M. D. C., and Jimenez, A. (2016) Active edible films: Current state and future trends. *Journal of Applied Polymer Science*. 133(2), 42631.
61. Kashiri, M., Cerisuelo, J. P., Domínguez, I., Lopez-Carballo, G., Muriel-Gallet, V., Gavara, R., et al. (2017) Zein films and coatings as carriers and release systems of Zataria Multiflora Boiss. Essential oil for antimicrobial food packaging. *Food Hydrocolloids*. 70, 260–268.
62. Gómez-Guillén, M. C., Giménez, B., López-Caballero, M. E., Montero, M. P. (2011) Functional and bioactive properties of collagen and gelatin from alternative sources: a review. *Food Hydrocolloids*. 25, 1813–1827.
63. Bonnaillie, L. M., Zhang, H., Akkurt, S., Yam, K. L., and Tomasula, P. M. (2014) Casein films: the effects of formulation, environmental conditions and the addition of citric pectin on the structure and mechanical properties. *Polymer*. 6, 2018–2036.
64. Dangaran, K., Tomasula, P. M., and Qi, P. (2009) Structure and function of protein based edible films and coatings. In: Kerry C. Huber, Milda E. Embuscado (eds) *Edible Films and Coatings for Food Applications*. New York: Springer. 25–56. doi: 10.1007/978-0-387-92824-1
65. Regalado, C., Pérez-Pérez, C., Lara-Cortés, E., and García-Almendarez, B. (2006) Whey protein based edible food packaging films and coatings. *Advances in Agricultural and Food Biotechnology*. 237–261.
66. Aleksandra, Nešić, Gustavo, Cabrera-Barjas, Dimitrijević-Branković 3, Suzana, Davidović, Sladjana, Radovanović, Neda, and Delattre, Cédric. (2020) Prospect of polysaccharide-based materials as advanced food packaging. *Molecules*. 25, 135.
67. Aydin, Furkan, Kahve, Halil Ibrahim, and Mustafa, Ardic. (2017) Lipid based edible films: a review. *Journal of Scientific and Engineering Research*. 4(9), 86–92.
68. Callegarin, F., Gallo J-A. Q., Debeaufort. F., and Voilley A. (1997) Lipids and biopackaging. *Journal of the American Oil Chemists' Society*. 74(10), 1183–1192.
69. Cha. D. S., and Chinnan. M. S. (2004) Biopolymer-based antimicrobial packaging: a review. *Critical Reviews in Food Science and Nutrition*. 44(4), 223–237.
70. Debaufort, F., and Voilley. A. (1995) Effect of surfactants and drying rate on barrier properties of emulsified films. *International Journal of Food Science and Technology*. 30(2), 183–190.
71. Stuchell. Y. M., and Krochta J. M. (1995) Edible coatings on frozen king salmon: effect of whey protein isolate and acetylated monoglycerides on moisture loss and lipid oxidation. *Journal of Food Science*. 60(1), 28–31.
72. Wong, D. W. S., Gastineau, F. A., Grogorski, K. S., Tillin, S. J., and Pavlath, A. E. (1992) Chitosan lipid films: microstructure and surface energy. *Journal of Agricultural and Food Chemistry*. 40(4), 540–544.

73. Dhall, R. K. (2013) Advances in edible coatings for fresh fruits and vegetables: a review. *Critical Reviews in Food Science and Technology*. 53(5), 435–450.
74. Hall, D. J. (2012) Edible coatings from lipids, waxes ad resins. In E. A. Baldwi, R. Hagenmaier, and J. Bai (Eds.) *Edible Coatings and Films to Improve Food Quality.* Boca Raton: CRC Press. 2, 79–101.
75. Velickova, E., Winkelhausen, E., Kuzmanova, S., Alves, V. D., and Moldao-Martins, M. (2013) Impact of chitosan-beeswax edible coatings on the quality of fresh strawberries (Fragariaananassa cv Camarosa) under commercial storage conditions. *LWT – Food Science and Technology*. 52(2), 80–92.
76. Galus, S., and Kadzinśka, J. (2015) Food applications of emulsion-based edible films and coatings. *Trends in Food Science & Technology*. 45(2), 273–283.
77. Ramos, Ó. L., Pereira, R. N., Silva, S. I., Fernandes, J. C., Franco, M. I., Lopes-da-Silva, J. A., Pintado, M. E., and Malcata, F. X. (2012) Evaluation of antimicrobial edible coatings from a whey protein isolate base to improve the shelf life of cheese. *Journal of Dairy Science*. 95(11), 6282–6292.
78. Debaufort, F., Quezada-Gallo, J. A., and Voilley A. (1998) Edible films and coatings: tomorrow's packagings: a review. *Critical Reviews in Food Science and Nutrition.* 38(4), 299–313.
79. Liu, I., Kerry. J. F., and Kerry, J. P. (2006) Effect of food ingredients and selected lipids on the physical properties of extruded edible films/casings. *International Journal of Food Science and Technology.* 41(3), 295–302.
80. Talens, P., Perez-Masia, R., Fabra, M. J., Vargas, M., and Chiralt, A. (2012) Application of edible coatings to partially dehydrated pineapple for use in fruit-cereal products. *Journal of Food Engineering*. 112(1–2), 86–93.
81. Gautam, R. K., Kakatkar, A. S., Karani, M. N. (2016) Development of protein-based biodegradable films from fish processing waste. *International Journal of Current Microbiology and Applied Sciences*. 5(8), 878–888.
82. Ramos, O. L., Fernandes, J. C., Silva, S. I., Pintado, M. E., and Malcata, F. X. (2012) Edible films and coatings from whey proteins: a review on formulation, and on mechanical and bioactive properties. *Critical Reviews in Food Science and Nutrition.* 52(6), 533–552.
83. Kaewprachu, P., and Rawdkuen, S. (2014) Mechanical and physico-chemical properties of biodegradable proteinbased films: a comparative study. *Food and Applied Bioscience Journal*. 2(1), 14–29.
84. Jooyandeh, H. (2011) Whey protein films and coatings: a review. *Pakistan Journal of Nutrition*. 10(3), 296–301.
85. Coltelli, M. B., Wild, F., Bugnicourt, E., Cinelli, P., Lindner, M., Schmid, M., et al. (2016) State of the art in the development and properties of protein-based films and coatings and their applicability to cellulose based products: an extensive review. *Coatings*. 6(1), 1–59.
86. Dobrucka, R., and Cierpiszewski, R. (2014) Active and intelligent packaging food – research and development: a review. *Polish Journal of Food and Nutrition Sciences*. 64(1), 7–15.
87. Kocira, Anna, Kozłowicz, Katarzyna, Panasiewicz, Katarzyna, Staniak, Mariola, Szpunar-Krok, Ewa, and Hortyńska, Paulina. (2021) Polysaccharides as edible films and coatings: characteristics and influence on fruit and vegetable quality – a review, *Agronomy.* 11, 813.
88. Varma, A., and Gupta, G. (2022) Optimization of starch-based biofilm formulation and biodegradability assessment. *Journal of Pharmaceutical Sciences and Research*. 14(2), 705–708.
89. Lourdin, D., and Colonna, P. (2006) *Les matériaux à base d'amidons et de leurs derives. La chimie verte*, Colonna, P. (Ed.), Paris: Tec & Doc.
90. Maynard, F. (1999) Edible film and method. *European Patent* 0935921.
91. Petersen, K., Nielsen, P. V., Bertelsen, G., Lawther, M., Olsen, M. B., Nilsson, N. H., and Mortensen, G. (1999) Potential of biobased materials for food packaging. *Trends in Food Science & Technology*. 10, 52.
92. De la Cruz, G. V., Torres, J. A., and Martín-Polo, M. O. (2001) Temperature effect on the moisture sorption isotherms for methylcellulose and ethylcellulose films. *Journal of Food Engineering*. 48, 91–94.
93. Gemili, S., Yemenicioglu, A., and Altinkaya, S. A. (2009) Development of cellulose acetate based antimicrobial food packaging materials for controlled release of lysozyme. *Journal of Food Engineering*. 90, 453–462.
94. Möller, H., Grelier, S., Pardon, P., and Coma, V. (2004) Antimicrobial and physicochemical properties of chitosan-HPMC-based films *Journal of Agricultural and Food Chemistry*. 52, 6585.

95. Aspinall, G. O. (1959) Structural chemistry of the hemicelluloses. *Advances in Carbohydrate Chemistry*, 14, 429.
96. Hartman, J., Albertsson, A. C., Lindblad, M. S., and Sjoberg, J. (2006) Oxygen barrier materials from renewable sources: Material properties of softwood hemicellulose-based films. *Journal of Applied Polymer*. 4, 2985.
97. Stipanovic, A. J., Haghpanah, J. S., Amidon, T. E., Scott, G. M., Barber, V., and Mishr, M. (2007) ACS symposium series: opportunities for hardwood hemicellulose. *Biodegradable Polymer Blends, Materials, Chemicals, and Energy from Forest Biomass*. 7, 107.
98. Choi, J. H., Choi, W. Y., Cha, D. S., Chinnan, M. J., Park, H. J., Lee, D. S., and Park, J. M. (2005) Diffusivity of potassium sorbate in κ-carrageenan based antimicrobial film. *LWT – Food Sciences and Technology*, 38, 417.
99. Marcos, B., Aymerich, T., Monfort, J. M., and Garriga, M. (2007) Use of antimicrobial biodegradable packaging to control Listeria monocytogenes during storage of cooked ham. *International Journal of Food Microbiology*. 117, 148.
100. Izawa, H., Nawaji, M., Kanako, Y., and Kadokawa, J. I. (2009) Preparation of glycogen-based polysaccharide materials by phosphorylase-catalyzed chain elongation of glycogen. *Macromolecular Bioscience*. 9, 1098.
101. Rousseau, C. F., and Gagnieu, C. H. (2002) In vitro cytocompatibility of porcine type I atelocollagen crosslinked by oxidized glycogen. *Biomaterials*, 23, 1503.
102. Crini, G., and Badot, P. M. (2008) Application of chitosan, a natural aminopolysaccharide, for dye removal from aqueous solutions by adsorption processes using batch studies: A review of recent literature. *Progress in Polymer Science*, 33, 399.
103. Özmeric, N., Özcan, G., Haytac, C. M., Alaaddinoglu, E. E., Sargon, M. F., and Senel, S. (2000) Chitosan film enriched with an antioxidant agent, taurine, in fenestration defects. *Journal of Biomedical Materials Research*. 51(500).
104. Reichert, Corina L., Bugnicourt, Elodie, Coltelli, Maria-Beatrice, Cinelli, Patrizia, Lazzeri, Andrea, Canesi, Ilaria, et al. (2020) Bio-based packaging: materials, modifications, industrial applications and sustainability, *Polymers*. 12(7), 1558. https://doi.org/10.3390/polym12071558
105. Yadav, Ajay, Mangaraj, S., Singh, Ranjeet, M. Naveen Kumar, and Arora, Simran. (2018) Biopolymers as packaging material in food and allied industry. *International Journal of Chemical Studies*. 6(2), 2411–2418.
106. Peelman, N., Ragaert, P., de Meulenaer, B., Adons, D., Peeters, R., Cardon, L., et al. (2013) Application of bioplastics for food packaging. *Trends in Food Science and Technology*. 32, 128–141.
107. Jiménez-Rosado, M., Zarate-Ramírez, L., Romero, A., Bengoechea, C., Partal, P., and Guerrero, A. (2019) Bioplastics based on wheat gluten processed by extrusion. *Journal of Cleaner Production*. 239, https://doi.org/10.1016/j.jclepro.2019.117994
108. Novamont GmbH. (2020) https://germany.novamont.com/page.php?id_page=2&id_first=2
109. Bio-Based News. (2020) Available online: http://news.bio-based.eu/globally-2014-a-good-year-for-renewables/.
110. GualapackGroup. (2020) Available online: https://gualapackgroup.com.
111. Nawab, A., Alam, F., Haq, M. A., Haider, M. S., Lutfi, Z., Kamaluddin, S., et al. (2018) Innovative edible packaging from mango kernel starch for the shelf life extension of red chili powder. *International Journal of Biological Macromolecules*. 114, 626–631.

# 7 Polymeric Biomaterial
## *As Corrosion Protector*

*Shumaila Masood, Fahmina Zafar, and Nahid Nishat*

**CONTENTS**



## 7.1 INTRODUCTION

Commercial sectors bear huge losses through corrosion on a daily basis which is very uneconomical because of the high maintenance cost of the machinery and other metal-based structures [1]. Corrosion, a metallic malice, is a spontaneous and natural phenomenon which weakens metallic surfaces, nuts and bolts and alloys through various oxidation and reduction reactions taking place between the metallic interface and its environment, which generally contains salt and water [2]. The failures caused by corrosion can be disastrous and extensively impactful, not only from an economic point of view but also from a human safety perspective [3]. The losses caused by corrosion directly affect the gross domestic product (GDP) of a country. A study on the role of corrosion management in industry and government impact on the US was conducted by the National Association of Corrosion Engineers (NACE) which estimated that the global cost of corrosion was US$2.5 trillion, which is approximately equal to 1–5% the GDP of each nation. The indirect costs caused by corrosion damage were observed to be more significant than the direct costs [4].

Indirect costs like environmental concerns and health hazards due to corroded materials have been widely studied. It is a well-known fact that corrosion cannot be fully eradicated, but it can overcome with the help of various means such as using alloys, keeping the environment clean, constant greasing, using inhibitors or using paints and coatings [5]. Among all the aforementioned methods, the application of paints and coatings has gained high potential for corrosion prevention since it is cost effective and easier to apply. These paints and coatings protect the metallic equipment by creating a barrier between the environment and the metal surface, thereby increasing the part life [6,7].

Protective coatings are applied to various metallic structures depending upon their exposure zones. The type of protection they provide strongly depends upon the requirements. Various types of protective coatings are acknowledged, such as antibacterial, antifungal, flame-retardant, water-repellant and anticorrosive, etc. [8–11]. These coatings are applied to metallic structures using a

DOI: 10.1201/9781003240884-7

**FIGURE 7.1** Various biobased materials.

number of methods such as a roller or moving belt. These protective coatings can be functionalized by organic polymer, hybrid, inorganic materials or metallic layers [12,13].

The most widely used anticorrosive coatings for metal protection are polymeric coatings. Polymeric coatings are being designed and formulated in such a manner that they help to achieve more than one protective mechanism. They have high anticorrosive performance along with less toxicities and higher chemical stabilities as compared to other non-polymeric coatings. They are widely used in infrastructure sectors (e.g., building, pipelines bridges) and transport (e.g., aircraft, ships and automobiles). Polymeric coatings prevent water and other corrosive elements from penetrating the metal surface by producing a dense barrier [14,15]. The commercially available polymeric coatings are generally produced from petroleum products, which are limited. Thus, these exhausting resources need to be replaced with more abundant materials. Researchers are constantly focusing on developing functionalized coatings from more reliable and abundant resources.

In recent years, considerable attention has been paid to the development of biobased polymeric coatings in order to reduce the environmental pollution to as great as extent as possible. Biobased materials have numerous benefits such as abundance, less toxicity, cost-effectiveness and environment-friendliness [16]. Various renewable resources like cellulose, chitin, chitosan, vegetable oil and cashew nut shell liquid (CNSL) contain ketones, aldehyde, multiple bonds, hydroxyl, esters and other functional groups (Figure 7.1) [17]. These functional groups help provide good adhesion with the substrate, and as a result provide excellent protection. Therefore, these renewable resources are constantly being applied in a number of polymeric industries after several modifications to increase their protection performance [18].

## 7.2 GREEN MATERIALS AS COATINGS

The process of synthesizing polymers from renewable resources fulfills the requirements of green chemistry such as pollution prevention, use of biobased feedstock energy efficiency and design for degradation, as a result of which these materials are called green materials [19]. Green polymeric

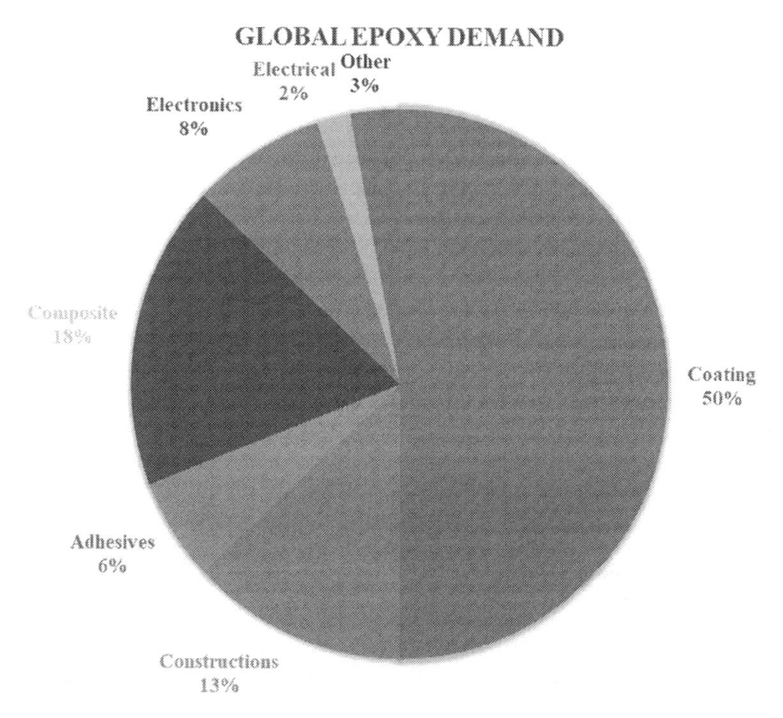


**FIGURE 7.2** Percentage distribution of epoxy applications.

materials have gained vast attention over the last few decades mainly due to increasing environmental concerns and limitations of petroleum feedstocks [20]. Four major categories of green materials extensively utilized in coating industries are *epoxy*, *polyurethane*, *polyesters* and *hybrid coatings* [21].

### 7.2.1 Epoxy

The global epoxy market size was found to be US$6617.7 million in 2021 and is expected to reach US$8027.7 million by the end of 2026 [22]. Figure 7.2 illustrates the percentage distribution of demand for epoxy in various fields at a global level. It can be observed that about 50% of the epoxy resins produced globally are utilized in the coating industry [23].

Epoxies are one of the most widely used materials in the coating industry because of their high weather resistance, excellent tensile strength, superior chemical, thermal and solvent resistance, etc. [24]. They are generally prepared by a condensation reaction between epichlorohydrin and a diphenyl propane derivative [25]. These resins are generally characterized by the presence of at least one three-membered ring in the chain (Figure 7.3). The three-membered ring in the resin is highly strained and thus very reactive, and allows chain extension without the removal of small molecules like water [26].

Bisphenol A (BPA), earlier used as chemical estrogen, is one of the most commonly used monomers for the synthesis of epoxy resin. However, the use of BPA has been restricted as it acts as an endocrine disruptor and can easily mimic the hormones present in the human body and therefore may cause severe health effects [27]. Bellido-Aguilar et Al., synthesized biobased and solvent-less epoxy for a coating application from cardanol by using furfurylamine (FA) or 1.8 diamino-p-methane (DAPM) as a curing agent. These fully biobased coatings were found to have anticorrosive and antiicing properties and thus eliminated the use of BPA [28].

Zheng et al., fabricated eco-friendly and hydrophobic epoxy coatings from isosorbide, a bio-epoxy with similar properties to BPA, by incorporating superhydrophobic $SiO_2$ nanoparticles. The

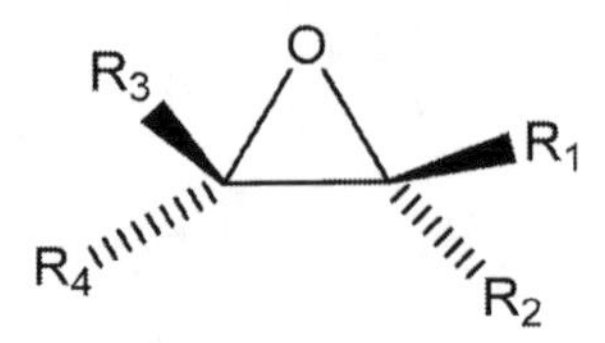


FIGURE 7.3 Structure of epoxy containing a three-membered ring.

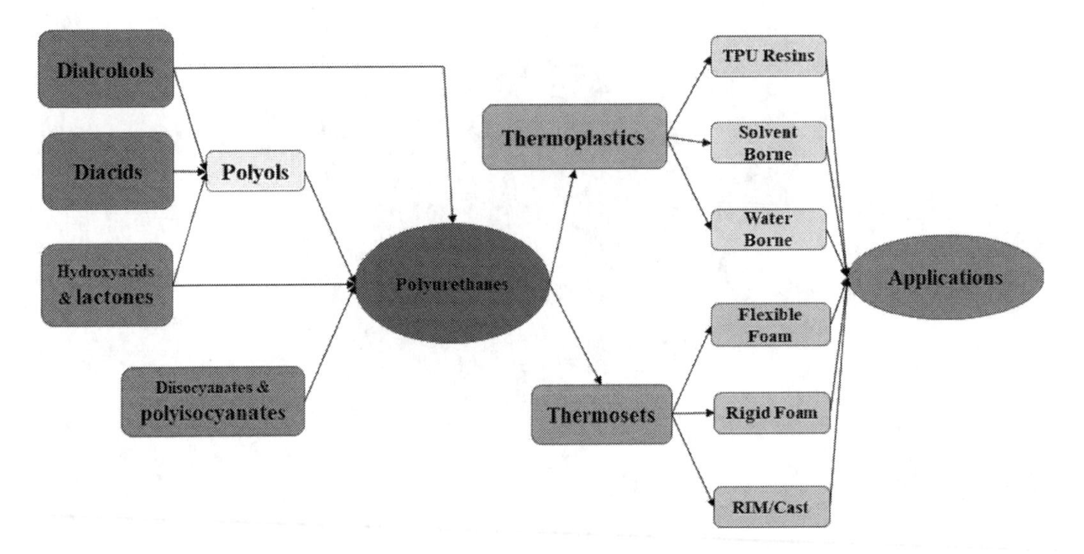


FIGURE 7.4 Various starting materials for PU synthesis and different types of PUs.

coatings thus formed possessed excellent anticorrosive properties and high contact angle values of up to 134 [29].

### 7.2.2 Polyurethane

Polyurethanes (PUs) are among the most rapidly increasing polymers used in the coating industry. These polymers are synthesized by reacting poly-isocyanates with polyols together with polyether/polyesters [30]. At first low-molecular-weight polymers are synthesized which are linked with each other with the help of chain extenders such as diamines [31] and diols [32]. The properties of PUs can be tailored by choosing different polyols such as by using various chain extenders [33], by adding a soft segment [34] or by changing the proportion of hard-to-soft segments [35]. Various PUs are conventionally used in a number of applications. Figure 7.4 shows the starting materials for PU synthesis and different types of PUs.

Polyols, the active hydrogen compounds, are the key factor which affects the sustainability of PUs. Apart from polyols, volatile organic solvents (VOCs) also contribute to the toxic nature of PUs [36]. Two methods need to be applied in order to make PUs completely green polymeric materials, the first is using a renewable resource-based raw material and the other is minimizing the use of VOCs and replacing them by green solvents [37].

#### 7.2.2.1 Using Renewable Resources

The rapid depletion of petroleum deposits and increasing environmental concerns caused by this have caused researchers to look for bioderived polyols for green PU synthesis. Renewable resource-based diols and diacids are being focused on for the preparation of green PUs. Figure 7.5 shows

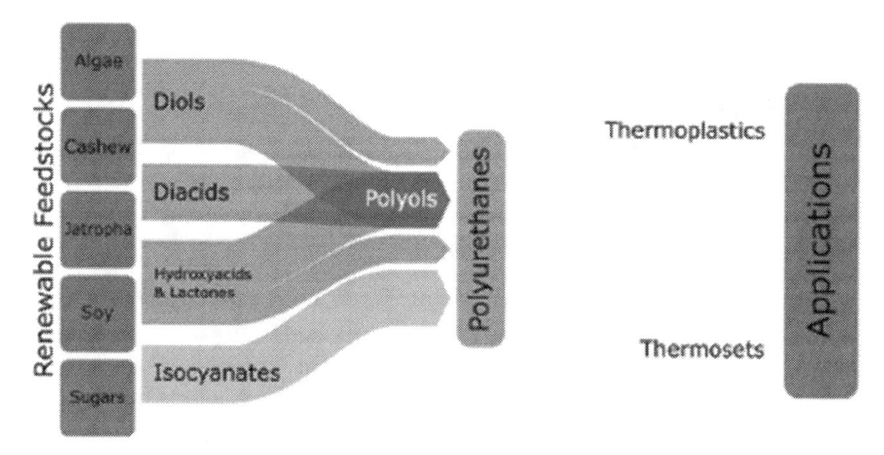


**FIGURE 7.5** Renewable feedstocks for bioderived polyols. (Reprinted with permission from ACS [38].)

some of the renewable feedstocks that account for bioderived polyols [38]. A number of biomasses, such as lignin, proteins, oils and carbohydrates like sugar, starch and cellulose have been extensively used as precursors for various PUs.

Anand et al. reported the synthesis of green PU coatings prepared from renewable resources like diacids, sorbitol and 1,4-butanediol. Furthermore, the authors compared the coating properties of these green PUs with those obtained from non-renewable resources. Sebacic acid-based coatings were found to be suitable according to their anticorrosive properties for pencil hardness and gloss [39].

Lignin, an aromatic polymer, is the second most abundant after cellulose [40]. It has a wide range of applications as it improves the mechanical properties of various materials due to its rigid amorphous polyphenol network. In recent years, lignin has proved to be a great corrosion resistor both as an inhibitor and as a coating [41]. The epoxies and PUs derived from industrial lignin have a low lignin content because of their poor solubility and lower reactivity. To increase the functionality and improve the solubility of these, industrial lignins are generally modified either by oxypropylation or liquification [42]. However, these methods are energy consuming. Therefore, other milder chemical modification methods need to be developed. One of these is the thiol-ene reaction initiated by free radicals. This method is widespread as it has low sensitivity to oxygen and gives a higher yield [43]. To increase the reactivity of these polyols Cao et al., carried out *in situ* conversion of polyols to primary aliphatic hydroxyl with the help of allylation followed by a thiol-ene reaction to prepare lignin-based polyols. These lignin-based polyols along with hexamethylene diisocyanate were later utilized to prepare thermosetting PUs without using any catalyst. The synthesized coatings were found to possess high corrosion resistance as compared to other reported biomodified coatings [44]. Griffini et al. synthesized lignin-based thermoset PUs with a high lignin content by reacting the 2-methyltetrahydrofuran soluble fraction with toluene diisocyanate-based polyisocyanate using different OH/NCO ratios (Figure 7.6). The results of these studies showed that unmodified lignin can also react directly with polyisocyanates. Therefore, lignin-based PU thermosets can be highly utilized in the field of high-performance coatings and adhesives [45].

Neem oil has been extensively used for decades as a biopesticide in almost every household. The neem tree has well-known antibacterial properties and is used in many ways. However, neem tree extracts as a corrosion inhibitor still remained unexplored and they have been recently investigated by Peter et al. These authors investigated the inhibition efficiency of neem extract obtained from leaves, stem and roots and found them to be well performing [46]. Further, Chaudhari et al. reported that neem oil can also be used for the preparation of Pus. They synthesized PUs from neem oil and cured the coatings at room temperature. The authors first converted neem oil into fatty amide

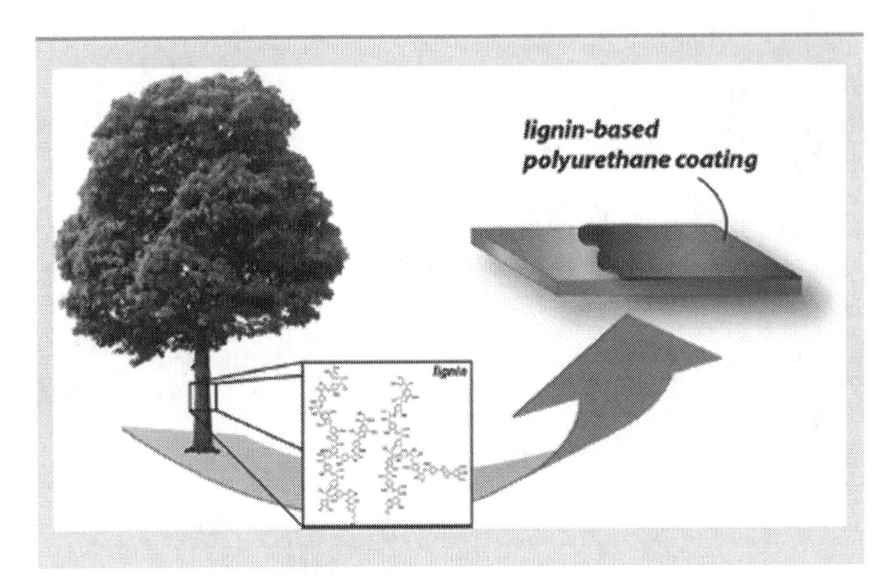


**FIGURE 7.6** Lignin-based PU. (Reprinted with permission from ACS [45].)

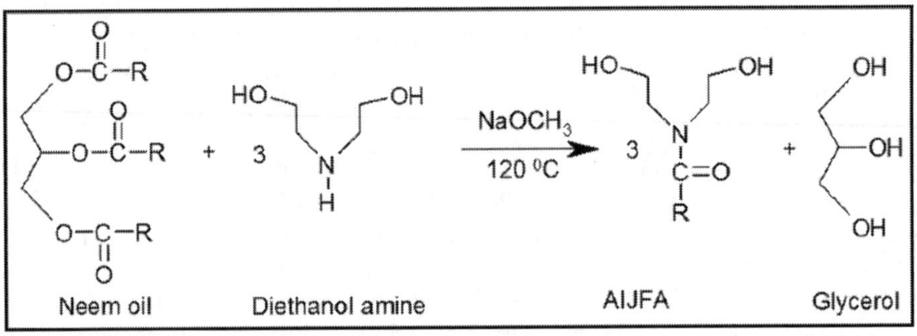

Neem oil + 3 Diethanol amine $\xrightarrow[120\,^{0}C]{NaOCH_3}$ 3 AIJFA + Glycerol

**FIGURE 7.7** Preparation of neem oil-based fatty amide. (Reprinted with permission from ACS [47].)

with the help of diethanolamine (Figure 7.7) and then prepared polyester amide with the help of phthalic anhydride (Figure 7.8). Further, this polyester amide was reacted to polyamidoamine-based polyurea to prepare self-healing green PUs (Figure 7.9). These self-healing coatings were found to have excellent anticorrosive behavior [47].

Melanin is a naturally occurring biomacromolecule that possesses good antibacterial activity, radical scavenging activity and photoprotection. For these properties, melanin is extensively used after several modifications to enhance its use as a polymeric material. The most amazing characteristic of melanin is its nanoparticle behavior. Wang et al. synthesized PU nanocomposites using the nanoparticle property of melanin (Figure 7.10).

These nanoparticles, used in small amounts, significantly improved the strength and toughness of PUs (Figure 7.11) [48].

#### 7.2.2.2 Removing VOCs

According to the United States Environmental Protection Agency (USEPA), VOCs are defined as carbon-containing compounds excluding carbon dioxide, carbon monoxide, metallic carbides and carbonates, carbonic acid and ammonium carbonate, which participate in atmospheric photochemical reactions [49]. These VOCs have high vapor pressure and low water solubility, and can easily

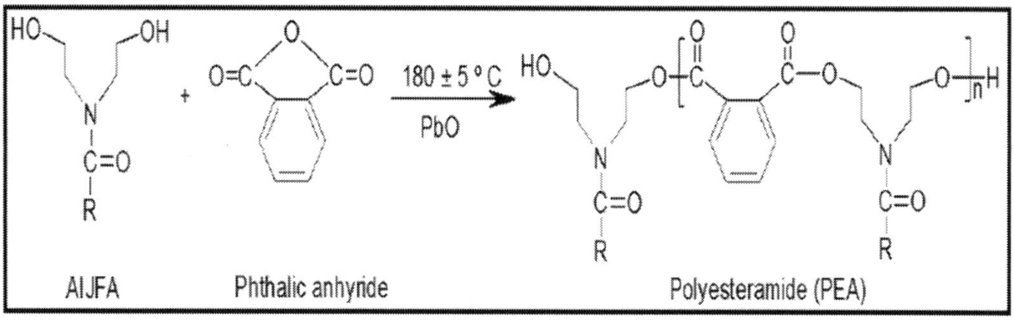

**FIGURE 7.8** Conversion of neem oil based fatty amide into polyester amide. (Reprinted with permission from ACS [47].)

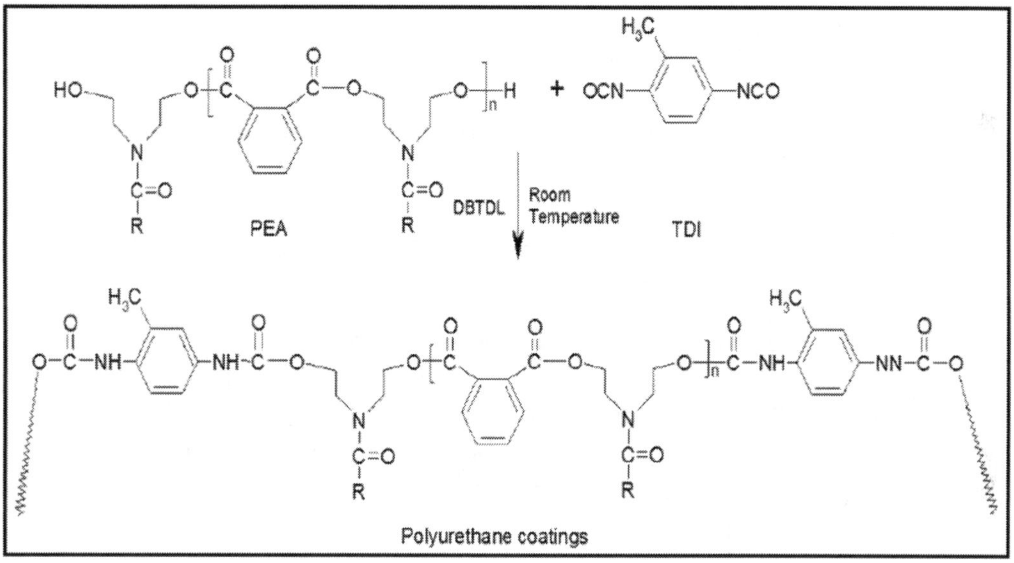

**FIGURE 7.9** Preparation of PU coating. (Reprinted with permission from ACS [47].)

evaporate at room temperature. VOCs are emitted in the form of gases from various solids and liquids and form photochemical oxidants, including ozone. They act as pollutants in the environment and adversely affect health, crops, forests and materials [50].

Previously, VOCs were extensively utilized in the preparation of solvent-borne PUs. However, due to their non-environmentally friendly behavior various agencies have issued guidelines to reduce their consumption [51]. In the search for ecofriendly polymeric materials, researchers switched to greener alternatives and invented waterborne PUs (WPUs) [52]. These WPUs offer many advantages over solvent-borne PUs and are widely applicable in the field of coatings, adhesives and inks. The processing of solvent-borne PUs involves the emission of organic solvents, thereby adding to the VOC content in the environment [53]. The synthesis of PUs from VOs is a hot topic of research [54]. However, these oils first need to be converted to the desired polyols, which increases the overall cost of the product.

Keeping in view the inexpensive and biobased requirements, Gogoi et al. synthesized biodegradable waterborne PUs having a hyperbranched structure from tannic acid. Tannic acid is a polyphenolic renewable resource with the structure shown in Figure 7.12.

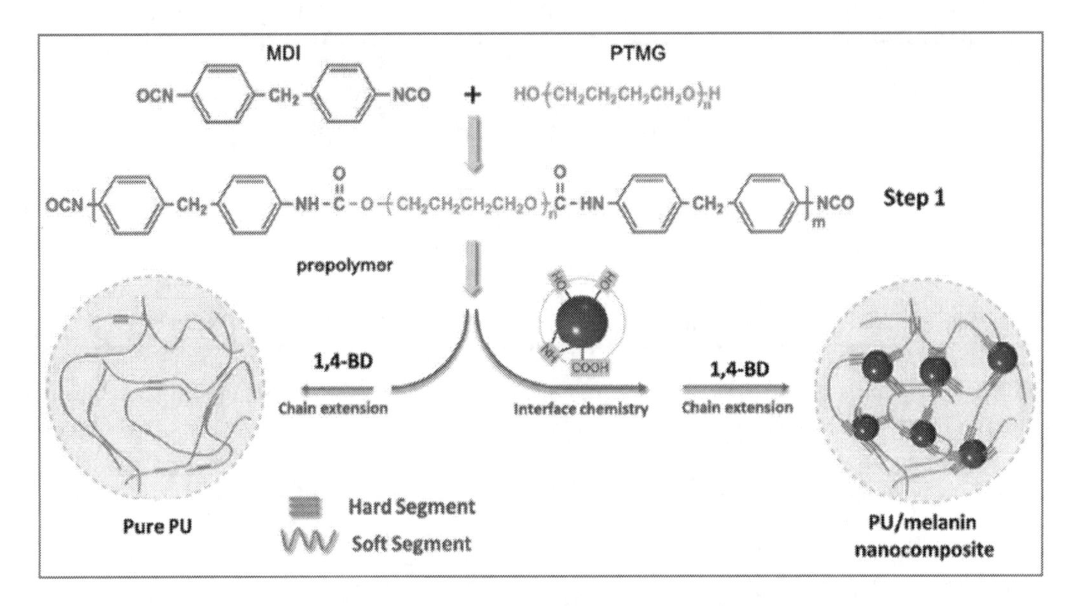


**FIGURE 7.10** Synthesis of PU nanocomposite using melanin. (Reprinted with permission from ACS [48].)

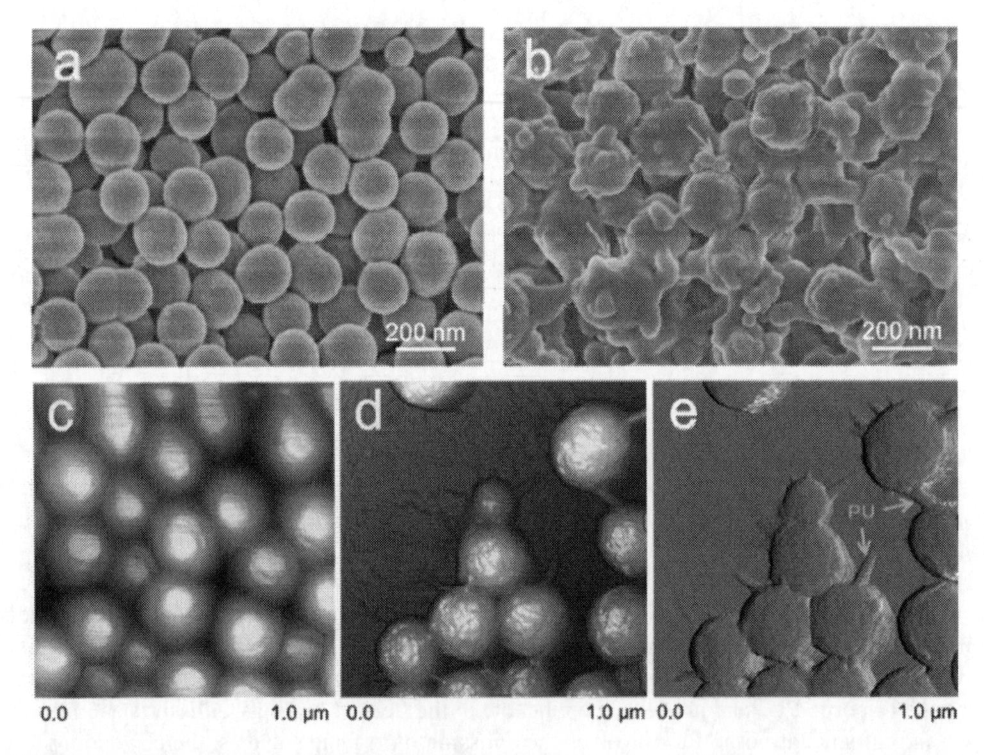


**FIGURE 7.11** Images showing SEM micrographs of (a) melanin; (b) melanin-PU nanocomposite and AFM height of (c) melanin; (d) melanin-PU nanohybrid and AFM of (e) melanin-PU nanohybrid. (Reprinted with permission from ACS [48].)

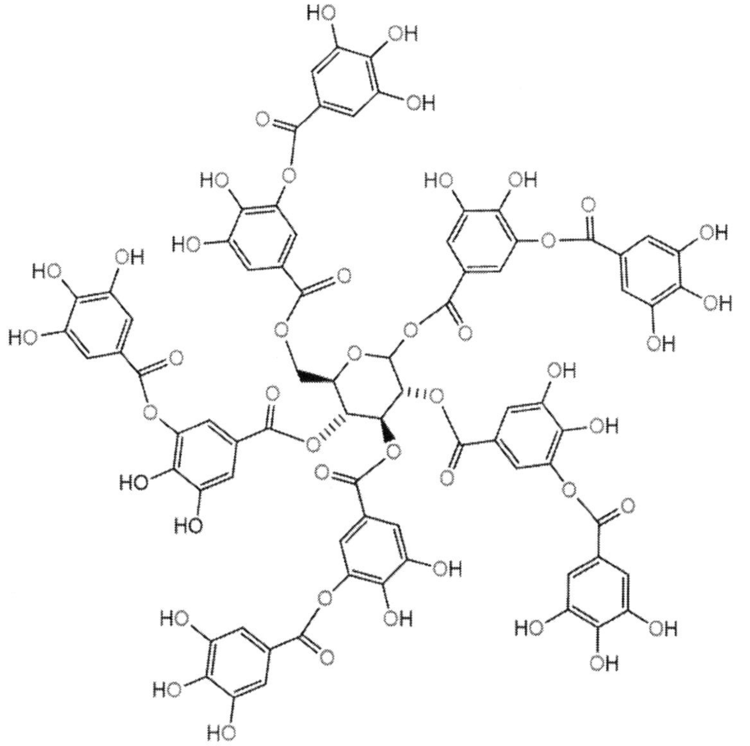

FIGURE 7.12 Structure of tannic acid.

The design of tannic acid-based polyurethane has significant bioactivity and biocompatibility (Figure 7.13). Since tannic acid readily acts as a redox material, the authors carefully examined its interaction with the biological system.

These waterborne PUs were thermally stable and showed mechanically desirable performance and also possess excellent antioxidant activity, along with great antibacterial properties [55].

The unique long, flexible fatty acid structure of plant oil makes it perfect to be used as a soft segment for PUs. Also, the triglyceride group and carbon–carbon double bond present in the moiety provide active functional sites for various modifications [56]. The triglyceride groups present in the plant oil contain varying compositions of fatty acid. These compositions may differ according to the type of plant, the crop, the season and the environmental conditions in which it is grown [57]. These plant oil-based PU coatings require higher temperature for curing and take a longer time for self-healing and reprocessing. Moreover, these coatings also exhibit poor thermal and mechanical properties. To overcome these problems researchers have suggested the incorporation of aromatic compounds containing dynamic bonds which can act as a hard segment and therefore can improve the mechanical properties of PUs [58].

In this context, the development of a highly cross-linked polymer network of castor oil-containing aromatic compound, soft and long molecular chains and a sufficient number of additives could offer a novel polymeric containing all the inherent properties of plant oil. Zhang et al. incorporated a phenyl ring containing S–S bonds as hard segments in a castor oil-based WPU system. They further tuned the ratio of hard and soft segments systematically. The films thus obtained were found to possess excellent mechanical properties. Furthermore, the simultaneous healing and outstanding processibility of plant-oil-based WPUs was reported for the first time [59].

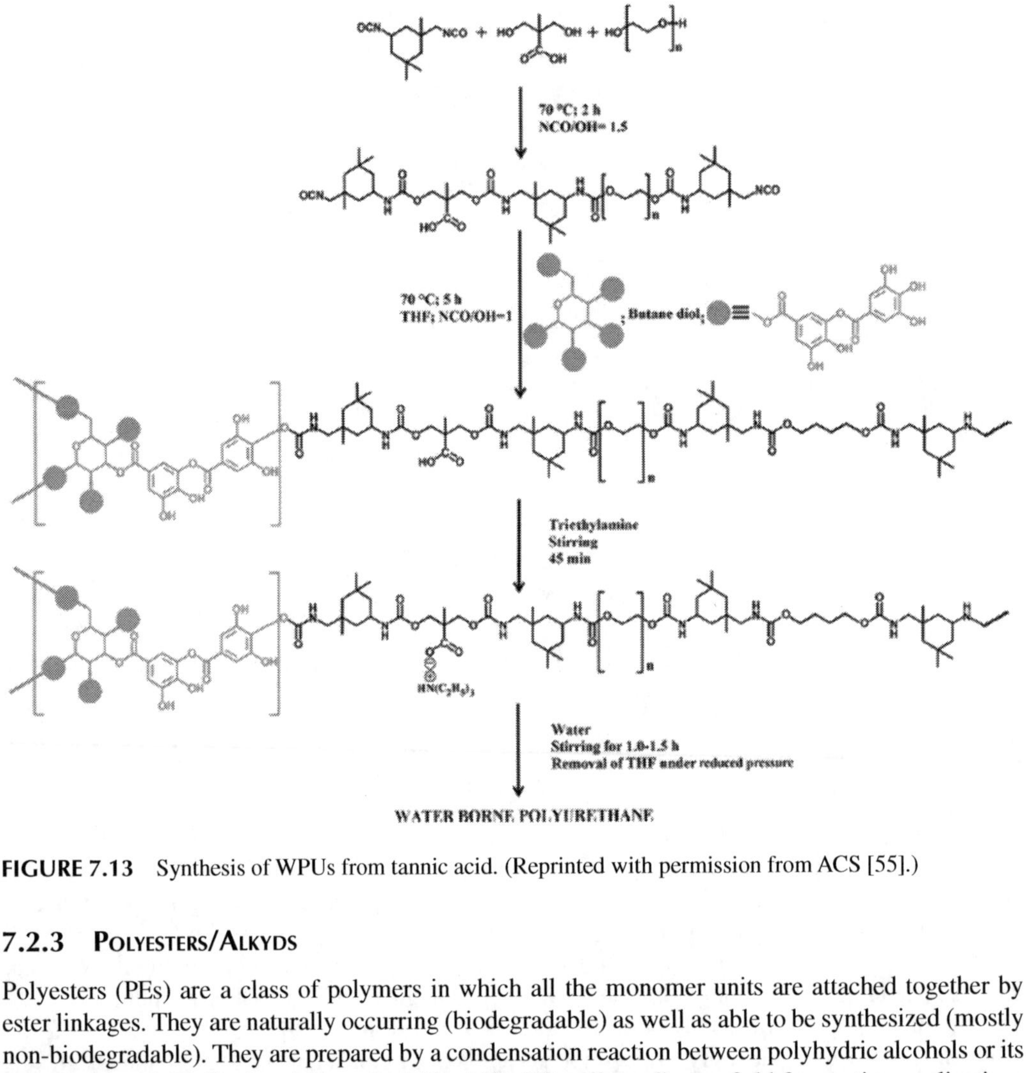

**FIGURE 7.13** Synthesis of WPUs from tannic acid. (Reprinted with permission from ACS [55].)

### 7.2.3 Polyesters/Alkyds

Polyesters (PEs) are a class of polymers in which all the monomer units are attached together by ester linkages. They are naturally occurring (biodegradable) as well as able to be synthesized (mostly non-biodegradable). They are prepared by a condensation reaction between polyhydric alcohols or its derivatives and di/tribasic acid or anhydrides [60]. PEs offer a diverse field for coating applications as the groups having ester linkages can vary over a broad range [61]. Carothers et al. synthesized aliphatic PEs by polycondensation of diols and dicarboxylic acid in 1930. The resulting aliphatic PE had a low melting point which restricted its application as a polymer [62]. Later, many tailored PEs were prepared using new monomers which increased the application of these polymers in coating industry. It was observed that the properties of PEs can be varied by varying the molar ratios of di/tri-functional monomers [63]. Apart from this, the incorporation aromatic rings also improve the mechanical properties of the PEs and decrease its hydrolytic stability. The aromatic diacids are known to improve the chemical resistance, hardness and $T_g$ of PEs [60]. However, aromatic PEs have been found to have better properties although they are prone to photo-oxidation and therefore turn yellow with time. On the other hand, aliphatic polyesters are found to perform better in this regard [64].

Like other polymers, earlier PEs were also prepared from petroleum feedstock. Later research concerning the replacement of these feedstocks with biobased materials was carried out. Biobased PEs offer unique properties such as biodegradability and fiber-forming ability, also, these PEs have been found to be biocompatible in some cases [65,66]. Various sources for the preparation of PEs are lignin, carbohydrates, VOs and sometimes suberin [67,68].

Carbohydrates have been a great source of aliphatic dicarboxylic acids such as succinic acid, itaconic acid and fumaric acid [69]. Gioia et al. synthesized polyester by combining chemically

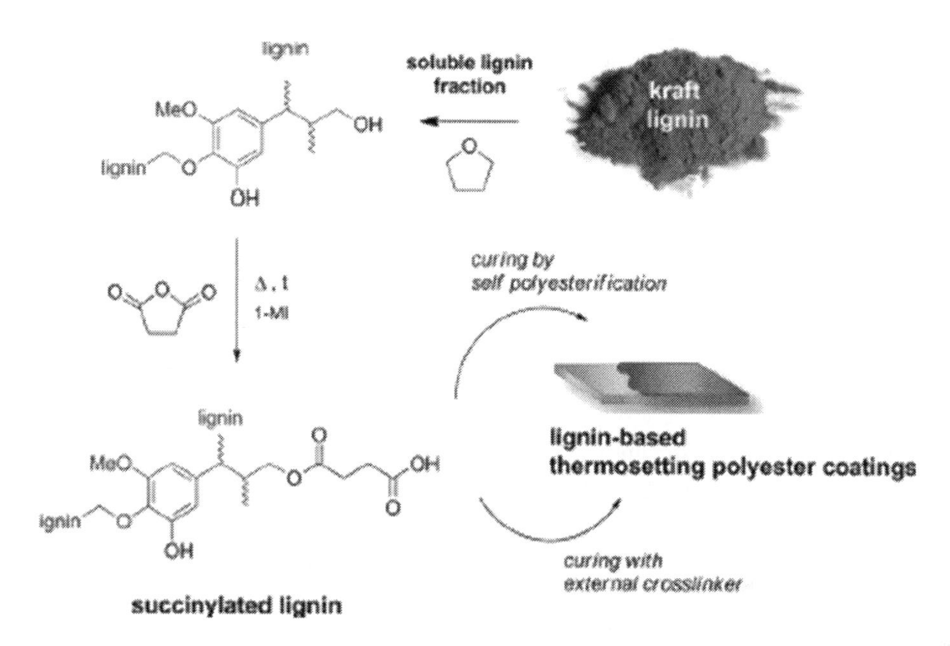


**FIGURE 7.14** Conversion of lignin into succinylated lignin. (Reprinted with permission from ACS [73].)

recycled poly(ethylene terephthalate) with isosorbide and succinic acid. Isosorbide, a renewable resource mainly extracted from sugar, is an interesting diol highly used in the preparation of PEs because of its rigid structure. PE thus formed was used to prepare powder coatings. These powder coatings were solvent free and were applied without any material loss by continuous recovery and reusing excess powder. The coatings thus produced were fully biobased and hence can be described as green [70].

Dai et al. synthesized unsaturated PE from itaconic acid and different polyols. Itaconic acid contains two carboxylic acids along with one carbon–carbon double bond. The carbon–carbon double bond is present as a pendant and showed high reactivity during radial polymerization. These green PE coatings were found to possess good hardness and excellent water- and solvent-resistant properties [71].

PEs can also be crafted from lignin because of the presence of the multiple hydroxyl groups in the structure of lignin [72]. Apart from succinic acid, succinic anhydride can also be utilized for the preparation of PEs. Succinic anhydride has been widely studied as a cross-linker for epoxy coatings (Figure 7.14).

However, little work has explored functionalizing succinic anhydride with lignin for PE preparation. To bridge this gap, Scarica et al. synthesized lignin-based PE thermosets with succinic anhydride for coating applications (Figure 7.15). These PE thermosets having high lignin content can be easily utilized as green coatings as well as adhesives [73].

### 7.2.4 Hybrid coatings

Hybrid coatings (HCs) are a class of material formulated by combining an organic segment (generally polymers) and an inorganic segment (a ceramic phase); they are also called organic–inorganic hybrid coatings [74]. These organic–inorganic components combine in a single phase which results in a hybrid network. The unique properties provided by these coatings are their most important characteristics. Through this method the organic and inorganic phases are bound with each other with the help of siloxane linkages [75]. Careful interaction results in new functionalities which can be achieved by adjusting the nature, type and proportion of both the components at their interface.

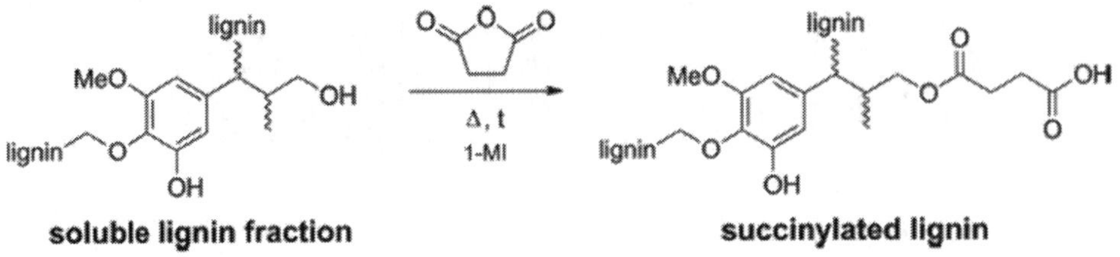

**FIGURE 7.15** Synthesis of succinylated lignin from soluble lignin. (Reprinted with permission from ACS [73].)

Among the various methods used in the preparation of HC, sol–gel is widely used as it is most convenient and versatile [76]. As compared to other methods which proceed at an elevated temperature, sol–gel is a low-temperature process. Material obtained through this method is of high purity and the particle size can be controlled in the nano-range. Therefore, these materials can be tailored depending upon the desired application [77].

The incorporation of renewable resource-based polymer matrix makes the HCs green. These polymers can be obtained from vegetable oils, vegetable seed oils, plant oils, chitosan, cellulose, etc. Uyama et al. synthesized acid-catalyzed oil-based hybrid materials from epoxidized triglycerides by using a modified clay, i.e., organophilic montmorillonite. The material thus produced exhibited high thermal stability [78]. Luca et al. synthesized hybrid films by reacting epoxidized castor oil and glycidoxypropyltrimethoxysilane and tetraethoxysilane. The hardness and tensile strength of these films were found to improve with an increase in the load of the inorganic phase [79]. Galia et al. prepared silicon hybrid PU from vegetable oil with excellent flame retardancy and reported its potential application in the biomedical field [80].

Most of the hybrid materials are synthesized by the addition of organosilane coupling groups in polymer matrix externally. Allauddin et al. reported the preparation of functional hybrid material by structural modifications of castor oil (Figure 7.16). In this study castor oil was first functionalized with alkoxy silane (3-glycidoxypropyltrimethoxy silane) and then it was reacted with different isocyanate ratios. The material thus formed was found to possess excellent properties [81].

## 7.3 METALLIC BIOMATERIALS AS ANTICORROSIVE AGENTS

The word *biomaterial* is used to define materials obtained through natural sources and to describe the materials implanted in the human body as pseudo-organs, tissue or function. Biomaterials obtained from bioresources are a great substitute for petroleum products because of their nontoxic behavior, cost-effectiveness and abundance. They enable thousands of modifications because of various functional groups such as hydroxyl, double bonds and aromatic rings, available in their backbones, which results in the formation of monomers and polymers [82–84]. Epoxy coatings have major importance in anticorrosive coating applications as they offer excellent solvent, moisture and chemical resistance, good adhesion and they are also cured easily. However, their drawbacks like brittleness, porosity and crack propagation remain a matter for concern. To remove these shortcomings various modifications have been made such as incorporating nanomaterials, pigmentation and utilizing flexible curing agents [85,86]. Aliphatic epoxies are known to have properties like flexibility, good adhesion and corrosion inhibition, and are commonly prepared by the epoxidation of polyunsaturated vegetable oils [87–89]. Hedge et al. reported on polymeric coatings incorporated with graphene oxide. Graphene is known to possess chemical inertness, and offer protection against various corrosive agents such as oxygen, water and other corrosive ions [90]. Therefore, the incorporation of graphene improves the corrosion performance of the coatings [91].

Sunflower oils contain 70–75% linoleic acid and hence are used in the formulation of various paints and coatings [92]. The highest content of polyol present in sunflower oil offers a large

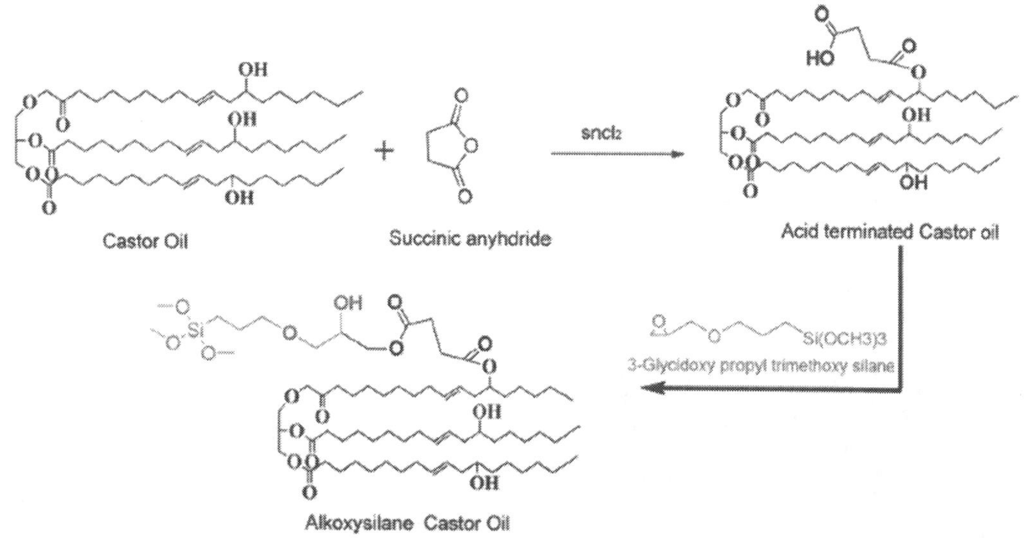

FIGURE 7.16 Synthesis of alkoxysilane castor oil. (Reprinted with permission from ACS [81].)

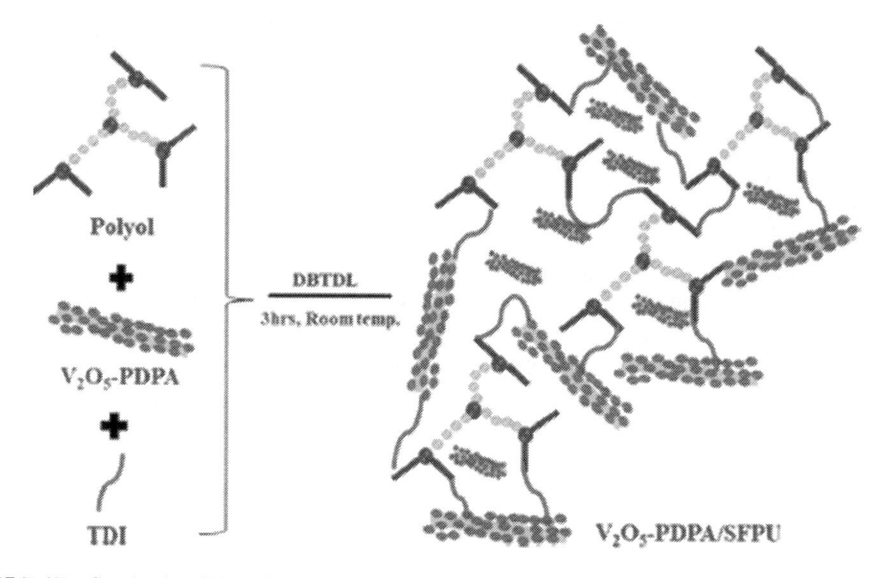


FIGURE 7.17 Synthesis of $V_2O_5$ incorporated PUs. (Reprinted with permission from ACS [93].)

number of hydroxyl groups. These hydroxyl groups help in the processing of a highly cross-linked polymer network. Despite various properties provided by these PUs they lack good anticorrosive performance. Khatoon et al. reported on the enhanced anticorrosive behavior of PUs originating from sunflower oil with the incorporation of $V_2O_5$ (Figure 7.17). As $V_2O_5$ has poor solubility in sunflower oil, the authors also reported the utilization of conducting polymers over the surface of particles to improve the electrical conductivity and solubility of $V_2O_5$ to obtain the desired properties (Figure 7.18) [93].

The incorporation of metal in the polymer matrix improves its properties such as flame retardancy, high tensile strength, thermal stability as well as chemical resistance. Metal-incorporated polymers better serve as adhesives, elastomers and resins. Chantarasiri et al. synthesized Ni- and

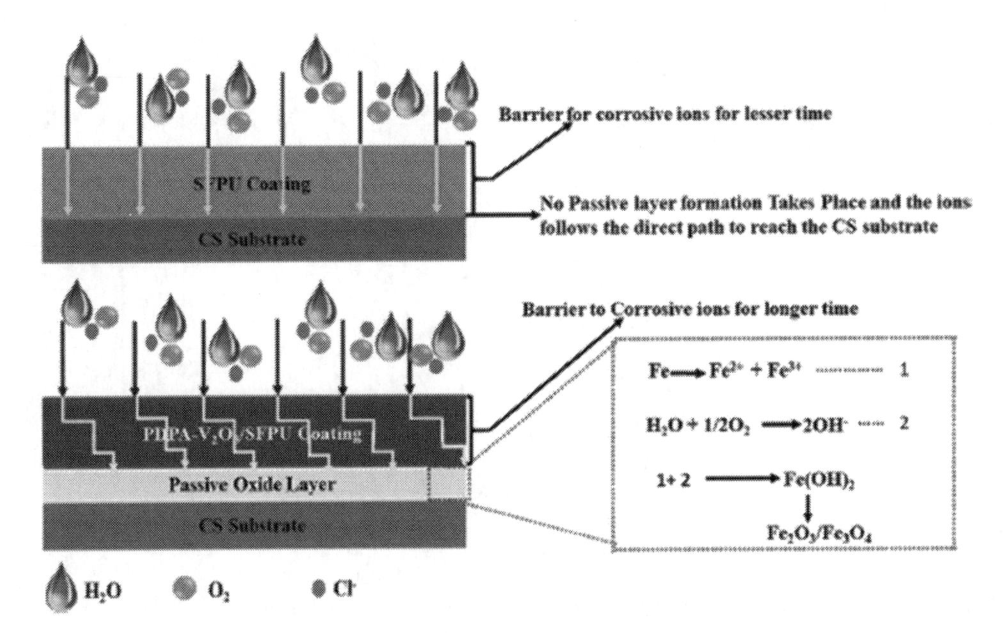


**FIGURE 7.18** Corrosion prevention mechanism of $V_2O_5$-incorporated sunflower oil-based green PUs. (Reprinted with permission from ACS [93].)

Zn-containing polyurethane-urea. The metal-containing polymer possessed excellent thermal stability and good flame retardancy [94]. Ahamad et al. synthesized metal-chelated epoxy polymer by polycondensation of a Schiff base metal complex with epichlorohydrin. The metal-chelated resin showed excellent anticorrosive properties along with good thermal behavior and antimicrobial properties [95]. Jayakumar et al. reported that the incorporation of metal in the polymeric chains of PUs improved its mechanical and thermal properties along with its flame retardancy. These PUs proved to be the best materials that can be applicable in various fields as anticorrosive and antimicrobial agents and as well as thickeners, etc. [96].

Biomaterials intended to interface with biological systems are used to enhance, treat or replace certain tissues, organs or functions in the human body [97]. Before introducing such materials to the market, they are inspected for various issues such as mechanical, pathobiological, biocompatibility, toxicology and the healing process, etc. [98]. The functionality and diversity of biomaterials along with their processing methods have also emerged, with the variability of hybrid, natural and synthetic materials currently being used in industry [99–101]. Highly mobile electron-containing metals are characterized by nondirectional metallic bonds. After the final shaping of metals, the required mechanical properties can be attained by mechanical or thermal processes. Metals have been enormously used in medical implants for centuries due to their mechanical properties and biocompatibility. Gold has been used in dentistry [102], α- and β-Ti alloys owing to their high strength and toughness [103] and Co–Cr alloys because of their high wear resistance and low friction coefficient [104] have been used as bone substitute materials. However, these metallic biomaterials have a tendency to oxidize in solution by losing their electron, causing them to corrode. This results in the release of toxic/cytotoxic, allergic and carcinogenic products like Ni, Al, V, Co and Cr in the biological environment [105]. The release of these products can be controlled by applying surface coatings. These coatings also provide protection against various in vivo microbes.

Xu et al. reported the development of multifunctional lysozyme–phosphate composite coatings for Ti (titanium)-based biomaterials. These coatings were prepared by soaking the substrate in polyphosphate solution and reduced lysozyme. These coatings were found to resist the growth of

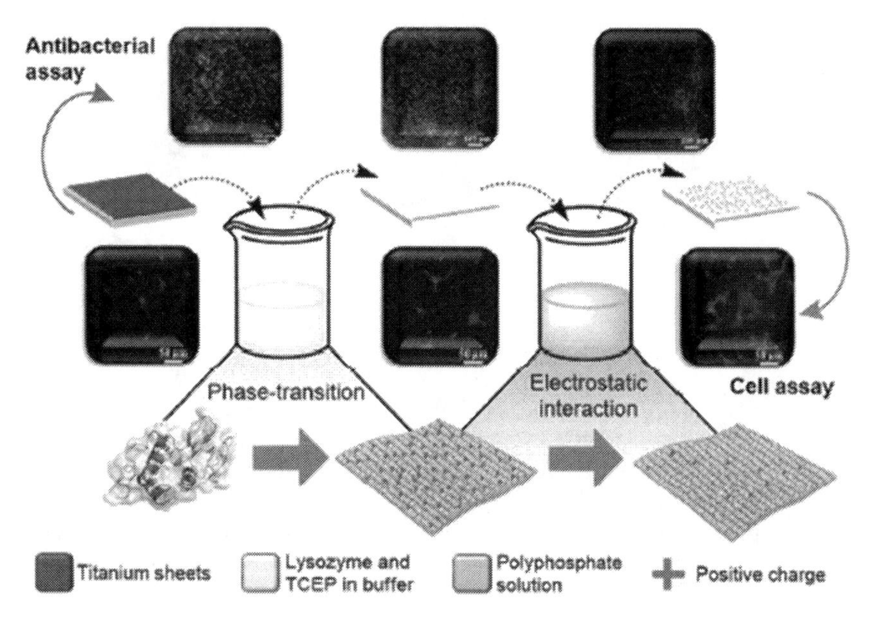


**FIGURE 7.19** Development of multifunctional lysozyme-phosphate composite coatings for titanium-based biomaterials. (Reprinted with permission from ACS [106].)

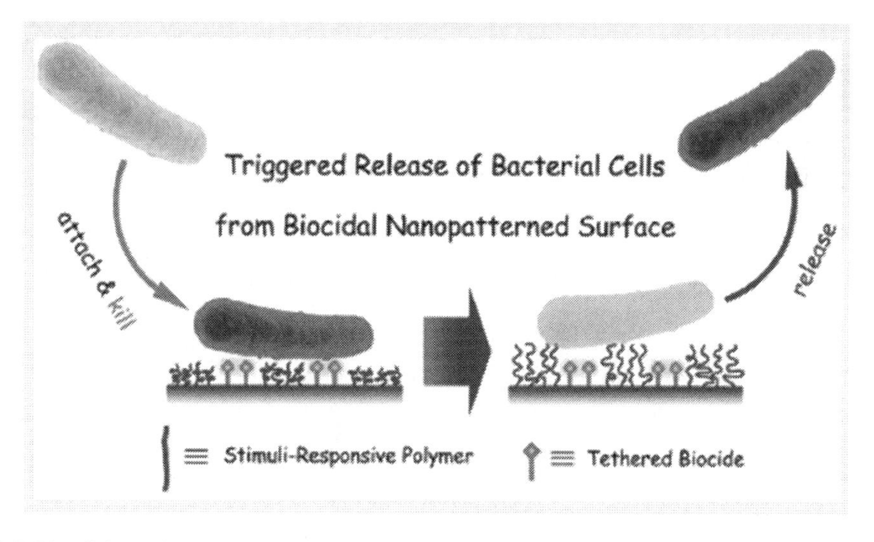


**FIGURE 7.20** Schematic representation of the kill-and-release strategy. (Reprinted with permission from ACS [107].)

Gram-negative bacteria (*E. coli*) effectively and enhanced cell adhesion, differentiation and proliferation (Figure 7.19). The coated biomaterials thus formed were found to perform better than pure Ti-based biomaterials [106].

However, the accumulation of dead bacteria on the surface may serve to help in bacterial adhesion and may also activate the inflammation system or trigger the immune response. Various intelligent techniques have been developed as an antibacterial surface by the kill-and-release strategy (Figure 7.20) [107] in which dead bacteria are released on demand through a change in pH, temperature or salt content [108–110].

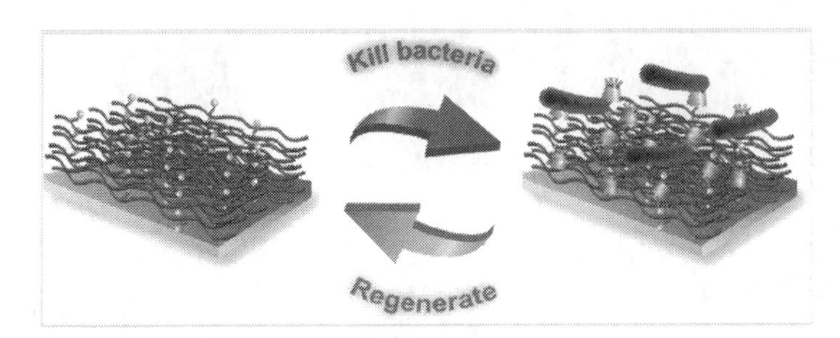


**FIGURE 7.21** Diagrammatic representation of the release of dead bacteria and surface regeneration. (Reprinted with permission from ACS [111].)

Wei et al. fabricated a multifunctional and regeneratable antibacterial surface by a universal strategy using a layer-by-layer technique and host–guest interactions (Figure 7.21). These coatings were developed from β-cyclodextrin modified with a quaternary ammonium salt group. The prepared coatings were capable of maintaining antibacterial effects for the long term. These coatings were able to keep the surface free from the accumulation of dead bacteria and debris [111].

Yavari et al. developed multifunctional coatings for the protection of metallic meta-biomaterial-based implants from various possible infections and also for the stimulation of bone tissue for its regeneration. The authors used the layer-by-layer technique to apply multiple layers of chitosan- and gelatin-based coatings containing either vancomycin or bone morphogenetic protein on the surface of the substrate. The developed coatings were found to have excellent mechanical properties, a well-connected porous structure, exceptional bone regenerating ability and the ability to protect the implant from infections [112].

## 7.4 SYNTHESIS AND APPLICATIONS

### 7.4.1 Synthesis

A significant amount of research has been carried out on the synthesis and applications of biomaterials over the past few decades. This area has proved to be a bridge between chemistry, material science, medicine and bioengineering. Biomaterials offer a wide range of the required physical, chemical, biological and mechanical properties and helps in the regeneration of damaged endogenous cells. Various organic systems have been used for the synthesis of biomaterials including polymers, biopolymers and peptides/proteins. These materials are either those systems which are based on covalent bonds or self-assembled molecular systems which convert into supramolecular architectures [113–115]. Self-assembled biomaterials have been proven to be best candidates for mimicking the extracellular matrix [116]. The superior ability of self-assembled material is their self-healing property, due to which they are injectable biomaterials that reassemble in vivo. Different biomaterial applications require different functional and structural requirements, therefore no one material can be suited for all purposes. Self-assembled biomaterials have considerably been explored in the field of tissue-engineering scaffolds and drug-delivery carriers. Hwang et al. reported the synthesis of self-assembled biomaterials from cholesteryl-(L-acid). These materials have been reported to be used easily for surface modification of porous three-dimensional tissue-engineering scaffolds. The authors concluded that through a periodic multilayer strategy these materials can be exposed repeatedly toward biological components of cells in a particular location [90].

While designing a biomaterial the degradation and depolymerization of the biomaterial after the completion of therapeutic activity is an important aspect to keep in mind. Since these biomaterials

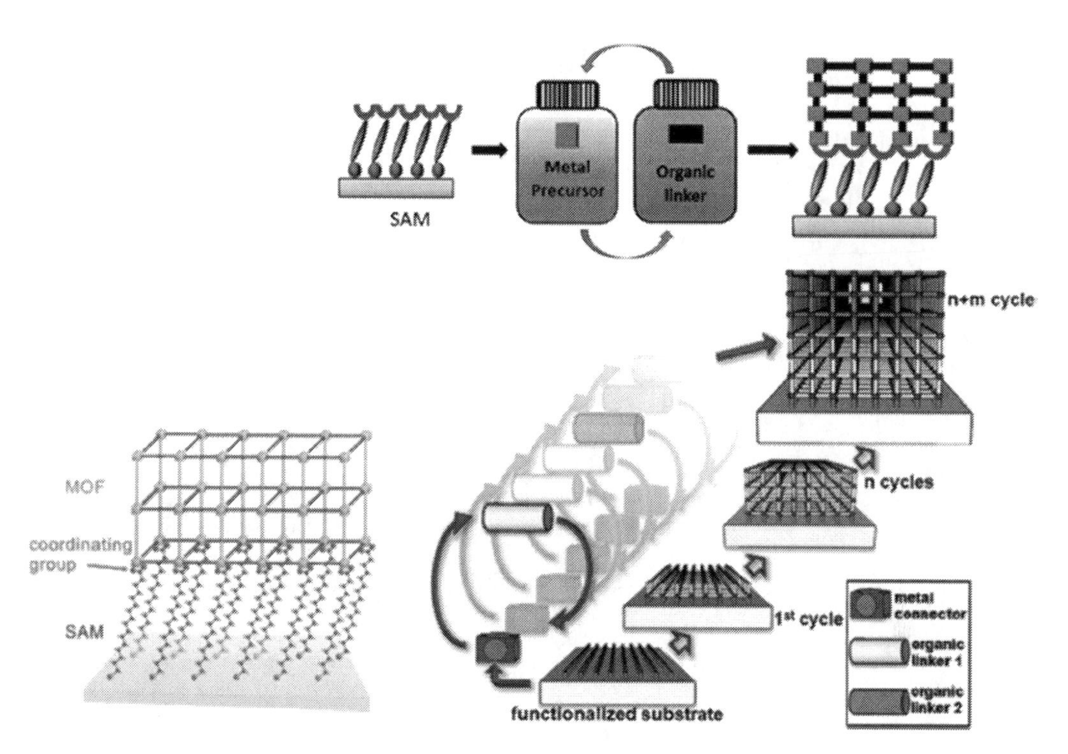


**FIGURE 7.22** Formation of metal organic framework-modified biomaterial. (Reprinted with permission from ACS [124].)

are made up of branched or hyperbranched polymers, they must contain degradable bonds and linkages like acetals, etc. The commonly applied bonds are of disulfide, which gets cleaved under reactive conditions [117].

Biomaterials are also used for the manufacturing of wound dressing materials. Wound healing has evolved from ancient times and is a complex process. Dressing of wounds is generally done to protect them from environmental irritants and maintain the moisture at the wound interface. Dressing materials from chitin and chitosan have been very well studied due to their antifungal and bactericidal properties. Chitin- and chitosan-based biomaterials like hydrogels, sponges, membranes, nanofibers, etc. have been widely accepted owing to their ease of processing [118–121]. However, chitin and chitosan derivatives from different polymers like hyaluronic acid, alginate, polyethylene glycol diacrylate, etc. have been reported to perform better [122]. Chitin-based fibrous materials are known to possess high durability and less toxicity. Chitosan membranes are used to hasten wound healing, whereas chitin and chitosan hydrogels are well known for their ability to contract and heal wounds [123].

Begum et al. synthesized a novel biomaterial by the incorporation of a porous crystalline framework in the soft polymer matrix (Figure 7.22).

These metal organic framework templated polymers have proved to be persistent modular materials and can be easily modified as per the required bio-functioning. The surface-anchored metal organic framework (SAMOF) is converted to surface-anchored polymeric gels (SAPG) by the extraction of metal ions for their application as thin films (Figure 7.23).

The conversion by functionalization enables the fabrication of polymers into the desired biofunction at the internal and external surface. The authors reported that the conversion of SAMOF into SAPG introduced antibacterial property to the thin film (Figure 7.24) [124].

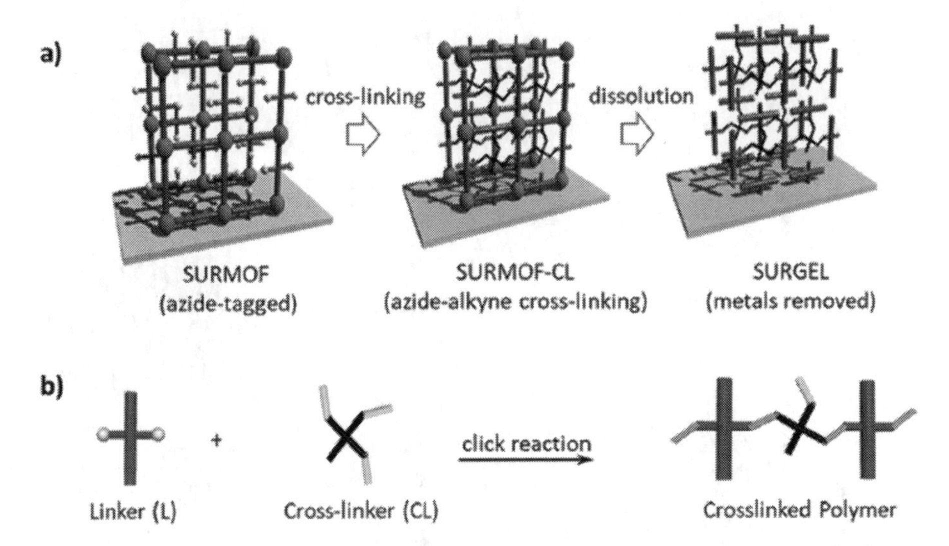


**FIGURE 7.23** Conversion of SAMOF into SAPG. (Reprinted with permission from ACS [124].)

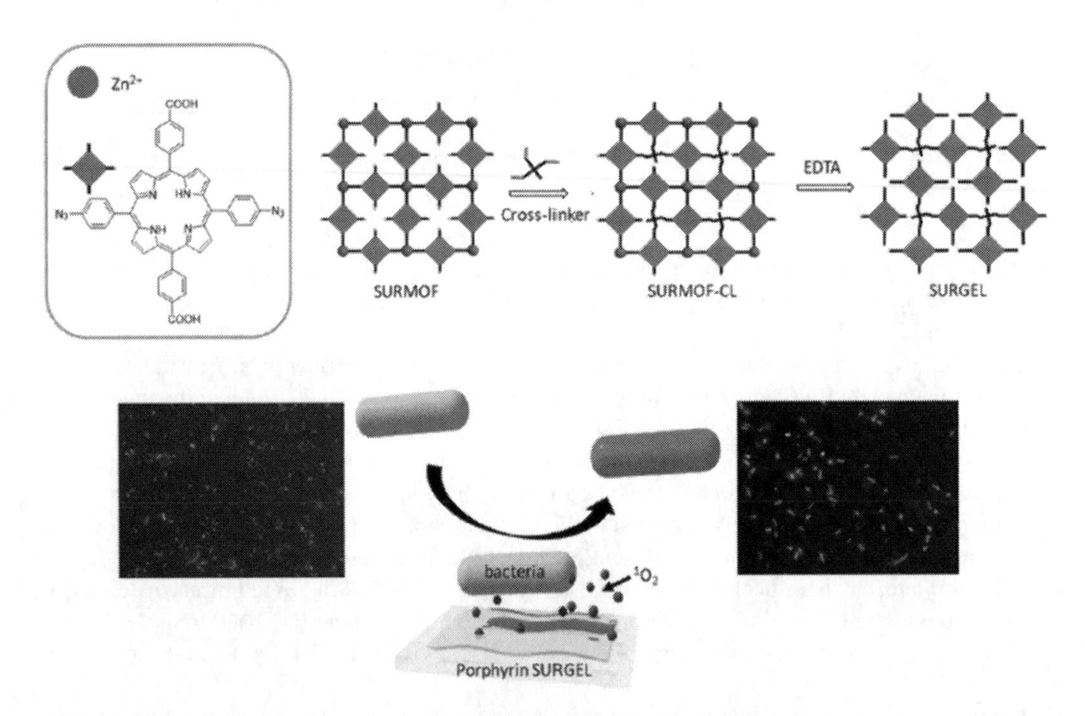


**FIGURE 7.24** SEM images showing (green) living bacteria and (red) dead bacteria after the application of SAPG thin film depicting its antibacterial activity. (Reprinted with permission from ACS [124].)

### 7.4.2 Applications

Biomaterials are widely used in medical and dentistry applications and are intended to interface with biological systems for replacement, treatment and augmentation of biological functions. The application of biomaterials depends upon their type. Table 7.1 presents different types of biomaterials used in different applications.

**TABLE 7.1**
**Types of Biomaterials Used in Various Applications**

| Biomaterials | Applications |
|---|---|
| **Metallic biomaterials** | |
| a) 316L stainless steel | Fracture fixation, surgical instruments and stents |
| b) Ni-Ti | Bone plates, stent, orthodontic wires |
| c) Co-Cr-Mo, Cr-Ni-Cr-Mo | Bone and joint replacement, dental implants |
| d) Gold alloys | Dental restorations |
| **Polymeric biomaterials** | |
| a) Ultra-high-molecular-weight polyethylene | Knee, hip, shoulder joints |
| b) Polymethyl methacrylate, polyphosphazenes | Bone cement, bone tissue engineering |
| c) Acetal, polyethylene, polyurethane | Heart pacemaker |
| d) Nylon, PVC, silicones | Gastrointestinal segments |
| e) Polyester, polytetrafluoroethylene, PVC | Blood vessels |
| **Ceramic and glass biomaterials** | |
| a) Alumina | Joint replacement, dental implants |
| b) Zirconia | Joint replacement |
| c) Porcelain | Dental restoration |
| d) Calcium phosphate | Bone repair and augmentation, surface coating on metal |
| e) Bioactive glass | Bone replacement |
| **Composite biomaterials** | |
| a) PMMA-glass fillers | Dental restoration (as dental cement) |
| b) BIS-GMA-quartz/silica filler | Dental restoration |
| c) Silver nanocomposite material | Combating implant related infections |

## REFERENCES


[1] S. Zheng, D.A. Bellido-Aguilar, J. Hu, Y. Huang, X. Zhao, Z. Wang, X. Zeng, Q. Zhang, and Z. Chen, Waterborne bio-based epoxy coatings for the corrosion protection of metallic substrates, *Prog. Org. Coatings*. 136 (2019) 105265. https://doi.org/10.1016/j.porgcoat.2019.105265

[2] D.E. Arthur, A. Jonathan, P.O. Ameh, and C. Anya, A review on the assessment of polymeric materials used as corrosion inhibitor of metals and alloys, *Int. J. Ind. Chem.* 4 (2013) 2. https://doi.org/10.1186/2228-5547-4-2

[3] A. Saikia, D. Sarmah, A. Kumar, and N. Karak, Bio-based epoxy/polyaniline nanofiber-carbon dot nanocomposites as advanced anticorrosive materials, *J. Appl. Polym. Sci.* 136 (2019) 47744. https://doi.org/10.1002/app.47744

[4] Y. Yons, *Corrosion Engineering: Principles and Practice* (n.d.). www.academia.edu/17546075/Corrosion_Engineering_Principles_and_Practice (accessed August 15, 2021).

[5] M. Abdullah Dar, A review: plant extracts and oils as corrosion inhibitors in aggressive media, *Ind. Lubr. Tribol.* 63 (2011) 227–233. https://doi.org/10.1108/00368791111140431

[6] H. Uhlig, and R. Revie, *Urlig's Corrosion Handbook* (2013) Wiley. 1296.

[7] P.A. Sørensen, S. Kiil, K. Dam-Johansen, and C.E. Weinell, Anticorrosive coatings: A review, *J. Coatings Technol. Res.* 6 (2009) 135–176. https://doi.org/10.1007/S11998-008-9144-2

[8] S. Hofacker, M. Mechtel, M. Mager, and H. Kraus, Sol–gel: a new tool for coatings chemistry, *Prog. Org. Coatings*. 45 (2002) 159–164. https://doi.org/10.1016/S0300-9440(02)00045-0

[9] A.J. Vreugdenhil, V.N. Balbyshev, and M.S. Donley, Nanostructured silicon sol-gel surface treatments for Al 2024-T3 protection, *J. Coatings Technol.* 73 (2001) 35–43. https://doi.org/10.1007/BF02730029

[10] S. Masood, F. Zafar, and N. Nishat, Green flame retardant material from cashew nut shell liquid, in: *Appl. Adv. Green Mater.*, Elsevier, 2021: pp. 663–679. https://doi.org/10.1016/B978-0-12-820484-9.00025-8

[11] L. Matějka, K. Dušek, J. Pleštil, J. Kříž, and F. Lednický, Formation and structure of the epoxy-silica hybrids, *Polymer (Guildf)*. 40 (1999) 171–181. https://doi.org/10.1016/S0032-3861(98)00214-6
[12] D. Akram, E. Sharmin, and S. Ahmad, Linseed polyurethane/tetraethoxyorthosilane/fumed silica hybrid nanocomposite coatings: Physico-mechanical and potentiodynamic polarization measurements studies, *Prog. Org. Coatings*. 77 (2014) 957–964. https://doi.org/10.1016/j.porgcoat.2014.01.024
[13] M. Toorani, and M. Aliofkhazraei, Review of electrochemical properties of hybrid coating systems on Mg with plasma electrolytic oxidation process as pretreatment, *Surfaces and Interfaces*. 14 (2019) 262–295. https://doi.org/10.1016/j.surfin.2019.01.004
[14] Y. Feng, and Y.F. Cheng, An intelligent coating doped with inhibitor-encapsulated nanocontainers for corrosion protection of pipeline steel, *Chem. Eng. J. C* 315 (2017) 537–551. https://doi.org/10.1016/J.CEJ.2017.01.064
[15] D.G. Shchukin, and H. Möhwald, Smart nanocontainers as depot media for feedback active coatings, *Chem. Commun*. 47 (2011) 8730. https://doi.org/10.1039/c1cc13142g
[16] J. Wisniak, Jojoba oil and derivates, *Prog. Chem. Fats Other Lipids (United Kingdom)*. 15:3 (1977) 167–218. https://doi.org/10.1016/0079-6832(77)90001-5
[17] J. Salimon, N. Salih, and E. Yousif, Industrial development and applications of plant oils and their biobased oleochemicals, *Arab. J. Chem*. 5 (2012) 135–145. https://doi.org/10.1016/j.arabjc.2010.08.007
[18] M. Alam, D. Akram, E. Sharmin, F. Zafar, and S. Ahmad, Vegetable oil based eco-friendly coating materials: A review article, *Arab. J. Chem*. 7 (2014) 469–479. https://doi.org/10.1016/j.arabjc.2013.12.023
[19] J. Peyrton, C. Chambaretaud, A. Sarbu, and L. Avérous, biobased polyurethane foams based on new polyol architectures from microalgae oil, *ACS Sustain. Chem. Eng*. 8 (2020) 12187–12196. https:// doi.org/10.1021/acssuschemeng.0c03758
[20] L. Yu, S. Petinakis, K. Dean, A. Bilyk, and D. Wu, Green Polymeric Blends and composites from renewable resources, *Macromol. Symp*. 249–250 (2007) 535–539. https://doi.org/10.1002/masy.200750432
[21] M. Faccini, L. Bautista, L. Soldi, A.M. Escobar, M. Altavilla, M. Calvet, A. Domènech, and E. Domínguez, Environmentally friendly anticorrosive polymeric coatings, *Appl. Sci*. 11 (2021) 3446. https://doi.org/10.3390/app11083446
[22] Epoxy Resins Market Share, Size, Growth Global Forthcoming Developments, Industry Updates, Leading Players, Future Business Prospects and Future Investments by Forecast to 2026 – MarketWatch (n.d.). www.marketwatch.com/press-release/epoxy-resins-market-share-sizegrowth-global-forthcoming-developments-industry-updates-leading-playersfuture-business-prospects-and-future-investments-by-forecast-to-2026-2021-10-07 (accessed November 18, 2021).
[23] S. Kumar, S. Krishnan, S. Mohanty, and S.K. Nayak, Synthesis and characterization of petroleum and biobased epoxy resins: a review, *Polym. Int*. 67 (2018) 815–839. https://doi.org/10.1002/pi.5575
[24] O. Dagdag, A. El Harfi, A. Essamri, M. El Gouri, S. Chraibi, M. Assouag, B. Benzidia, O. Hamed, H. Lgaz, and S. Jodeh, Phosphorous-based epoxy resin composition as an effective anticorrosive coating for steel, *Int. J. Ind. Chem*. 9 (2018) 231–240. https://doi.org/10.1007/S40090-018-0152-5/FIGURES/13
[25] S. Remanan, T.K. Das, and N.C. Das, Graphene as a reinforcement in thermoset resins, in: *Polym. Nanocomposites Contain. Graphene*, Elsevier, 2022: pp. 317–341. https://doi.org/10.1016/B978-0-12-821639-2.00012-4
[26] J.A. Brydson, Epoxide Resins, in: *Plast. Mater*., Elsevier, 1999: pp. 744–777. https://doi.org/10.1016/B978-075064132-6/50067-X
[27] O.E. Ohore, and S. Zhang, Endocrine disrupting effects of bisphenol A exposure and recent advances on its removal by water treatment systems: A review, *Sci. African*. 5 (2019) e00135. https://doi.org/10.1016/j.sciaf.2019.e00135
[28] D.A. Bellido-Aguilar, S. Zheng, Y. Huang, X. Zeng, Q. Zhang, and Z. Chen, Solvent-free synthesis and hydrophobization of biobased epoxy coatings for anti-icing and anticorrosion applications, *ACS Sustain. Chem. Eng*. 7 (2019) 19131–19141. https://doi.org/10.1021/acssuschemeng.9b05091
[29] S. Zheng, D.A. Bellido-Aguilar, Y. Huang, X. Zeng, Q. Zhang, and Z. Chen, Mechanically robust hydrophobic bio-based epoxy coatings for anti-corrosion application, *Surf. Coatings Technol*. 363 (2019) 43–50. https://doi.org/10.1016/j.surfcoat.2019.02.020

[30] Y. Guo, J. Guo, S. Li, X. Li, G. Wang, and Z. Huang, Properties and paper sizing application of waterborne polyurethane emulsions synthesized with TDI and IPDI, *Colloids Surfaces A Physicochem. Eng. Asp*. 427 (2013) 53–61. https://doi.org/10.1016/j.colsurfa.2013.03.017
[31] B.C. Chun, T.K. Cho, and Y.-C. Chung, Enhanced mechanical and shape memory properties of polyurethane block copolymers chain-extended by ethylene diamine, *Eur. Polym. J*. 42 (2006) 3367–3373. https://doi.org/10.1016/j.eurpolymj.2006.08.013
[32] D.-K. Lee, H.-B. Tsai, R.-S. Tsai, and P.H. Chen, Preparation and properties of transparent thermoplastic segmented polyurethanes derived from different polyols, *Polym. Eng. Sci*. 47 (2007) 695–701. https://doi.org/10.1002/pen.20742
[33] A. Eyvazzadeh Kalajahi, M. Rezaei, F. Abbasi, and G. Mir Mohamad Sadeghi, The effect of chain extender type on the physical, mechanical, and shape memory properties of poly(ε -Caprolactone)-based polyurethane-ureas, *Polym. Plast. Technol. Eng*. 56 (2017) 1977–1985. https://doi.org/10.1080/03602559.2017.1298797
[34] N. Yoshihara, M. Enomoto, M. Doro, Y. Suzuki, M. Shibaya, and H. Ishihara, Effect of soft segment components on mechanical properties at low temperatures for segmented polyurethane elastomers, *J. Polym. Eng*. 27 (2007) 291–312. https://doi.org/10.1515/POLYENG.2007.27.4.291
[35] M.A. Pérez-Limiñana, F. Arán-Aís, A.M. Torró-Palau, C. Orgilés-Barcel, and J.M. Martín-Martínez, Influence of the hard-to-soft segment ratio on the adhesion of water-borne polyurethane adhesive, *J. Adhes. Sci. Technol*. 21 (2007) 755–773. https://doi.org/10.1163/156856107781362635
[36] Y. Liu, H. Liang, S. Li, D. Liu, Y. Long, G. Liang, and F. Zhu, Preparation of waterborne polyurethane with high solid content and elasticity, *J. Polym. Res*. 26 (2019) 146. https://doi.org/10.1007/s10965-019-1795-4
[37] A. Das, and P. Mahanwar, A brief discussion on advances in polyurethane applications, *Adv. Ind. Eng. Polym. Res*. 3 (2020) 93–101. https://doi.org/10.1016/j.aiepr.2020.07.002
[38] T.A.P. Hai, M. Tessman, N. Neelakantan, A.A. Samoylov, Y. Ito, B.S. Rajput, N. Pourahmady, and M.D. Burkart, Renewable polyurethanes from sustainable biological precursors, *Biomacromolecules*. 22 (2021) 1770–1794. https://doi.org/10.1021/ACS.BIOMAC.0C01610
[39] A. Anand, R.D. Kulkarni, C.K. Patil, and V.V. Gite, Utilization of renewable bio-based resources, viz. sorbitol, diol, and diacid, in the preparation of two pack PU anticorrosive coatings, *RSC Adv*. 6 (2016) 9843–9850. https://doi.org/10.1039/C5RA17202K
[40] Q. Fan, T. Liu, C. Zhang, Z. Liu, W. Zheng, R. Ou, and Q. Wang, Extraordinary solution-processability of lignin in phenol–maleic anhydride and dielectric films with controllable properties, *J. Mater. Chem. A*. 7 (2019) 23162–23172. https://doi.org/10.1039/C9TA06665A
[41] M.M. El-Deeb, Evaluation of the modified extracted lignin from wheat straw as corrosion inhibitors for aluminum in alkaline solution, *Int. J. Electrochem. Sci*. 13 (2018) 4123–4138. https://doi.org/10.20964/2018.05.49
[42] J.P.S. Aniceto, I. Portugal, and C.M. Silva, Biomass-based polyols through oxypropylation reaction, *ChemSusChem*. 5 (2012) 1358–1368. https://doi.org/10.1002/cssc.201200032
[43] C.E. Hoyle, and C.N. Bowman, Thiol–Ene click chemistry, *Angew. Chemie Int. Ed*. 49 (2010) 1540–1573. https://doi.org/10.1002/ANIE.200903924
[44] Y. Cao, Z. Liu, B. Zheng, R. Ou, Q. Fan, L. Li, C. Guo, T. Liu, and Q. Wang, Synthesis of lignin-based polyols via thiol-ene chemistry for high-performance polyurethane anticorrosive coating, *Compos. Part B Eng*. 200 (2020) 108295. https://doi.org/10.1016/j.compositesb.2020.108295
[45] G. Griffini, V. Passoni, R. Suriano, M. Levi, and S. Turri, Polyurethane coatings based on chemically unmodified fractionated lignin, *ACS Sustain. Chem. Eng*. 3 (2015) 1145–1154. https://doi.org/10.1021/acssuschemeng.5b00073
[46] Azadirachta Indica Extracts as Corrosion Inhibitor for Mild Steel in Acid Medium (PDF) (n.d.). www.researchgate.net/publication/266448594_Azadirachta_Indica_Extracts_as_Corrosion_Inhibitor_for_Mild_Steel_in_Acid_Medium (accessed September 16, 2021).
[47] A.B. Chaudhari, P.D. Tatiya, R.K. Hedaoo, R.D. Kulkarni, and V.V. Gite, Polyurethane prepared from neem oil polyesteramides for self-healing anticorrosive coatings, *Ind. Eng. Chem. Res*. 52 (2013) 10189–10197. https://doi.org/10.1021/ie401237s
[48] Y. Wang, T. Li, X. Wang, P. Ma, H. Bai, W. Dong, Y. Xie, and M. Chen, Superior performance of polyurethane based on natural melanin nanoparticles, *Biomacromolecules*. 17 (2016) 3782–3789. https://doi.org/10.1021/acs.biomac.6b01298

[49] What are volatile organic compounds (VOCs)?|US EPA (n.d.). www.epa.gov/indoor-air-quality-iaq/what-are-volatile-organic-compounds-vocs (accessed September 3, 2021).
[50] R. Tisserand, and R. Young, The respiratory system, in: *Essent. Oil Saf.*, Elsevier, 2014: pp. 99–110. https://doi.org/10.1016/B978-0-443-06241-4.00006-0
[51] B.K. Kim, T.K. Kim, and H.M. Jeong, Aqueous dispersion of polyurethane anionomers from H12MDI/IPDI, PCL, BD, and DMPA, *J. Appl. Polym. Sci.* 53 (1994) 371–378. https://doi.org/10.1002/app.1994.070530315
[52] K.-L. Noble, Waterborne polyurethanes, *Prog. Org. Coatings.* 32 (1997) 131–136. https://doi.org/10.1016/S0300-9440(97)00071-4
[53] B.K. Kim, T.K. Kim, and H.M. Jeong, Aqueous dispersion of polyurethane anionomers from H12MDI/IPDI, PCL, BD, and DMPA, *J. Appl. Polym. Sci.* 53 (1994) 371–378. https://doi.org/10.1002/APP.1994.070530315
[54] Y. Xia, and R.C. Larock, Vegetable oil-based polymeric materials: synthesis, properties, and applications, *Green Chem.* 12 (2010) 1893. https://doi.org/10.1039/c0gc00264j
[55] S. Gogoi, and N. Karak, Biobased biodegradable waterborne hyperbranched polyurethane as an eco-friendly sustainable material, *ACS Sustain. Chem. Eng.* 2 (2014) 2730–2738. https://doi.org/10.1021/sc5006022
[56] I. Singh, S.K. Samal, S. Mohanty, and S.K. Nayak, Recent advancement in plant oil derived polyol-based polyurethane foam for future perspective: A review, *Eur. J. Lipid Sci. Technol.* 122 (2020) 1900225. https://doi.org/10.1002/ejlt.201900225
[57] F.D. Gunstone, J.L. Harwood, and J.L. Harwood, *The Lipid Handbook with CD-ROM* (2007). CRC Press. https://doi.org/10.1201/9781420009675
[58] D.K. Chattopadhyay, and K.V.S.N. Raju, Structural engineering of polyurethane coatings for high performance applications, *Prog. Polym. Sci.* 32 (2007) 352–418. https://doi.org/10.1016/J.PROGPOLYMSCI.2006.05.003
[59] C. Zhang, H. Liang, D. Liang, Z. Lin, Q. Chen, P. Feng, and Q. Wang, Renewable castor-oil-based waterborne polyurethane networks: Simultaneously showing high strength, self-healing, processability and tunable multishape memory, *Angew. Chemie Int. Ed.* 60 (2021) 4289–4299. https://doi.org/10.1002/ANIE.202014299
[60] K.V.S.N. Raju, and D.K. Chattopadhyay, Polyester coatings for corrosion protection, in: *High-Performance Org. Coatings*, Elsevier, 2008: pp. 165–200. https://doi.org/10.1533/9781845694739.2.165
[61] J. Argyropoulos, P. Popa, G. Spilman, D. Bhattacharjee, and W. Koonce, Seed oil based polyester polyols for coatings, *J. Coatings Technol. Res.* 6 (2009) 501–508. https://doi.org/10.1007/s11998-008-9154-0
[62] W.H. Carothers, Studies on polymerization and ring formation. I. An introduction to the general theory of condensation polymers, *J. Am. Chem. Soc.* 51 (1929) 2548–2559. https://doi.org/10.1021/ja01383a041
[63] S. Lundmark, M. SjÖling, and A.-C. Albertsson, Polymerization of oxepan-2,7-dione in solution and synthesis of block copolymers of oxepan-2,7-dione and 2-oxepanone, *J. Macromol. Sci. Part A – Chem.* 28 (1991) 15–29. https://doi.org/10.1080/00222339108052083
[64] B.A.J. Noordover, R. Duchateau, R.A.T.M. van Benthem, W. Ming, and C.E. Koning, Enhancing the functionality of biobased polyester coating resins through modification with citric acid, *Biomacromolecules.* 8 (2007) 3860–3870. https://doi.org/10.1021/bm700775e
[65] J.M.G. Cowie, and Valeria Arrighi, *Polymers Chemistry and Physics of Modern Materials*, Third Edition (2007). CRC Press.
[66] U. Edlund, and A.-C. Albertsson, Polyesters based on diacid monomers, *Adv. Drug Deliv.* Rev. 55 (2003) 585–609. https://doi.org/10.1016/S0169-409X(03)00036-X
[67] G.-Q. Chen, and M.K. Patel, Plastics derived from biological sources: Present and future: A technical and environmental review, *Chem. Rev.* 112 (2012) 2082–2099. https://doi.org/10.1021/cr200162d
[68] K. Yao, and C. Tang, Controlled polymerization of next-generation renewable monomers and beyond, *Macromolecules.* 46 (2013) 1689–1712. https://doi.org/10.1021/ma3019574
[69] V. Menon, and M. Rao, Trends in bioconversion of lignocellulose: Biofuels, platform chemicals & biorefinery concept, *Prog. Energy Combust. Sci.* 38 (2012) 522–550. https://doi.org/10.1016/j.pecs.2012.02.002

[70] C. Gioia, M. Vannini, P. Marchese, A. Minesso, R. Cavalieri, M. Colonna, and A. Celli, Sustainable polyesters for powder coating applications from recycled PET, isosorbide and succinic acid, *Green Chem.* 16 (2014) 1807–1815. https://doi.org/10.1039/C3GC42122H
[71] J. Dai, S. Ma, X. Liu, L. Han, Y. Wu, X. Dai, and J. Zhu, Synthesis of bio-based unsaturated polyester resins and their application in waterborne UV-curable coatings, *Prog. Org. Coatings.* 78 (2015) 49–54. https://doi.org/10.1016/j.porgcoat.2014.10.007
[72] C.A. Cateto, M.F. Barreiro, A.E. Rodrigues, M.C. Brochier-Salon, W. Thielemans, and M.N. Belgacem, Lignins as macromonomers for polyurethane synthesis: A comparative study on hydroxyl group determination, *J. Appl. Polym. Sci.* 109 (2008) 3008–3017. https://doi.org/10.1002/APP.28393
[73] C. Scarica, R. Suriano, M. Levi, S. Turri, and G. Griffini, Lignin functionalized with succinic anhydride as building block for biobased thermosetting polyester coatings, *ACS Sustain. Chem. Eng.* 6 (2018) 3392–3401. https://doi.org/10.1021/acssuschemeng.7b03583
[74] S.V. Harb, A. Trentin, R.F.O. Torrico, S.H. Pulcinelli, C.V. Santilli, and P. Hammer, Organic-inorganic hybrid coatings for corrosion protection of metallic surfaces, *New Technol. Prot. Coatings* (2017) 19–51. https://doi.org/10.5772/67909
[75] R. Zandi-zand, A. Ershad-langroudi, and A. Rahimi, Organic–inorganic hybrid coatings for corrosion protection of 1050 aluminum alloy, *J. Non. Cryst. Solids.* 351 (2005) 1307–1311. https://doi.org/10.1016/j.jnoncrysol.2005.02.022
[76] M.D. Soucek, Z. Zong, and A.J. Johnson, Inorganic/organic nanocomposite coatings: The next step in coating performance, *J. Coatings Technol. Res.* 3 (2006) 133–140. https://doi.org/10.1007/s11998-006-0016-3
[77] S. Amiri, and A. Rahimi, Hybrid nanocomposite coating by sol–gel method: a review, *Iran. Polym. J.* 25 (2016) 559–577. https://doi.org/10.1007/s13726-016-0440-x
[78] H. Uyama, M. Kuwabara, T. Tsujimoto, M. Nakano, A. Usuki, and S. Kobayashi, Organic-inorganic hybrids from renewable plant oils and clay, *Macromol. Biosci.* 4 (2004) 354–360. https://doi.org/10.1002/mabi.200300097
[79] M.A. de Luca, M. Martinelli, and C.C.T. Barbieri, Hybrid films synthesised from epoxidised castor oil, γ-glycidoxypropyltrimethoxysilane and tetraethoxysilane, *Prog. Org. Coatings.* 65 (2009) 375–380. https://doi.org/10.1016/j.porgcoat.2009.03.002
[80] G. Lligadas, J.C. Ronda, M. Galià, and V. Cádiz, Novel silicon-containing polyurethanes from vegetable oils as renewable resources. *Synthesis and Properties, Biomacromolecules.* 7 (2006) 2420–2426. https://doi.org/10.1021/bm060402k
[81] S. Allauddin, R. Narayan, and .V.S.N. Raju, Synthesis and properties of alkoxysilane castor oil and their polyurethane/urea–silica hybrid coating films, *ACS Sustain. Chem. Eng.* 1 (2013) 910–918. https://doi.org/10.1021/sc3001756
[82] M.Y. Shah, and S. Ahmad, Waterborne vegetable oil epoxy coatings: Preparation and characterization, *Prog. Org. Coatings.* 3 (2012) 248–252. https://doi.org/10.1016/J.PORGCOAT.2012.05.001
[83] K. Wazarkar, and A. Sabnis, Cardanol based anhydride curing agent for epoxy coatings, *Prog. Org. Coatings.* 118 (2018) 9–21. https://doi.org/10.1016/j.porgcoat.2018.01.018
[84] F. Mustata, N. Tudorachi, and D. Rosu, Curing and thermal behavior of resin matrix for composites based on epoxidized soybean oil/diglycidyl ether of bisphenol A, *Compos. Part B Eng.* 42 (2011) 1803–1812. https://doi.org/10.1016/j.compositesb.2011.07.003
[85] O. ur Rahman, M. Kashif, and S. Ahmad, Nanoferrite dispersed waterborne epoxy-acrylate: Anticorrosive nanocomposite coatings, *Prog. Org. Coatings.* 80 (2015) 77–86. https://doi.org/10.1016/j.porgcoat.2014.11.023
[86] S. Ahmad, A.P. Gupta, E. Sharmin, M. Alam, and S.K. Pandey, Synthesis, characterization and development of high performance siloxane-modified epoxy paints, *Prog. Org. Coatings.* 54 (2005) 248–255. https://doi.org/10.1016/J.PORGCOAT.2005.06.013
[87] G. Kurt Çömlekçi, and S. Ulutan, Acquired self-healing ability of an epoxy coating through microcapsules having linseed oil and its alkyd, *Prog. Org. Coatings.* 129 (2019) 292–299. https://doi.org/10.1016/j.porgcoat.2019.01.022
[88] K. Thanawala, N. Mutneja, A. Khanna, and R. Raman, Development of self-healing coatings based on linseed oil as autonomous repairing agent for corrosion resistance, *Materials (Basel).* 7 (2014) 7324–7338. https://doi.org/10.3390/ma7117324

[89] A.H. Navarchian, N. Najafipoor, and F. Ahangaran, Surface-modified poly(methyl methacrylate) microcapsules containing linseed oil for application in self-healing epoxy-based coatings, *Prog. Org. Coatings*. 132 (2019) 288–297. https://doi.org/10.1016/j.porgcoat.2019.03.029

[90] M.B. Hegde, K.N.S. Mohana, K. Rajitha, and A.M. Madhusudhana, Reduced graphene oxide-epoxidized linseed oil nanocomposite: A highly efficient bio-based anti-corrosion coating material for mild steel, *Prog. Org. Coatings*. 159 (2021) 106399. https://doi.org/10.1016/j.porgcoat.2021.106399

[91] Y. Zhang, Y.-W. Tan, H.L. Stormer, and P. Kim, Experimental observation of the quantum Hall effect and Berry's phase in graphene, *Nature*. 438 (2005) 201–204. https://doi.org/10.1038/nature04235

[92] L. Fan, and N.A.M. Eskin, The use of antioxidants in the preservation of edible oils, *Handb. Antioxidants Food Preserv.* (2015) 373–388. https://doi.org/10.1016/B978-1-78242-089-7.00015-4

[93] H. Khatoon, and S. Ahmad, Vanadium pentoxide-enwrapped polydiphenylamine/polyurethane nanocomposite: High-performance anticorrosive coating, *ACS Appl. Mater. Interfaces*. 11 (2018) 2374–2385. https://doi.org/10.1021/ACSAMI.8B17861

[94] N. Chantarasiri, D. Teerawat, J. Wannipa, S. Duangruthai, and S. Suebphan, Synthesis, characterization and thermal properties of metal-containing polyurethane-ureas from hexadentate Schiff base metal complexes, *Eur. Polym. J.* 40 (2004) 1867–1874. https://jglobal.jst.go.jp/en/detail?JGLOBAL_ID=200902270418212372 (accessed October 5, 2021).

[95] T. Ahamad, and S.M. Alshehri, Thermal, microbial, and corrosion resistant metal-containing poly(Schiff) epoxy coatings, *J. Coatings Technol. Res.* 9 (2012) 515–523. https://doi.org/10.1007/s11998-011-9393-3

[96] R. Jayakumar, S. Nanjundan, and M. Prabaharan, Metal-containing polyurethanes, poly(urethane–urea)s and poly(urethane–ether)s: A review, React. *Funct. Polym.* 3 (2006) 299–314. https://doi.org/10.1016/J.REACTFUNCTPOLYM.2004.12.008

[97] R.L. Reis, 2nd Consensus conference on definitions on biomaterials science, *J. Tissue Eng. Regen. Med.* 14 (2020) 561–562. https://doi.org/10.1002/TERM.3016

[98] Buddy D. Ratner, Composites in biomaterials science, *J. Biomater. Sci.* 1 (1996) 94–105.

[99] J.Z. Lu, L.J. Wu, G.F. Sun, K.Y. Luo, Y.K. Zhang, J. Cai, C.Y. Cui, and X.M. Luo, Microstructural response and grain refinement mechanism of commercially pure titanium subjected to multiple laser shock peening impacts, *Acta Mater.* 127 (2017) 252–266. https://doi.org/10.1016/j.actamat.2017.01.050

[100] D. Hong, D.-T. Chou, O.I. Velikokhatnyi, A. Roy, B. Lee, I. Swink, I. Issaev, H.A. Kuhn, and P.N. Kumta, Binder-jetting 3D printing and alloy development of new biodegradable Fe-Mn-Ca/Mg alloys, *Acta Biomater*. 45 (2016) 375–386. https://doi.org/10.1016/j.actbio.2016.08.032

[101] P. Trivedi, K.C. Nune, R.D.K. Misra, S. Goel, R. Jayganthan, and A. Srinivasan, Grain refinement to submicron regime in multiaxial forged Mg-2Zn-2Gd alloy and relationship to mechanical properties, *Mater. Sci. Eng. A*. 668 (2016) 59–65. https://doi.org/10.1016/j.msea.2016.05.050

[102] David F. Williams, Robert W. Cahn, and Michael B. Bever, *Concise Encyclopedia of Medical & Dental Materials* (1990) 412. Springer.

[103] D.M. Brunette, P. Tengvall, M. Textor, and P. Thomsen, *Titanium in Medicine* (2001). https://doi.org/10.1007/978-3-642-56486-4

[104] A.Marti, Cobalt-base alloys used in bone surgery, *Injury*. 31 (2000) D18–D21. https://doi.org/10.1016/S0020-1383(00)80018-2

[105] M. Spector, Biomaterial failure, *Orthop. Clin. North Am.* 23 (1992) 211–217. https://doi.org/10.1016/S0030-5898(20)31732-6

[106] X. Xu, D. Zhang, S. Gao, T. Shiba, Q. Yuan, K. Cheng, H. Tan, and J. Li, Multifunctional biomaterial coating based on bio-inspired polyphosphate and lysozyme supramolecular nanofilm, *Biomacromolecules*. 19 (2018) 1979–1989. https://doi.org/10.1021/ACS.BIOMAC.8B00002

[107] Q. Yu, J. Cho, P. Shivapooja, L.K. Ista, and G.P. López, Nanopatterned smart polymer surfaces for controlled attachment, killing, and release of bacteria, *ACS Appl. Mater. Interfaces*. 5 (2013) 9295–9304. https://doi.org/10.1021/am4022279

[108] T. Wei, Q. Yu, W. Zhan, and H. Chen, A smart antibacterial surface for the on-demand killing and releasing of bacteria, *Adv. Healthc. Mater.* 5 (2016) 449–456. https://doi.org/10.1002/adhm.201500700

[109] Z. Cao, L. Mi, J. Mendiola, J.-R. Ella-Menye, L. Zhang, H. Xue, and S. Jiang, Reversibly switching the function of a surface between attacking and defending against bacteria, *Angew. Chemie Int. Ed.* 51 (2012) 2602–2605. https://doi.org/10.1002/anie.201106466

[110] G. Cheng, H. Xue, Z. Zhang, S. Chen, and S. Jiang, A switchable biocompatible polymer surface with self-sterilizing and nonfouling capabilities, *Angew. Chemie Int. Ed.* 47 (2008) 8831–8834. https://doi.org/10.1002/anie.200803570

[111] T. Wei, W. Zhan, L. Cao, C. Hu, Y. Qu, Q. Yu, and H. Chen, Multifunctional and regenerable antibacterial surfaces fabricated by a universal strategy, *ACS Appl. Mater. Interfaces.* 8 (2016) 30048–30057. https://doi.org/10.1021/acsami.6b11187

[112] S. Amin Yavari, M. Croes, B. Akhavan, F. Jahanmard, C.C. Eigenhuis, S. Dadbakhsh, H.C. Vogely, M.M. Bilek, A.C. Fluit, C.H.E. Boel, B.C.H. van der Wal, T. Vermonden, H. Weinans, and A.A. Zadpoor, Layer by layer coating for bio-functionalization of additively manufactured metabiomaterials, *Addit. Manuf.* 32 (2020) 100991. https://doi.org/10.1016/J.ADDMA.2019.100991

[113] J.D. Hartgerink, E. Beniash, and S.I. Stupp, Self-assembly and mineralization of peptide-amphiphile nanofibers, *Science* 294. (2001) 1684–1688. https://doi.org/10.1126/science.1063187

[114] J.J. Hwang, S.N. Iyer, L.-S. Li, R. Claussen, D.A. Harrington, and S.I. Stupp, Self-assembling biomaterials: Liquid crystal phases of cholesteryl oligo(L-lactic acid) and their interactions with cells, *Proc. Natl. Acad. Sci.* 99 (2002) 9662–9667. https://doi.org/10.1073/pnas.152667399

[115] T. Aida, E.W. Meijer, and S.I. Stupp, Functional supramolecular polymers, *Science* 335 (2012) 813–817. https://doi.org/10.1126/science.1205962

[116] S.I. Stupp, Self-Assembly and biomaterials, *Nano Lett.* 10 (2010) 4783. https://doi.org/10.1021/NL103567Y

[117] R.W. Graff, X. Wang, and H. Gao, Exploring self-condensing vinyl polymerization of inimers in microemulsion to regulate the structures of hyperbranched polymers, *Macromolecules.* 48 (2015) 2118–2126. https://doi.org/10.1021/ACS.MACROMOL.5B00278/SUPPL_FILE/MA5B00278_SI_001.PDF

[118] H. Nagahama, T. Kashiki, N. Nwe, R. Jayakumar, T. Furuike, and H. Tamura, Preparation of biodegradable chitin/gelatin membranes with GlcNAc for tissue engineering applications, *Carbohydr. Polym.* 3 (2008) 456–463. https://doi.org/10.1016/J.CARBPOL.2007.12.011

[119] H. Nagahama, N. Nwe, R. Jayakumar, S. Koiwa, T. Furuike, and H. Tamura, Novel biodegradable chitin membranes for tissue engineering applications, *Carbohydr. Polym.* 73 (2008) 295–302. https://doi.org/10.1016/j.carbpol.2007.11.034

[120] H. Tamura, T. Furuike, S.V. Nair, and R. Jayakumar, Biomedical applications of chitin hydrogel membranes and scaffolds, *Carbohydr. Polym.* 84 (2011) 820–824. https://doi.org/10.1016/j.carbpol.2010.06.001

[121] K.T. Shalumon, N.S. Binulal, N. Selvamurugan, S.V. Nair, D. Menon, T. Furuike, H. Tamura, and R. Jayakumar, Electrospinning of carboxymethyl chitin/poly(vinyl alcohol) nanofibrous scaffolds for tissue engineering applications, *Carbohydr. Polym.* 77 (2009) 863–869. https://doi.org/10.1016/j.carbpol.2009.03.009

[122] M.A. Brown, M.R. Daya, and J.A. Worley, Experience with chitosan dressings in a civilian EMS system, *J. Emerg. Med.* 37 (2009) 1–7. https://doi.org/10.1016/j.jemermed.2007.05.043

[123] R. Jayakumar, M. Prabaharan, P.T. Sudheesh Kumar, S.V. Nair, and H. Tamura, Biomaterials based on chitin and chitosan in wound dressing applications, Biotechnol. *Adv.* 29 (2011) 322–337. https://doi.org/10.1016/j.biotechadv.2011.01.005

[124] S. Begum, Z. Hassan, S. Bräse, C. Wöll, and M. Tsotsalas, Metal–organic framework-templated biomaterials: Recent progress in synthesis, functionalization, and applications, *Acc. Chem. Res.* 52 (2019) 1598–1610. https://doi.org/10.1021/ACS.ACCOUNTS.9B00039

# 8 New Emerging Trends in Polymeric Biomaterials in Textile Industries

*Priyanka Dhingra, Shashank Sharma, Vikas Bhardwaj, Anila Dhingra, Yogita Madan, Divya Bajpai Tripathy, and Anjali Gupta*

**CONTENTS**



DOI: 10.1201/9781003240884-8

## 8.1 INTRODUCTION

Nanotechnology is a new science and technology with potential applications in a wide variety of sectors, including material sciences, material-processing technology, mechanics, electronics, optical sciences, medicines, energy and aerospace, plastics, and textiles. Despite the fact that this technology is still in an early stage, it is already a helpful tool for increasing textile performance and so is garnering international attention. Nanotechnology's unique application in textiles provides a wider range of qualities with the possibility for better and novel product applications [1]. Nanotechnologies can deliver new or enhanced features by changing or improving their properties. However, the materials offered reflect present worldwide research in nanotextiles and commercial operations. In all disciplines of science and technology, the usage of nano-techniques and nanomaterial-based processes is rapidly increasing. Nanotechnology also benefits the textile sector, and has many uses therein. Various products derived from nanotechnology combined with textiles, ranging from nanocomposites and nanofibres to innovative polymeric coatings, have been successful in high-performance applications and traditional textiles to bring new functionality and increased performance. Nanotechnological developments in textiles provide increased repeatability, dependability, robustness, and other benefits. Nanoparticle functionalisation throughout different textile manufacturing steps such as dyeing, finishing, and coating improves product performance and gives previously unattainable functionality.

Nano-Tex, a subsidiary of Burlington Industries in the United States, studied nanotechnology in textiles [2]. Following them, other textile firms across the globe began to invest in nanotechnology development in other areas. Coating, for example, is a typical method of applying nanoparticles to fabrics. Nanoparticles, a surfactant, chemicals, and carrier media are commonly used in coatings that can affect the surface of textiles [3]. The uses of sprays, transfer print, wash, rinse, and cushioning are some methods for coating textiles. Padding is the most commonly used of these methods [4,5]. With the use of a padder set to the correct pressure and speed, nanoparticles are attached to the fabrics, then dried and cured. Water repellence, dirt resistance, wrinkle resistance, antibacterial, antistatic, and UV protection, flame retardation, dye-ability enhancement, and other features have been added to textiles thanks to nanotechnology. Despite the fact that nanotechnology has several potential uses in the textile industry, only a few of the notable features imparted by nano-treatment are critically reviewed in this work.

## 8.2 EMERGING BIOPOLYMERS

### 8.2.1 Soybean Fibre

Soybean fibre is a regenerated protein fibre created by combining soybean protein with PVA. It is biodegradable, allergy-free, and antimicrobial material. Clothing made of soy fibre is less durable, but it has a soft, elastic feel about it. Soybean protein is a globular protein that must be denatured (by alkali, heat, or enzymes) and degraded before being converted into a spinnable dope. Soy fibre is utilised in nutrition-based items and some pastry kitchen goods. Research shows that soy fibre enables significant medical advances generally connected with dietary fibre, including further developed laxation and a cholesterol-reduction capacity [4].

### 8.2.2 Poly(alkylenedicarboxylate)

Petroleum-derived (i.e., non-renewable) or biomass-derived (i.e., renewable) monomers can be used to make aliphatic APDs, with the former being the most common approach. Although both may be made to the same level of purity, the latter is more expensive. Figure 8.1 depicts common dicarboxylic and diol monomers found in APDs. Succinic acid (SA), adipic acid (AA), ethylene glycol (EG), and 1,4-butanediol (1,4BD) are among them. PBS (polybutylene succinate) is an aliphatic polyester with characteristics comparable to PET. Succinic acid and 1,4-butanediol are combined to make PBS.

PBS is a semicrystalline polyester that has a higher melting point than PLA. The crystal structure and degree of crystallinity determine its mechanical and thermal characteristics. The melting temperature is around 115°C, while the $T_g$ is approximately 32°C. PBS is harder than PLA; however, it has a lower rigidity and Young's modulus [19].

Bio-succinic acid (SA) is manufactured directly by bioengineered yeast and *E. coli* fermentation. Bio-succinic acid may be catalytically hydrogenated to yield 1,4-butanediol, which can also be done by fermentation. Bioethylene glycol is made from bioethylene, a byproduct of the catalytic dehydration of ethanol obtained from fermentation. A variety of fermentation-based techniques can create bio-adipic acid. Polyurethanes are used in coatings, adhesives, and foams, flexible packaging, agricultural films, biodegradable bags, and blends and composites with other biobased polymers that all benefit from aliphatic poly(alkylenedicarboxylates) [20]. It is used in cleaning electronics items, including cleaning of devices and a lot more [6].

### 8.2.3 Biodegradable Polyurethanes (PURs)

Toughness, durability, biocompatibility, and biostability are all attributes of PURs. Polyether-based PURs, unlike polyester derivatives, are resistant to microbial degradation. A diisocyanate, a diol, and a chain-extension agent are used to make biodegradable PURs that are used as thermoplastics. The interaction of a cyclic carbonate with an amine to form the urethane bond is the first example of avoiding diisocyanate. Specifically, in the presence of a strong base, there is a polyaddition process between L-lysine and a bifunctional five-membered cyclic carbonate. Enzymatic production of PEUs via enzymatic polyesterification has been described [21].

#### 8.2.3.1 Clothing

When scientists observed that polyurethane could be used to manufacture better strings, they put it together with nylon to make lighter, stretchable pieces of clothing. Over the long term, polyurethanes have been improved and shaped into polyurethane coatings, spandex fibres, and thermoplastic

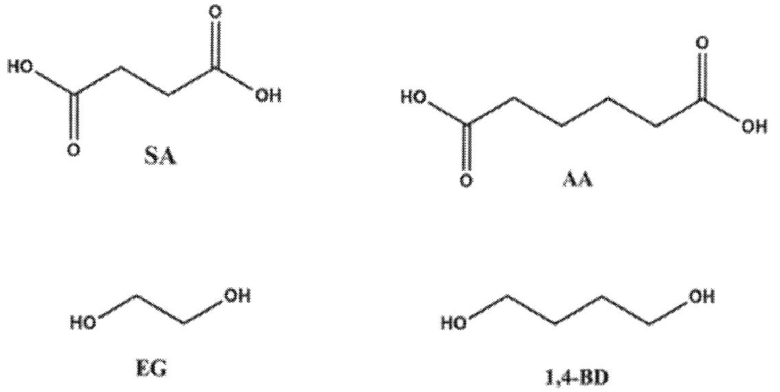

**FIGURE 8.1** Dicarboxylic and diol monomers found in APDs.

elastomers. With the current advances in the polyurethane systems, producers can be helped to make a wide variety of polyurethane clothing from artificial skins and cow hides, which can be used for cloths, dynamic garments, and a variety of ornaments [10].

#### 8.2.3.2 Apparatuses

Polyurethanes are a critical part in huge machines that customers use regularly. The most broadly recognised use for polyurethanes is inflexible foams for the heat protection system of a cooler. These polyurethane foams are an important and pragmatic substance which are useful for meeting energy requirements in customers' coolers. The incredible warmth-safeguarding features of non-stretchable polyurethane foams result from the mixture of a fine, closed cell foam plane and cell gases that prevent heat movement [15].

#### 8.2.3.3 Vehicles

Polyurethanes are utilised throughout vehicles. In addition, they make vehicle seats more comfortable, and are used in guards, inside "feature" roof segments, the vehicle frame, spoilers, entryways, and windows. All of these are made with the use of polyurethanes. They also enable producers to allow more car "mileage" by reducing the weight while also expanding efficiency, quietness, protection, and sound assimilation [17].

#### 8.2.3.4 Building and Construction

Modern homes require superior designs and products. This requires solid materials which are lightweight so that they are convenient and can be easily moved. The modern house needs flexible and stylish goods. Also, it can also help to protect the climate by reducing energy use. With its fantastic lightweight feature, it offers strength and flexibility. This is why it can be utilised in buildings and other developments. Polyurethane is utilised all around buildings including on floors and adaptable cushions. In the rooftop, intelligent plastic covers using polyurethane foam can block light and heat, helping the house to remain cool while reducing energy utilisation. Polyurethane building substances are suitable for new homes and rebuilding projects. The material offers a wide assortment of shadings and profiles for dividers and rooftops, while the foam can be used in passage entryways which are accessible in various areas and styles [19].

#### 8.2.3.5 Blended Wood

Polyurethanes play a significant role in modern materials like composite wood. Polyurethane-based sheets are used in blended wood items. It enables natural items to be used. It is used in medium-thickness fibreboard, and many more materials [20].

#### 8.2.3.6 Devices

Often known as "plotting compounds," non-foam polyurethanes are very frequently used in electrical and equipment undertakings to seal and safeguard fragile, pressure-delicate, microelectronic parts, and printed circuit boards. Polyurethane compounds are widely used by architects to meet a number of physical, warmth, and electrical features. By providing excellent dielectric and concrete properties, as well as amazing dissolvable, water, and extreme temperature impediments, they can safeguard devices [23].

#### 8.2.3.7 Flooring

Either as a foam or on top as a covering, polyurethane can be used to make the floors on which we walk more robust, while also being more elegant. By using versatile polyurethane foam as a mat underlay in private or business uses future floor coverings can be fabricated that protect the surfacing, give more comfort and support, while also reducing any sound. Polyurethanes are used to cover floors from wood and parquet to cement. This material is scratch and solvent resistant, and is easy to clean [23].

#### 8.2.3.8 Products

Polyurethane, by and large as a versatile foam, is possibly the most widely used substance in house designs, for instance, it is used in furniture, floors, and bed coverings. As a cushioning substance, versatile polyurethane foam is used to make furniture more strong, aesthetically pleasing, and consistent [24].

#### 8.2.3.9 Marine

Many people enjoy boating, though it has some risks. The increasing popularity of boating and its safety are due to boating technology. And polyurethane has made some great contributions towards this [21]. Polyurethane epoxy resins seal boat structures from water, the environment and disintegration, and are used in parts that develop drag, impact hydrodynamics, and can reduce strength. Boaters now can have the comforts of home on the water, on account of versatile polyurethane foam. Moreover, rigid polyurethane foam protects boats from noise and high temperatures, gives tear resistance, and increases the heavy load-breaking point while adding insignificant weight. Thermoplastic polyurethane is similarly incredible for use in the ocean business. It is adaptable, durable, and an easy to use substance that is suitable for fittings for wire and connection coatings, belts, engine tubing, and even boat framing [28].

#### 8.2.3.10 Clinical

Polyurethanes are regularly used in different clinical areas, including medical tubes, emergency bedding, surgical drapes, wound dressings, and a variety of devices. The most common use is in organ implants. Polyurethane use in clinical trials can be more cost effective and can provide greater longevity and strength to parts [13].

#### 8.2.3.11 Bundling

Polyurethane packaging foam (PPF) can give more intelligent, perfectly sized padding that can extraordinarily and safely protect items that need to remain securely set up during travel. PPF is widely used to securely safeguard and transport numerous items, such as electronic devices, fragile crockery, and huge modern parts. An adaptable answer for some packaging difficulties, PPF can save time and money by providing the best container with every shipment [23].

### 8.2.4 Polylactic Acid (PLA)

PLA has been around since 1845, but it did not become popular until the early 1990s. It is the only melt-processable fibre made from annual renewable natural resources like maize starch (in the US), tapioca (roots, chips, or starch, particularly in Asia), or sugar cane (in the rest of the world). It is a synthetic polyethylene terephthalate-like thermoplastic aliphatic polyester (PET). The steps in the manufacturing process are as follows: D- and L-lactic acid monomer production; fabrication of polymer-modified fibres, films, plastics, and bottles; production of D-, L-, and meso-lactide polymer (PLA); corn starch fermentation (unrefined dextrose); D- and L-lactic acid monomer production; polymerisation of D-, L-, and meso-lactides (PLA); and alteration of polymers (Figure 8.2).

PLA is strong, drapes well, resists wrinkles, and is UV light-resistant. It melts at 170 degrees Celsius and has a density of 1.25 grams per cubic centimetre. The PP is higher and the limiting oxygen index is 25 points higher than for PET. PLA has lower flammability and contains fewer flame retardants as a result. Compared to PET and PP, water absorption is slightly higher (0.4–0.6%). In a range of situations, it possesses a high level of resilience.

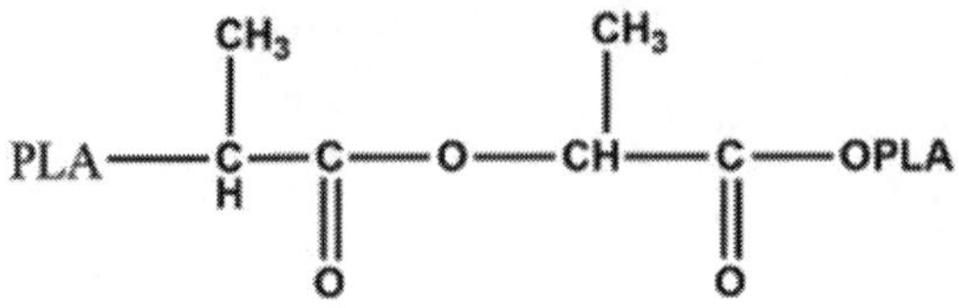

FIGURE 8.2 Structure of PLA fibre polymer.

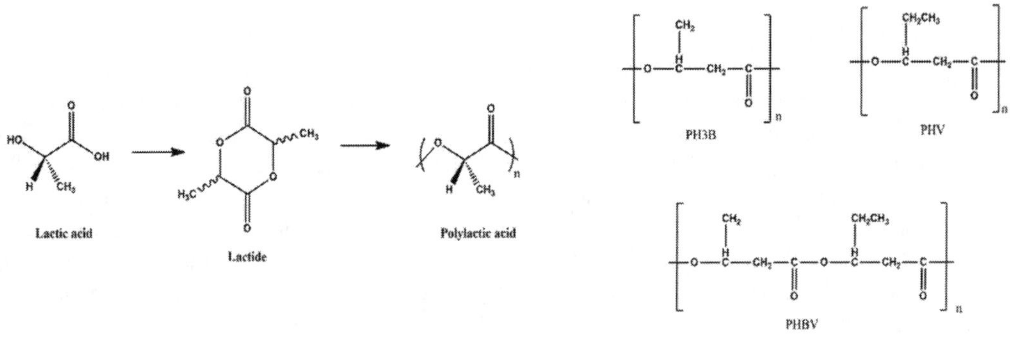

FIGURE 8.3 Chemical structure of PH3B and its copolymer PHBV.

Woven shirts (with ironability), microwavable trays, hot-fill applications, and even engineered plastics are examples of its applications. Sutures, stents, dialysis media, and medicine-delivery devices are examples of its biomedical applications. PLA can be used to make stiff thermoforms, films, labels, and bottles, but it is not suitable for hot-fill containers or gaseous liquids like beer or soda.

It is widely utilised for structure improvement. Because of its high strength it has been used as a better material for ligament replacements or as a device to replace non-biodegradable fibres, for example, Dacron. An injectable type of PLA has recently been approved for the restoration of facial fat loss[17].

Because of better wrap and superb wrinkle obstruction, filaments can be utilised in horticultural areas, wipes, nappies, geotextiles, and so forth. Fibre has good UV opposition and predominant versatility that enables its application in areas such as floor covering tiles, modern divider boards, and vehicle outfitting. By controlling the proportion and dissemination of D- and L-isomers in the polymer chain, it is feasible to prompt different glasslike liquefying points during melt handling. This component offers particular advantages in fabricating bicomponent filaments [26].

### 8.2.5 Bacterial Polyesters

Microorganisms synthesise the bacterial polyesters polyhydroxyalkanoates (PHAs) with poly-(R)-3-hydroxybutyrate (P3HB) as the first homologue (Figure 8.3). Zeneca has created a bacterial storage substance, polyhydroxybutyrate copolymer (PHBV), known as "Biopol." Bioproducts are made by fermenting PH3B and copolymerising it with PHV. It is a high-molecular-weight polyester that is also thermo-plastic (melting at 1800°C) and it may be melt spun into biocompatible, biodegradable surgical fibres.

Production from 100% renewable resources, rapid and total biodegradability, and exceptional stamina and stiff nature are only a few of its benefits. Expensive thermal degradability, brittleness, and high price are among its drawbacks [22].

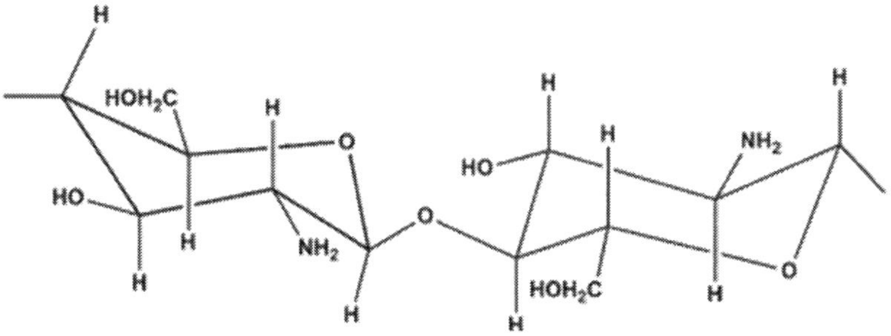

FIGURE 8.4 Two monomers of alginic acid.

Bacterial polyesters have been found to have useful features for use as thermoplastics, elastomers, and adhesives, and are biodegradable and biocompatible. Poly(3-hydroxyalkanoates) (PHAs) and poly(β-malate) are the most representative polyesters synthesised from microorganisms [19].

### 8.2.6 Sodium Alginate Fibre

Sodium alginate is a polymeric acid made up of two monomers: L-guluronic acid (G) and D-mannuronic acid (M) (Figure 8.4). It is non-irritating and non-toxic. Alginate fibre is utilised for wound dressings because it creates a wet healing environment. Aqueous calcium chloride is mixed with aqueous sodium alginate to make calcium alginate.

Alginate has numerous food source and biomedical uses, because of its biocompatibility, low harmfulness, somewhat minimal expense, and gentle gelation [5]. In the food industry, alginate is utilised as a thickening specialist, gelling specialist, emulsifier, stabiliser, and surface improver. Currently, alginate is added to various types of food, for example, frozen yoghurt, jam, corrosive milk drinks, dressings, instant noodles, beverages, and so on [6]. Alginic corrosive has drug uses, and it is added to tablets as a transporter to speed up tablet crumbling for quicker arrival of the restorative part, and also in beauty care products because of its usefulness as a thickener and dampness retainer. For instance, alginate holds the shade of lipstick on the lip surface by framing the gel organisation.

As they are primarily similar to the macromolecular-based parts of the frame and can frequently be delivered into the frame through minimally intrusive organization, alginate-based hydrogels are a profoundly encouraging possibility for use as medication conveyance frameworks [7] and as biomedical inserts [8,9].

Sodium alginate gels are dynamically utilised as a suitable structure for mammalian cell cultures in biomedical examinations. They can be immediately changed in accordance with the fill in either 2-D or even more physiologically significant three-layered culture structures. The shortfall of mammalian cell receptors for alginate, together with lower protein adsorption to alginate gels, makes it possible for these substances to be used in various areas as a proper clean substance, then it is possible to combine highly precise and numerical kinds for cell grab. Furthermore, fundamental revelations uncovered during in vitro examinations can be instantly deciphered in vivo, in view of the biocompatibility and straightforward analysis of alginate [11].

### 8.2.7 Chitin and Chitosan

Chitin and chitosan are both cellulose derivatives, however chitin has an acetamido group, while chitosan has an α-amino group instead of the C-2 hydroxyl activity. Shrimp shells are now the only commercial source of chitin. Crab and lobster shells, on the other hand, contain the polymer (Figure 8.5).

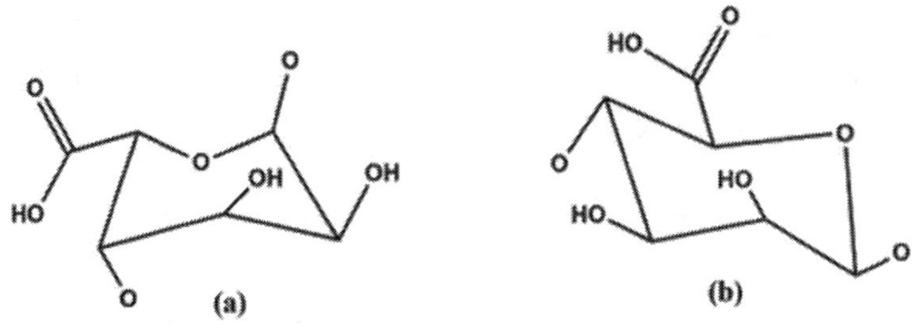

FIGURE 8.5 The structure of chitosan.

Chitin derivatives have been utilised to give fabrics antistatic and soil-repellent properties. Chitosan, on the other hand, can remove colours from discharge water and is utilised in printing and finishing processes. Both have made significant contributions to the field of medical textile sutures, threads, and fibres. They are useful in the quest for a new generation of biomaterials with the incorporation of nanomaterials.

#### 8.2.7.1 Chitosan in Antibacterial Finishing of Substances

Substances promote and provide room for related microbes, which promotes the spread of disease and jeopardises human prosperity.To avoid this, antibacterial treatment of substances is essential. Chitosan has wide reach antibacterial activity, including against for *Escherichia coli*, *Staphylococcus aureus*, and *Bacillus subtilis* [5]. There are essentially two methods for producing antibacterial surfaces using chitosan. The chief methodology is finishing the surface by coating it. Aloson et al. [6] used citrus extricate and sodium dihydrogen phosphate as the driving force to join it with cotton substances under UV light, imparting practical antibacterial features [6]. The resulting procedure is to design antibacterial strains, then weave these into antibacterial fibres. Japan Fuji Textile Co., Ltd. used a stable ultrafine chitosan powder with a particle size of ca. 5 μm, that was added to a thick mixture for blending. Finally, strands for use in chiropody with high antibacterial features have been produced [7]. It is against defective material finishing. Many consumers are drawn to textile products made from common fibres like silk, cotton, and hemp.

These surfaces have generally low flexibility, tending to cause wrinkling and irreversible distortion, consequently adversely affecting the style and comfort of products. These defect on the surface can be additionally corrected using fibre change, changing the surface design, fibre blending, or wrapping. A helpful and effective method for eroding the surface's antiwrinkle execution is to completely finish the chitosan [18].

#### 8.2.7.2 Chitosan in Substance Colouring and Wrapping Up

Chitosan was first used as a dye in the manufacture of textiles. It is actually a cationic polymer. It is also suitable for fixing anionic tones. The protonation of the free amino group on the chitosan molecule in an acidic environment also contributes to the generating force.

The positive charge of the fibre is extended whenever the surface is decreased in the chitosan design, which reduces the shock force between the fibre and the anionic tones. Chitosan is used as an antistatic coating for materials . Due to their inclination for erosion-based power, polyester surfaces can cause garments to cling to the skin. It is expected that participants in modern sectors requiring delicate electronic devices would wear antistatic clothing [10]. Therefore, in the material industry, antistatic surface finishing is becoming increasingly popular. Additionally, the chitosan molecules are very hygroscopic due to the abundance of strong polar social interactions on the chitosan particles, such as hydroxyl and amino gatherings, which outline a persistent water film on the fiber's outer layer. The $CO_2$ in the air and the electrolytes in the strands can be partially separated by this water layer, which also indirectly increases the surface's electrical conductivity and gives the polyester surface its antistatic properties [11].

### 8.2.8 Sorona

A fibre made of the biopolymer Sorona (R) presents an excellent possibility to replace spandex. Stretch is made possible by a synthetic fibre made of nylon and elastane, which are both petrochemical byproducts that may be used to create textures. The stretch qualities of spandex degrade with time, it cannot be colored or tinted during printing, and it cannot be recycled (unfastened). A cellulose polymer developed from corn sugar is called Sorona. When obtained ethically, it provides a naturally safe solution for a wide range of applications, such as sportswear, athletic wear, performance clothes, apparel, and even coverings. It is extremely likely that throughout the recycling process, a produced fibre will be separated from the cellulose polymer and removed [15].

A careful review of the textile industry is very essential, and bio-based products are crucial. This is not to say that we should completely replace current filaments; rather, it means that adding biocatalysts and biopolymers to their assembly cycle or DNA might improve the ecological qualities of the textiles we produce and use on a daily basis [30].

Sorona is a cellulose polymer that has a wide range of other qualities in addition to stretch. It provides exceptional shading power, is protective, sturdy, versatile, resistant to defects, quickly dries, feels incredibly delicate, and doesn't lose its distinctive shape. A few other facts about Sorona include the fact that 37% of the polymer is produced using annual sustainable plant-based ingredients. It converts a formerly petrochemical process into an eco-efficient organic one as a progressive bio-PDOTM particle. When compared to the production of Nylon 6, Sorona uses 30% less energy and emits 63% fewer ozone-damaging substances (Spandex is a blend of elastane and nylon). Sorona is OEKO-TEX Standard 100 certified and is a USDA-verified bio-based product [23].

The circular strategy demands that we currently incorporate end-of-life considerations into the products we produce. Samit and his team focused on the development of bio-based fibres in the future and the manageability attributes of Dupont products [24]. These developments will alter future fabrics so that they may be used to create safe textures that reinstate biodiversity. In the past, manufactured textiles were designed to last forever, but because nature is regenerative, future engineered textiles should develop as nature intended [13].

The components and advancements of the mixed fabrics we regularly use will change as reuse technology progresses. Despite the fact that billions of pounds of investment are being provided, corporate reuse on the scale required for the circular plan to be implemented to meet global demand is still in its infancy. Most textiles are being recycled by businesses, either because rPET degraded into plastic pellets that may be recycled and recovered to create rPET polyester textures or because most textiles are currently destroyed as garbage before being separated by colour using optical sorting technology [21]. When constructed of various cellulose yarns, maybe polyester blends, or blended strands of petrochemical origin, mixed fabrics are difficult to reuse because of unfastening (parting) of the woven filaments [22].

The denim industry could perhaps unfasten denim by employing catalysts to break down and remove the cellulose strands from the manufactured stretch yarns that are added for comfort. Research and improvements are advancing quickly to provide the necessary infrastructure for reuse using various methods. However, the technology and energy utilised in the reuse system should be worthwhile from an ecological and economic standpoint also [34].

Future improvements to polyester will include managing the way that small strands shed. Utilising current separation advancements, it is possible to catch the tiny microglobules before they reach the streams. For material science engineers, making textiles that degrade when exposed to specific microbes (as in the biodiversity of nature) is a top priority [39].

Materials research is started by a scientific specialist and accelerated with industrial support. We are approaching the beginning of a new era for the textile industry, and just as the digitisation of the print sector will moderate the world's valuable assets for a considerable period of time, so too will the continuous advancement of fibre developments – the futures of both textiles and computerised printing are always intertwined [10].

### 8.2.9 Cellulose Nanocomposites

The usefulness of cellulose and cellulose derivatives-based nanocomposites for food packaging applications has been demonstrated by a number of scientists [24,25]. Ghaderi et al. [27] used a low-value agricultural by-product, sugarcane bagasse, to create a high-performance cellulose nanocomposite film with a high tensile strength (140 MPa). They claimed that the produced film could be used for food packaging [27].

Due to their high crystallinity and aspect ratio, large specific surface area, abundance of surface hydroxyl groups to form hydrogen bonds, and eco-friendly nature, cellulose nanostructures, i.e. cellulose nanofibres (CNFs), have recently attracted increasing attention as an organic nano-reinforcement or nano-filler. Many articles have recently been published on biopolymer-based nanocomposite films reinforced with CNFs, and the addition of CNF-enhanced mechanical, barrier, and thermal characteristics, according to those studies [25,26]. Xiao et al. [27] created a CNF-containing nanocomposite film based on soy protein isolate. Tensile strength, oxygen and water vapour barrier characteristics, water resistance, and thermal stability were all improved in the produced films [27]. Similarly, better mechanical and barrier characteristics were reported for a whey protein isolate and polydextrose-based nanocomposite film incorporating CNFs [28].

### 8.2.10 Starch Nanocomposites

A bionanocomposite film was created utilising starch as the biopolymer matrix, pre-oxidised sucrose as the cross-linker, and CNFs as the reinforcing filler in a recent work [29]. The film's barrier qualities (reduced water vapour transmission and oxygen permeability), mechanical performance (increased tensile strength and elongation), and functionality (reduced water absorption, diffusion, permeability, and swelling) were improved as a result of the cross-linking [29]. Esterification or etherification of starch occurs when the hydroxyl groups of the starch are replaced by other chemical groups, resulting in a reduction in water molecule attraction [30–32]. The use of starch in combination with other biopolymers has been found to improve moisture resistance and reduce brittleness [33]. The use of a starch and gelatine blend has been shown to increase the mechanical strength of composite films considerably [33]. Fakhouri et al. [39] created a hybrid edible film and coating from modified maize starch and gelatine, which they utilised to store Red Crimson grapes in a refrigerator. When compared to the uncoated control, the coated grapes had a better look after 21 days of storage [39].

### 8.2.11 Chitosan Nanocomposites

Chitosan is a water-soluble chitin that is used in a variety of sectors, including the food industry. It is a hydrocolloid having antibacterial, antifungal, and antioxidant properties [35]. Because of its highly reactive amino groups, chitosan is positively charged at acidic pH, whereas most hydrocolloids are neutral or negatively charged. Because of its particular benefits, such as solubility in mild acids, strong film-forming abilities, superior antibacterial activity, and selective permeability to gases ($CO_2$ and $O_2$), chitosan has been widely used in nanocomposite films and coatings [35]. However, its employment in food packaging is limited due to its weak mechanical and water vapour barrier qualities [36].

Several attempts have been made to overcome these drawbacks, including the use of nanomaterials such as nanoclays, metal and metal oxide nanoparticles, carbon nanotubes, and so on to strengthen chitosan [35]. MMT is a multilayer silicate-based natural inorganic clay that is widely recognised for its strong cation exchange capacity and excellent aqueous dispersibility. Because of the excellent dispersion of MMT throughout the biopolymer via intercalation of cationic polymer chains into MMT interlayers, MMT incorporation into the chitosan matrix improves its mechanical and barrier characteristics [36,37]. Antimicrobial and antioxidant capabilities are built into chitosan. To increase its antibacterial and antioxidant activities, many nanomaterials and plant-based active components have been utilised. The nanoparticles also increase the chitosan nanocomposites' physicochemical characteristics, such as mechanical, barrier, and optical properties. ZnO nanoparticles are favoured because of their excellent antibacterial properties, nutritional value, and GRAS (Generally Recognised as Safe) status by the US Food and Drug Administration [38].

### 8.2.12 Polylactic Acid (PLA) Nanocomposites

PLA is a biodegradable thermoplastic polyester with high mechanical qualities that could be used in the future. PLA's inherent brittleness, limited heat stability, and low impact strength, on the other hand, limit its applicability. Blending with nanomaterials (such as clay, nanocellulose, and other biopolymers), elastomers, plasticisers, and other biopolymers has been done to increase its physicochemical and functional qualities. Swaroop and Shukla ([39] used solution-casting methods to create PLA and MgO nanoparticle-based composite films for food packaging applications. PLA was used to strengthen MgO nanoparticles (up to 4 wt.%) in this work. When compared to unreinforced PLA film, they discovered that 2 wt.% reinforced PLA film enhanced tensile strength and oxygen barrier properties by 29% and 25%, respectively. The coatings that were created were translucent, possessed antimicrobial qualities, and could protect food from UV rays. The authors have recently generated blown PLA-MgO nanocomposite film via an extrusion technique using an industrial-scale melt-processing apparatus, continuing their prior work [39]. The tensile strength and plasticity of the extruded film with 2 wt.% MgO reinforcement increased by about 22% and 146%, respectively. For a 1 wt.% formulation, the oxygen and water vapour barrier characteristics were increased by about 65% and 57%, respectively. They also indicated that the blowing technique may be utilised to produce PLA/MgO nanocomposites with better physicochemical and antibacterial characteristics for food packaging applications on a wide scale [40]. As a result, ZnO nanoparticles were added to the PLA matrix to increase the mechanical, water vapour barrier, UV-light barrier, and antibacterial characteristics of the nanocomposites [41]. Also, cellulose nanofibrils were recognised as an appropriate reinforcing material for PLA nanocomposites due to their high aspect ratio, large specific surface area, and superior mechanical strength/weight performance. PLA is usually combined with another biopolymer, such as poly-hydroxybutyrate (PHB), for food packaging applications. PLA, PHB, and limonene ternary blends were produced by Arrieta and coworkers [48]. They found that combining PLA with PHB (3:1) increased the oxygen barrier, surface water resistance, and mechanical characteristics [48].

## 8.3 CHARACTERISATION METHODS

Biomaterials' chemical properties and molecular structures must frequently be determined and analysed for various purposes. In this section, we focus on those methods which are used for analysing biopolymeric textiles. In the textile industry, biopolymers can be utilised as fibres or as finishing agents. Therefore, the end result is a textile fibre that has to be thoroughly examined. Characterisation studies may be classified into two parts: studies that characterise the biopolymer to be utilised and studies that analyse the completed textile item.

## 8.3.1 Characterisation of Biopolymers

### 8.3.1.1 Morphology

The microscale structure of the polymer chain is connected to its morphology. It is influenced by the interactions between the amorphous and crystalline sections, as well as other elements like branching, folding, and cross-linking. The morphology of a polymer is extremely important, as it influences its characteristics, therefore it should be properly studied. One of the most frequent methods for photographing the microstructure and morphology of materials is to use a scanning electron microscope (SEM). Transmission electron microscopy (TEM) is often used to acquire direct information about the morphology, crystal structural, and composition of biomaterials to higher resolution than what is usually achievable with SEM. A scanning electron microscope (SEM) scans the surface of a sample with a focused beam of electrons, whereas a transmission electron microscope (TEM) studies an ultrathin layer of a sample by sending a beam of electrons through the specimen. The scattered intensity of an X-ray beam impacting a sample is measured by X-ray diffraction, a non-destructive analytical method. The incident and scattered angles, polarisation, wavelength, and energy collected are then utilised to investigate the crystal structure, chemical composition, and physical characteristics of materials [5,7].

### 8.3.1.2 Molecular Mass Measurements

Polymer type, source, cultivation conditions, extraction procedures, and other factors all influence the physical qualities of a biopolymer. To establish the correct use for the biopolymer, it is necessary to first calculate its molecular mass. The number average molecular weight, weight average molecular weight, and polydispersity all contribute to molecular mass distribution. Molecular mass measurement techniques can be classified in five sections, i.e., number average molecular weight determination (osmometry techniques), weight average molecular weight determination (light-scattering technique, higher average molecular weight determination (gel permeation chromatography, sedimentation technique), viscosity average molecular weight (Mŋ) determination (viscometry), and new/emerging methods [matrix-assisted laser desorption/ionisation mass spectroscopy (MALDI-MS), diffusion-ordered NMR spectroscopy] [5,7,42].

### 8.3.1.3 Examination of Molecular Structure

The most important step in defining a biopolymeric product is to determine its molecular structure. Infrared (IR) spectroscopy is a widely used visual method for identifying chemicals and analysing their chemical structures. Because different chemical functional groups absorb IR radiation at different frequencies, IR spectroscopy may infer the presence or absence of certain chemical functional groups in a molecule. This non-destructive approach may be used to examine gaseous, liquid, or solid materials. The energy of IR photons is insufficient to trigger valence electron transitions; however, IR radiation excites vibrational and rotational movements in molecules. UV spectroscopy is also used for quantitative analysis of molecules, depending on the absorbed light at visible or infrared regions of the electromagnetic spectrum. In addition, Raman spectroscopy is a technique for determining vibrational modes of molecules, and gas chromatography is generally used for separating and analysing compounds, etc.

Li et al. [25] reported the first accurately measured average molecular weight of polymers by DOSY. This technique tends to separate the NMR signals of different species based on their diffusion coefficient. On the assumption that linear correlation exists between the logarithm of diffusion coefficient (log D) and the logarithm of molecular weight (log Mw) according to this equation, external calibration curves are made from which the molecular weight of narrow polydispersity polymers are predicted.

$$\text{LogD} = \text{df} \log \text{Mw} + \text{logC}$$

where df is the fractal dimension of molecules and C is a constant.

## 8.4 APPLICATIONS OF BIOPOLYMERS IN MAKING TEXTILES FOR VARIOUS INDUSTRIES

Biopolymers are widely employed in the textile industry for a variety of applications. Biopolymeric materials can be used as raw materials (fibres) to make woven or nonwoven textiles when fibre synthesis is possible. More current biopolymeric fabrics, such as feather-based nonwovens [6,7], are examples of traditional natural-based materials, such as cotton, silk, and so on. Biopolymers have also been used in the manufacture of various fabrics. They have been utilised as a binding material, such as chitosan, and levelling agents, such as cellulose ethers [8] to aid dyeing, resulting in sharper and more uniform colour in final fabrics [8]. When the rheologic qualities of flow are crucial, such as during the spinning process, they can be utilised as a viscosity modifier, such as alginate [9]. There have also been reports of using biopolymers, such as gum arabic, for textile sizing [9].

Biopolymers such as guar and other gums have also been widely used in textile colouring and dyeing, where dyes and pigments are fixed into the textile components [10]. Carrageenan and other biopolymers have been used as thickening agents in fabric printing [11]. Thickeners are used in textile printing to provide print paste stickiness and plasticity, which may be achieved using these compounds. As a result, the usage of biopolymers maintains printing quality and sharpness by preventing colour leakage [11].

Biopolymers have also been widely employed in textile finishing, such as with antibacterial [12], durable-press, antistatic [13], and deodorant properties. Wool manufacturing has also employed biopolymers, such as chitosan to alleviate the felting and washing shrinkage problem [14]. Biopolymers can also be used in the production of functioning textiles. Cellulosic or feather-based textiles have been developed when the textile's protective qualities are critical [15]. Fabricating valuable smart textiles for wound dressing, drug delivery, and medical applications [16,17] are some other uses for these polymers.

Biopolymers are employed in a variety of industries, including food packaging, cosmetics, and medicine [17]. They can be used to replace petroleum-based polymers in a variety of applications. Some biopolymers have also been used in applications where traditional plastics would be ineffective, such as the manufacture of artificial tissue. These applications may necessitate biocompatible and biodegradable materials that are sensitive to pH, physicochemical, and temperature changes [15]. In comparison to synthetic polymers, biopolymers frequently have poor tensile characteristics, chemical resistance, and processability. They can be strengthened with fillers to increase their qualities and make them more appropriate for certain applications. Biopolymer composites are biopolymers that have been reinforced in this way. An overview of several typical biopolymer composites, their qualities, and the industries in which they are already widely utilised may be found in Table 8.1.

## 8.5 NEED FOR BIODEGRADABLE POLYMERS IN THE TEXTILE INDUSTRY

Textiles have a huge part to play in the re-examination of the print business. What we print onto is similarly vital to the innovation that works with the enriching surface-printed item. In the last 85 years, new aspects like polyester (created by Dupont in 1935) have become progressively more used – solid, reasonable and accessible in any amount. These filaments and their subsidiaries are fundamentally refined from unrefined petroleum and have numerous uses. The main objective is to make textiles without impacting the climate. We have made textiles that keep going without harming the environment. Engineered strands add to the enterprises' carbon impression at assembly and at end of utilisation. Basically, at present, the produced strands are not biodegradable. They do corrupt (after many years), yet as they do they pollute the climate with trillions of tiny particles; and as these small globules degenerate, they diminish in size. Polyester has now been found on the planet's most distant seas and is present at any part of the planet.

Saying this does not imply that cellulose textures are without shortcomings either. At the same time, normal filaments are biodegradable – soil disintegration because of the overcultivation of

**TABLE 8.1**
**Summary of Biopolymer Composites Production Methods, Properties, and Applications**

| Matrix/Filler | Production Method | Properties | Applications |
|---|---|---|---|
| PLA/PEG/chit | Extrusion | Low stiffness/high flexibility | Bone and dental implants, food packaging |
| PLA/potato pulp | Extrusion/injection | Low stiffness and ductility, good processability | Food packaging |
| PHB/wood sawdust fibres | Extrusion | Improved degradation in soil | Agriculture or plant nursery |
| Nanocellulose/CNT | Cast moulding | Good electrical conductivity | Supercapacitors, sensors |
| Potato starch/wheat gluten | Compression moulding | Improved maximum stress and extensibility | Development of biobased plastics |
| PLA/cellulose | Extrusion/injection | Improved rigidity and biodegradability | Packaging, automotive |
| PLA/MgO | Solution casting | Improved stability and bioactivity | Medical implants, tissue engineering, orthopaedic devices |

*Source:* A. M. Díez-Pascual, 2019.

cotton, pesticides, GM yields, and cottons' gigantic water requirement during development and handling additionally have an immense part to play in the harmful impacts of the textile business.

The filaments of things in future should be rethought to convey ecological insurance and an answer to the impractical utilisation of textiles that sully our planet and destroy our living space. Textile science should be reconsidered. Dupont's Bio-material group is trying to give the answers to the business's most significant problems. These advancements will change the filaments of the past to offer manageable answers for the present and future improvement of the global textile industry.

### 8.5.1 Specific Advantages of Biopolymers

- They are completely biodegradable.
- The amount of "oil (petroleum)" required for production is much lower.
- They emit less greenhouse emissions throughout the manufacturing process. Other polymers need 60% more greenhouse emissions and 50% more non-renewable energy than Ingeo® [polylactic acid (PLA) from Natureworks] [18].

## 8.6 CONCLUSION

Biopolymers derived from plants, such as starches, cellulosic substances, agar, and carnauba; biopolymers derived from animals, such as gelatine, casein, whey protein, and beeswax; and biopolymers derived from microorganisms, such as dextrans, xanthan, pullulan, bacterial cellulose, and polylactic acids, have all been extensively researched and used as biodegradable alternatives. These films, as well as their mixtures or hybrids, are most typically developed using the solution-casting process. Numerous studies have found that adding nanomaterials like CNFs, nano-MMT, ZnONPs, and AgNPs to these biopolymers increases their physico-chemical, mechanical, and barrier characteristics, while also adding functions including antibacterial and antioxidant activity. Fresh food has been packaged using biopolymer-based nanocomposites. According to some studies, depending on the variety of whole fruits and vegetables and storage circumstances, postharvest life can range from 2 to 4 months. The shelf life of cut fruits coated with nanocomposite coatings ranged from 14 to 40 days, depending on the kind of fruit. More research is needed to scale up and commercialise biopolymer-based nanocomposite films and coatings such that they are economical and simple to use by farmers and handlers.

## REFERENCES


1. Musante, C., and J. C. White (2012). Toxicity of silver and copper to Cucurbita pepo: differential effects of nano and bulk-size particles, *Environmental Toxicology* 27(9): 510–17.
2. Santos-Moriano, P., L. Fernandez-Arrojo, M. Mengibar, E. Belmonte-Reche, P. Peñalver, et al. (2018). Enzymatic production of fully deacetylated chitooligosaccharides and their neuroprotective and anti-inflammatory properties. *Biocatal Biotransformation* 36(1): 57–67.
3. Kim, S. K., S. Abe, K. Ishihara, M. Adachi, B. N. Ahn, et al. (2018). Antioxidant effects of phlorotannins isolated from Ishigeokamurae in free radical mediated oxidative systems. In: Kim, S. K., Abe, S., Ishihara, K., Adachi, M., Ahn, B.N. (eds) *Healthcare Using Marine Organisms*. Springer, pp. 1–6.
4. Joshi, M. (2005). Nanotechnology: opportunities in textiles. *Indian Journal of Fibre and Textile Research* 30: 477–9.
5. Jahandideh, Arash, Ashkani Mojdeh, and Nasrin Moini (2021). *Biopolymers in Textile Industries*. doi: 10.1016/B978-0-12-819240-5.00008-0
6. Thangavelu, K., and K. B. Subramani (2016). Sustainable biopolymer fibres production, properties, and applications. In: *Sustainable Fibres for Fashion Industry*. Springer, p. 109e140.
7. Omidi Meisam, Atena Fatehinya, Masomeh Farahani, Zahra Akbari, Saleheh Shahmoradi, Fatemeh Yazdian, Mohammadreza Tahriri, Keyvan Moharamzadeh, Lobat Tayebi, and Daryoosh Vashaee (2017). Characterization of biomaterials. In: Lobat Tayebi, and Keyvan Moharamzadeh (Eds.), *Biomaterials for Oral and Dental Tissue Engineering*. Woodhead, pp. 97–115, ISBN 9780081009611, https://doi.org/10.1016/B978-0-08-100961-1.00007-4.
8. Bansal, G., and V. Singh (2016). Review on chicken feather fibre (CFF) a livestock waste in composite material development. *International Journal of Waste Resources* 6: 4.
9. Klemm, D., B. Heublein, H.-P. Fink, and A. Bohn (2005). Cellulose: fascinating biopolymer and sustainable raw material, Angew. *Chemie International* 44: 3358e3393.
10. Hamedi, H., S. Moradi, S. M. Hudson, and A. E. Tonelli (2018). Chitosan based hydrogels and their applications for drug delivery in wound dressings: a review. *Carbohydrate Polymers* 199: 385–400.
11. Boroff, K. E., and A. Boroff (2018). Performance management-making a difference? *Case Journal* 14(1): 25–53.
12. Zia, K. M., S. Tabasum, M. Nasif, N. Sultan, N. Aslam, A. Noreen, and M. Zuber (2017). A review on synthesis, properties, and applications of natural polymer-based carrageenan blends and composites. *International Journal of Biological Macromolecules* 96: 282e301.
13. Li, J., J. He, and Y. Huang (2017). Role of alginate in antibacterial finishing of textiles. *International Journal of Biological Macromolecules* 94: 466e473.
14. Yegappan, R., V. Selvaprithiviraj, S. Amirthalingam, and R. Jayakumar (2018). Carrageenan based hydrogels for drug delivery, tissue engineering and wound healing. *Carbohydrate Polymers* 198: 385–400.
15. Altomare, L., L. Bonetti, C. E. Campiglio, L. De Nardo, L. Draghi, F. Tana and S. Farè (2018). Biopolymer-based strategies in the design of smart medical devices and artificial organs. *International Journal of Artificial Organs* 41(6): 337–59.
16. Moattari, M., H. M. Kouchesfehani, G. Kaka, S. H. Sadraie, and M. Naghdi (2018). Evaluation of nerve growth factor (NGF) treated mesenchymal stem cells for recovery in neurotmesis model of peripheral nerve injury. *Journal of Craniomaxillofac Surgery* 46(6): 898–904.
17. Hassan, M. E., J. Bai, and D. Dou (2019). Biopolymers; definition, classification and applications. *Egyptian Journal of Chemistry* 62(9): 1725–37.
18. Díez-Pascual, A. M. (2019). Synthesis and applications of biopolymer composites. *International Journal of Molecular Sciences* 20: 2321.
19. Younes, B. (2017). Classification, characterization, and the production processes of biopolymers used in the textiles industry. *Journal of the Textile Institute* 108: 674e682.

20. Andhare, P., K. Chauhan, M. Dave, and H. Pathak (2014). Microbial exopolysaccharides: advances in applications and future prospects. *Biotechnology* 3: 25.
21. Hirano, S. (1996). Chitin biotechnology applications. In: M. Raafat El-Gewely (ed.) *Biotechnology Annual Review.* Elsevier, p. 237e258.
22. Sarah, Ditty (2013). *7 Biopolymer Eco-fabrics You Need to Know About.* http://source.ethicalfashionforum.com/, August 12.
23. Gotro, J. (2013). Aliphatic poly(alkylenedicarboxylate) polyesters: organic or not organic? *Polymer Innovation Blog*. http://polymerinnovationblog.com/, April 15.
24. Lee, E. J., Huh B. K., Kim S. N., Lee J. Y., Park C. G., et al. (2017). Application of materials as medical devices with localized drug delivery capabilities for enhanced wound repair. *Progress in Materials Science* 89: 392–410.
25. Li, W., H. Chung, C. Daeffler, J. A. Johnson, and R. H. Grubbs (2012). Application of 1H DOSY for facile measurement of polymer molecular weights. *Macromolecules* 45(24): 9595–603.
26. Vidyavathi, M., S. K. M. Farhana, and A. Sreedevi (2018). Design and evaluation of lentil seed extract loaded bio scaffolds for wound healing activity. *Biomedical and Pharmacology Journal* 11(1): 503–11.
27. Ghaderi, M., M. Mousavi, H. Yousefi, and M. Labbafi (2014). All-cellulose nanocomposite film made from bagasse cellulose nanofibres for food packaging application. *Carbohydrate Polymers* 104: 59–65. doi: 10.1016/j.carbpol.2014.01.013
28. Quintero, R. I., F. Rodriguez, J. Bruna, A. Guarda, and M. J. Galotto (2013). Cellulose acetate butyrate nanocomposites with antimicrobial properties for food packaging. *Packaging Technology and Science* 26 (5): 249–65. doi: 10.1002/pts.1981
29. Rodrıguez, F. J., A. Torres, A. Penaloza, H. Sepúlveda, M. J. Galotto, A. Guarda, and J. Bruna (2014). Development of an antimicrobial material based on a nanocomposite cellulose acetate film for active food packaging. *Food Additives & Contaminants. Part A, Chemistry, Analysis, Control, Exposure & Risk Assessment* 31(3): 342–53. doi: 10.1080/19440049.2013.876105
30. Pelissari, F. M., M. M. Andrade-Mahecha, P. J. A. Sobral, and F. C. Menegalli (2017). Nanocomposites based on banana starch reinforced with cellulose nanofibres isolated from banana peels. *Journal of Colloid and Interface Science* 505: 154–67. doi: 10.1016/j.jcis.2017.05.106
31. Tibolla, H., F. M. Pelissari, J. T. Martins, E. M. Lanzoni, A. A. Vicente, F. C. Menegalli, and R. L. Cunha (2019). Banana starch nanocomposite with cellulose nanofibres isolated from banana peel by enzymatic treatment: in vitro cytotoxicity assessment. *Carbohydrate Polymers* 207:169–79. doi: 10.1016/j.carbpol.2018.11.079
32. Xiao, Y., Y. Liu, S. Kang, K. Wang, and H. Xu (2020). Development and evaluation of soy protein isolate-based antibacterial nanocomposite films containing cellulose nanocrystals and zinc oxide nanoparticles. *Food Hydrocolloids* 106: 105898. doi: 10.1016/j.foodhyd.2020.105898
33. Karimi, N., A. Alizadeh, H. Almasi, and S. Hanifian (2020). Preparation and characterization of whey protein isolate/polydextrose-based nanocomposite film incorporated with cellulose nanofibre and *L. Plantarum*: a new probiotic active packaging system. *LWT* 121: 108978. doi: 10.1016/j.lwt.2019.108978
34. Balakrishnan, P., M. S. Sreekala, V. G. Geethamma, N. Kalarikkal, V. Kokol, T. Volova, and S. Thomas (2019). Physicochemical, mechanical, barrier, and antibacterial properties of starch nanocomposites crosslinked with pre-oxidised sucrose. *Industrial Crops and Products* 130: 398–408. doi: 10.1016/j.indcrop.2019.01.007
35. Saliu, O. D., G. A. Olatunji, A. I. Olosho, A. G. Adeniyi, Y. Azeh, F. T. Samo, D. O. Adebayo, and O. O. Ajetomobi (2019). Barrier property enhancement of starch citrate bioplastic film by an ammonium-thiourea complex modification. *Journal of Saudi Chemical Society* 23(2): 141–9. doi: 10.1016/j.jscs.2018.06.004
36. Tian, H., G. Guo, X. Fu, Y. Yao, L. Yuan, and A. Xiang (2018). Fabrication, properties and applications of soy-protein-based materials: a review. *International Journal of Biological Macromolecules* 120(Pt A): 475–90. doi: 10.1016/j.ijbiomac.2018.08.110
37. Chen, S., and A. Nussinovitch (2000). The role of xanthan gum in traditional coatings of easy peelers. *Food Hydrocolloids* 14(4): 319–26. doi: 10.1016/S0268-005X(00)00008-4
38. Noorbakhsh-Soltani, S. M., M. M. Zerafat, and S. Sabbaghi (2018). A comparative study of gelatin and starch-based nanocomposite films modified by nano-cellulose and chitosan for food packaging applications. *Carbohydrate Polymers* 189: 48–55. doi: 10.1016/j.carbpol.2018. 02.012

39. Fakhouri, F. M., S. M. Martelli, T. Caon, J. I. Velasco, and L. H. I. Mei. (2015). Edible films and coatings based on starch/gelatin: film properties and effect of coatings on quality of refrigerated Red Crimson grapes. *Postharvest Biology and Technology* 109: 57–64. doi: 10.1016/j.postharvbio.2015.05.015
40. Kumar, S., A. Mukherjee, and J. Dutta (2020). Chitosan based nanocomposite films and coatings: emerging antimicrobial food packaging alternatives. *Trends in Food Science & Technology* 97: 196–209. doi: 10.1016/j.tifs.2020.01.002
41. Sharma, R., S. M. Jafari, and S. Sharma (2020). Antimicrobial bio-nano composites and their potential applications in food packaging. *Food Control* 112: 107086. doi: 10.1016/j.foodcont.2020.107086
42. Umoren, Saviour, and Solomon, Moses (2016). *Polymer Characterization: Polymer Molecular Weight Determination*. Polymer Science: research advances, practical applications and educational aspects, Formatex Research Center SL, 1, 412–419.

# 9 Water and Soil Remediation

*Nidhi Puri, Neha Dhingra, and Indu Bhushan*

## CONTENTS



## 9.1 INTRODUCTION

A clean environment is a basic need and right of every human being for survival. Environmental remediation broadly covers the treatment of soil, groundwater, surface water, and air. Groundwater is a natural resource which is fundamentally important for humankind. Pure groundwater is very much required for human consumption for good health and a disease-free environment. Moreover, agriculture and industry also need pure water as well as soil for their operation and development. Therefore, the importance of pure water is recognized for a healthy life throughout the world [1]. Going by the fact a healthy mind lives only in a healthy body, for that we need to have a clean environment all around us. The thoughtless disposal of waste in water as well as soil is spoiling our environment and causing various health issues. Therefore, a prime requirement is the removal of pollutants from both water and soil to clean and safeguard our environment.

Technological advancements and consumption demand have increased dramatically over the past few decades, resulting in a lack of natural resources preservation. Groundwater and soil contamination have drastically modified the physical, chemical, and biological nature, leading to a change in physicochemical considerations too, including taste, odour, hardness, colour, and even foaming. These unwanted problems can be resolved however; water contamination through harmful toxic chemicals has become a serious cause of concern. These contaminants of water and soil are associated with industrial advancements and growth where the wastes are disposed directly into the environment. Several sources of industrial wastes include wastewater coming from the electroplating, biochemical, mining and petroleum industries, and metal alloy-based industries where the heavy metal ions of Zn, Cu, Cr, Ni, Hg, etc. contaminate water and soil. Therefore, the proper treatment of these contaminated water sources and soil is required.

The cost-effectiveness, recyclability of used materials, cleaning efficiency, regeneration ability, and green eco-friendly behaviour of nanomaterials and nanoparticles (NPs) are some of the main

DOI: 10.1201/9781003240884-9

points which one needs to consider while working upon novel technologies for water and soil remediation. Nanotechnology has emerged as an effective method in different fields of research [2,3]. Nanomaterials are widely used for environmental remediation purposes. The most important methods involved in water treatment are adsorption, photo-degradation, disinfection, and filtration. Among these, nanoadsorption and the use of nanomembrane are commonly used methods for environment remediation. Many researchers have investigated organic/inorganic pollutants removal by utilizing nanomaterials [4–7]. With the need to have a complete and broad look on the base topic of water and soil remediation and its different aspects, techniques always require to be updated with the latest advancements. In this chapter, we have focused on the background of water and soil remediation requirements and presented an overview to the various current active practices using green eco-friendly biodegradable nanomaterials, the mechanisms involved, and the future directions.

## 9.2 CARBON-BASED NANOCOMPOSITES

Carbonaceous nanocomposites are those nanomaterials composed of carbon nanotubes (CNTs), graphene (G), nanosheets, and activated carbon materials. These nanocomposites have higher abilities for pollutant removal from wastewater and soil compared to conventional materials. The exceptionally large mechanical strength, high surface to volume ratio, great hydrophobicity, and high thermal features are some of the impressive properties of these materials which favour their use in the environment remediation process. Carbonaceous materials are used widely in medicines, biosensing applications, nano- and microelectronics, water and soil remediation, as well as their real-life applications. These materials have also proved to be excellent adsorbents in the removal of heavy metal ions-based pollutants and dyes from wastewater and soil.

### 9.2.1 Graphene-based Nanocomposites

Graphene, a 2D carbon allotrope, has been widely used in numerous scientific researches with wide applications in different fields [3,8]. The chemically functionalized form of graphene is graphene oxide (GO) which has outstanding hydrophilicity, making it promising in the ongoing adsorption process for pollutant removal. The nanocomposites made up of graphene/GO provide high surface area, favouring adsorption *via* $\pi$–$\pi$ conjugation; capable of accepting electrons to impede photo-catalytic activity; and prevention of graphene aggregation from agglomeration, resulting in enhanced applications in environmental remediation. Recently, Hammad and group reported the large-scale fabrication of iron oxide NPs/graphene (IO NPs/G)-based hybrid nanomaterials for the photo Fenton reaction to degrade organic pollutants such as methylene blue (MB), rhodamine B (RhB), acid orange 7 (AO 7), and phenol with an enhanced removal efficiency attributed to relatively large surface area and the rapid transfer of photo electrons between the constituent iron oxide and graphene [9]. Nhlane and group have also demonstrated the use of RGO in the effective removal of Hg(II) from contaminated water samples [10]. In another article, manganese dioxide ($MnO_2$) NPs along with zirconium (Zr) decorated over functionalized reduced GO (RGO) ternary nanocomposite (Zr-$MnO_2$@RGO) was fabricated *via* a facile chemical route for the removal of As(V) from industrial and groundwater samples [11]. It has been observed that the Zr and $MnO_2$ NPs enhanced the overall adsorption affinity and specific surface area of RGO, resulting in enhanced stability, metal ion removal efficiency, and reusability, even after five consecutive adsorption–desorption cycles. $MnO_2$ particles have also shown high adsorption capacity when incorporated into RGO functionalized with black cumin seeds to prepare a highly amorphous RGO-$MnO_2$/BC hybrid composite for the adsorptive removal of MB and As metal ions from wastewater [12]. The thiol-functionalized GO/Fe-Mn composite (SGO/Fe-Mn) has also been used to effectively remove Hg from contaminated soil, showing the potential applications of functionalized/modified GO in environmental remediation [13]. Table 9.1 gives a comparative description of the applications of water and soil remediation using carbon-based nanomaterials.

**TABLE 9.1**
**Comparative Study on the Applications of Water and Soil Remediation Using Carbon-based Nanomaterials**

| Target Pollutant | Matrix | Maximum Adsorption Capacity (mg/g) | Conditions/Key Findings | References |
|---|---|---|---|---|
| MB<br>RhB<br>AO 7<br>Phenol | IO NPs/G | 99% | • Optimum time: 60 min; concentration: 60 mg/mL; UV-A irradiation: 1.6 mW/cm$^2$; pH: 3.0 to 9 | [9] |
| Hg(II) | RGO nanosheets | >15% | • Optimum pH 4<br>• GO nanosheets prepared *via* modified Hummer's method | [10] |
| As(V) | Zr-$MnO_2$@RGO | 98.5–99.3% | • Pseudo 2nd order kinetic model<br>• Optimum pH 4<br>• Zr and $MnO_2$ NPs enhanced the adsorption affinity and specific surface area of RGO | [11] |
| MB<br>As | RGO-$MnO_2$/BC hybrid | 232.5<br>14.7 | • Hybrid composite could be regenerated *via* acid–base treatment<br>• Optimal conditions: 75 min, 7.0 pH, 27 °C | [12] |
| Hg | SGO/Fe-Mn | — | • Hg-polluted soil treatment<br>• Thiol-functionalized GO/Fe-Mn composite was used | [13] |
| TL | $Fe_3O_4$@CNT | 98.1% | • Optimum conditions: 20 °C, pH=7<br>• Recyclable up to 5 cycles | [15] |
| As(III) | Fe/$Fe_3C$@CNTs/CW | - | • Recyclable up to 5 cycles<br>• Optimum time 90 min<br>• 3D hierarchically porous hybrid monolith electrodes | [16] |
| $BrO_3^-$ | Fe/MWNTs | ~85% | • 65–85% remediation within 5 min<br>• Ultra-high-performance liquid chromatography was utilized to detect the adsorption process | [17] |
| BZ | CdS-CNTs/Nafion/GCE | — | • Linearity of sensor 0.1 nM–1.0 mM<br>• Sensitivity 5.43 μAμ/M/cm$^2$<br>• Detection limit 2.2 ± 0.1 pM (S/N = 3) | [18] |
| Oil | PP/CNTs/sorbitol | 40 g/g | • 97.2% recovery efficiency<br>• Excellent mechanical and thermal-insulating performance | [19] |
| Dye<br>Oil | CNT-PAA/MIL101(Fe)@Pt membrane | 98.5%<br>98.8% | • Separation throughput 11,000 L/m$^2$/h/bar<br>• Stable mechanical property<br>• Excellent anticorrosion ability | [20] |

### 9.2.2 Carbon Nanotube-based Nanocomposites

CNTs, the one-dimensional hollow cylindrical rolled form of graphene sheets having a very small diameter (~1 nm), are an important carbon allotrope used in various fields of research and development. Broadly, they are classified into two different types: SWNTs (single-walled carbon nanotubes) and MWNTs (multi-walled carbon nanotubes), consisting of a single layer and multiple

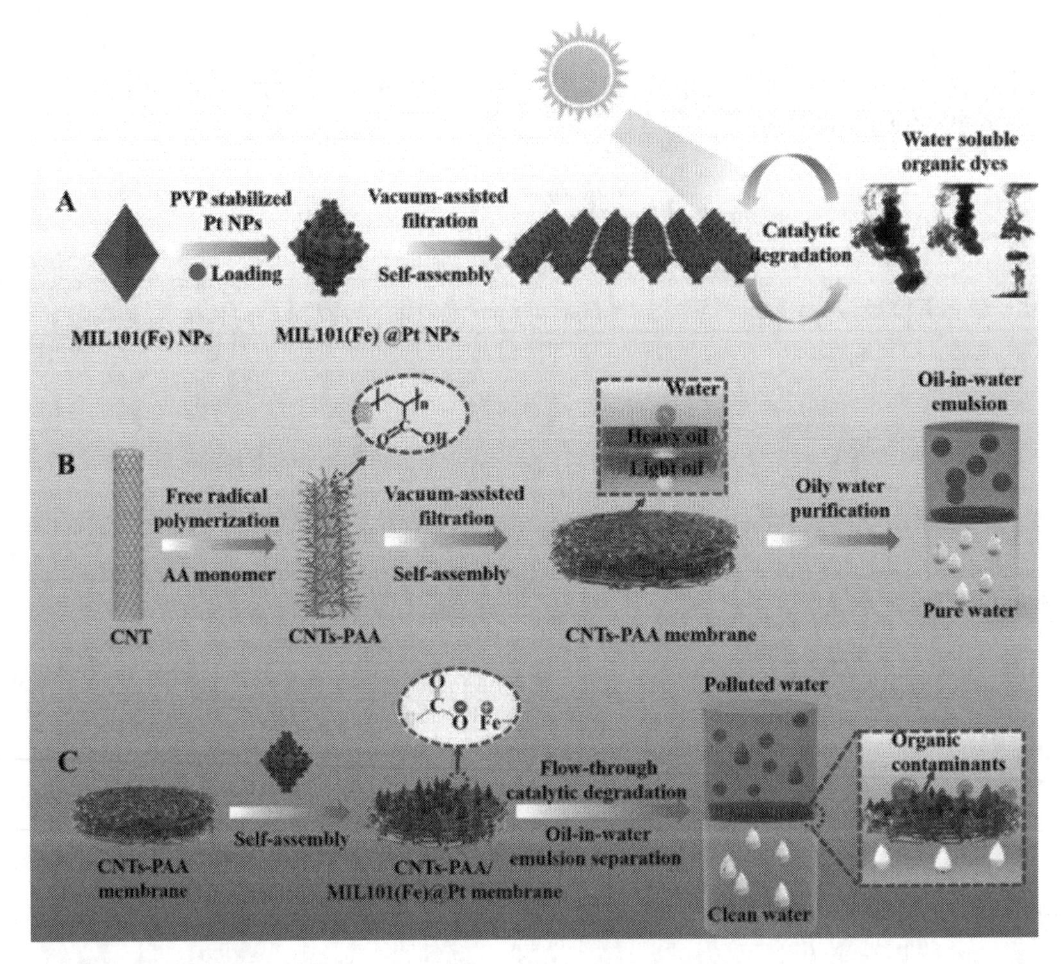


**FIGURE 9.1** Schematic presentation of (A) the preparation process of MIL101(Fe)@Pt NPs for the catalytic degradation of dye molecules; (B) the preparation of CNTs-PAA nanomembrane for the oil–water separation; and (C) the CNTs-PAA/MIL101 (Fe)@Pt composite nanomembrane assembling for both dye degradation and oil–water separation. (Data reprinted with permission of Elsevier [20].)

layers of rolled cylindrical graphene sheets, respectively. These materials are widely used in water and soil remediation processes. Many researchers have reviewed the rich science underpinning the applications of CNTs owing to their large aspect ratios, excellent thermal and electrical conductivity, high surface area, and superior optical characteristics [14]. The non-noble metal chainmail catalysts, consisting of metal oxide NPs on CNT-based carriers, are an important novel strategy which researchers are using to design a highly efficient and stable method for the water remediation process. Recently, $Fe_3O_4$ and MWNTs have been used to prepare magnetic nanocomposite $Fe_3O_4$@CNT to degrade tetracycline (TL) in wastewater [15]. Arsenite As(III), being a hyper-toxic material, always remains a challenge to remove from the environment. A group of researchers have developed a novel electrochemical integrated system to remove it from wastewater by converting toxic As(III) into As(V) [16]. For this purpose, a special carbonized wood (CW) framework was prepared on which clusters compressed in N,O-codoped CNTs were immobilized and then Fe/$Fe_3C$ NPs were encapsulated to form 3D porous hybrid (Fe/$Fe_3C$@CNTs/CW) monolith electrodes. The oxidation process of As(III) into the As(V) was promoted by the electro-Fenton reaction on the cathode which effectively removed As(III) within 90 min by electro-sorption exhibiting superior structural stability and satisfactory recycling capability up to five cycles. Alsohaimi et al. also synthesized a Fe/MWNTs nanocomposite for the remediation of $BrO_3^-$ ions from synthetic and commercial water samples [17].

CNTs decorated with different metal structures have also been used significantly in environmental remediation. Cadmium sulphide (CdS) NP-decorated CNT film-based nanocomposite material (CdS-CNTs) was prepared on glassy carbon electrode (GCE) by a facile solution method to develop a benzaldehyde (BZ) electrochemical sensor probe for removing water contaminants [18]. Superb sensitivity along with an ultra-low detection limit by the electrochemical technique indicated the promising results of utilizing a CNT-based nanocomposite in sensor applications for environment remediation. Zhao and group proposed a novel approach to fabricate microcellular polypropylene (PP) and CNTs/sorbitol-based nanocomposites (PP/CNTs/sorbitol) for oil absorption and oil/water separation utilizing a green, facile simple microcellular foaming technology [19]. The as-prepared nanocomposite possessed the unique hierarchical porous structure on cell walls, providing 97.2% recovery efficiency along with 40 g/g absorption capacity for applied oil–water solution. Hence, the high-efficiency, advantageous, and environment-friendly conventional polymer adsorbent-based nanocomposite foams have shown excellent performance in the treatment of oil/water separation. Liu et al. have also reported an effective strategy for the preparation of multifunctional polyacrylic acid (PAA) functionalized super hydrophilic and underwater super oleophobic CNTs and MIL101(Fe) at a Pt NP-based composite nanomembrane called CNTs-PAA/MIL101(Fe)@Pt for the effective deprivation of dyes and oil removal from water [20]. Figure 9.1 illustrates the construction of MIL101(Fe)@Pt NPs, CNTs-PAA membrane, and the CNTs-PAA/MIL101 (Fe)@Pt composite nanomembrane for the catalytic degradation of involved dye pollutants and further removal of an oil-in-water suspension.

## 9.3 CELLULOSE-BASED NANOCOMPOSITES

Cellulose is the most copious renewable, biodegradable natural long-chain polymer on Earth, and consists numerous of carbon, hydrogen, and oxygen atoms in its molecular structure. In recent years, it has become a most interesting research field because of its versatile properties which make its wider possibility of usage in advanced functional applications such as veterinary foods, textiles, packaging wood and paper, fibres and clothes, pharmaceutical industries, and water as well as soil remediation [21,22]. Nanocellulose, i.e. nanofibres and nanocrystals excavated from cellulose fibres, has opened up a new horizon in the wide innovative research for the scientific community because of its challenging conversion process from cellulose to nanocellulose. Nanocelluloses possess various important properties of cellulose including hydrophilicity, high surface area, and broad chemical modification capacity, enabling its potential usage in wastewater and soil contaminants removal, such as metal ions, organic and inorganic contaminants, bacteria, viruses, and oil. Nanocellulose can broadly be divided into three major parts: (1) cellulose nanocrystals (CNCs), needle-shaped 150–300 nm lengthened crystalline fibrils having diameters of 5–10 nm; (2) cellulose nanofibrils (CNFs), nanosized (5–50 nm wide) flexible fibrils, having both crystalline and amorphous domains at the micron scale; and (3) bacterial cellulose (BC), CNFs obtained from bacteria.

### 9.3.1 CELLULOSE-BASED NANOMATERIALS AS ADSORBENTS

Cellulose-based nanocomposites are highly produced biobased nanomaterials whose production has reached an industrial scale. These materials working as nanoadsorbents have gained wide and growing attention because of their remarkable surficial properties such as biodegradability, surface reactivity, and non-hazardous nature. The high aspect ratio of nanocellulose provides overall mechanical strength to its different networks, such as CNCs and CNFs. The high surface area along with high mechanical strength help in increasing the overall strength and adsorption capacity of nanocomposites. Moreover, the high surface reactivity provides various possibilities for surface functionalization, enabling its potential use in the adsorption of water and soil pollutants. Table 9.2 describes earlier studies in the field of water and soil remediation using cellulose-based nanoadsorbents and nanomembranes.

#### 9.3.1.1 Heavy Metal Ion Adsorption

A renewable porous cellulose adsorbent was prepared *via* a γ-initiated grafting process of poly(glycidyl methacrylate) (PGMA) grafted from cellulose extract for the effective removal of metal pollutants [Pb(II), Cd (II), and Cu(II) ions] from an aqueous solution [23]. The PGMA-grafted cellulose was modified with chelating species as iminodiacetic acid (IDA) to achieve natural-based and effective porous cellulose adsorbent for the effective chemisorb removal of toxic metal ions. Cellulosic ionic liquid [carboxymethyl-dimethyl ammonium ethyl cellulose (CMDEAEC)] has also been proved to be a highly selective and efficient low-cost environmentally friendly adsorbent for Au(III) [24]. A group of researchers have reported a novel nanocomposite, cellulose/ $HO_7Sb_3$, made up of organic poly-cellulose and inorganic $Sb(OH)_5$ by the sol–gel method for the effective adsorption of radionuclides Co(II), La(III), and Cs(I) from an aqueous solution [25]. Figure 9.2 schematically presents the proposed reaction mechanism for nanocomposite formation. It has been observed that the average 21-nm spherical-shaped $HO_7Sb_3$ particles were properly dispersed in the poly-crystalline cellulose matrix. The as-prepared cellulose/$HO_7Sb_3$ nanocomposite showed highly selective adsorption of radionuclides in the order La (III) < Co(II) < Cs(I).

In a recent report, cellulose biochar was utilized to synthesize magnetic iron-carbon-based nanoadsorbents for the effective removal of heavy metals ions of Cr(VI), Cu(II), As(V), and Pb(II) [26]. It has been observed that the cellulose biochar provided a good carbon support for the nanocomposite, enabling the efficient removal of heavy metal ions. These carbon compressed iron nanoparticles (CEINPs)-based nanocomposites retained their heterogeneous surface and showed adsorption of metal ions broadly in two main steps of diffusion and intraparticle diffusion at different pH. The adsorption mechanism is schematically shown in Figure 9.3. As per the proposed adsorption mechanism, Pb(II) removal was based on an additional stage of Pb(II) adsorption onto the internal pores of nanoadsorbent. The Cu(II), Cr(VI), and Pb(II) removal was performed *via* a redox reaction mechanism. An additional inner sphere bidentate complex was also formed in the removal of Cu(II) and As(V).

#### 9.3.1.2 Dye Adsorption

Biodegradable CNCs have worked remarkably well as green bioadsorbents in the elimination of dyes from water and soil. Wood-based colloidal CNCs in the given anionic forms of sulphated and

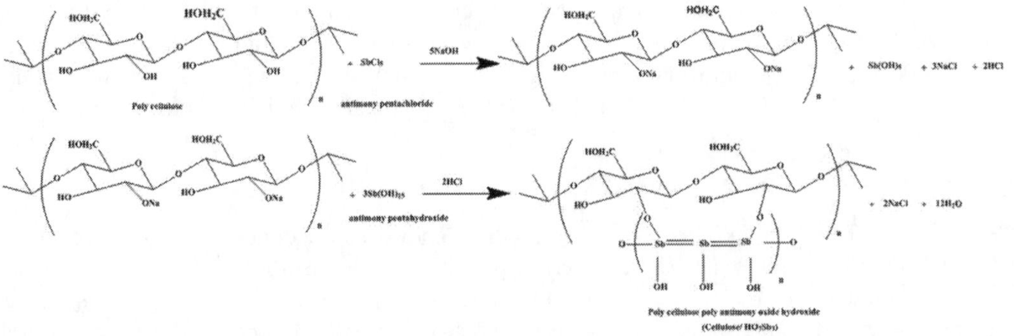

**FIGURE 9.2** Schematic representation of proposed reaction mechanism for cellulose/$HO_7Sb_3$ formation. (Data reprinted with permission of Elsevier [25].)

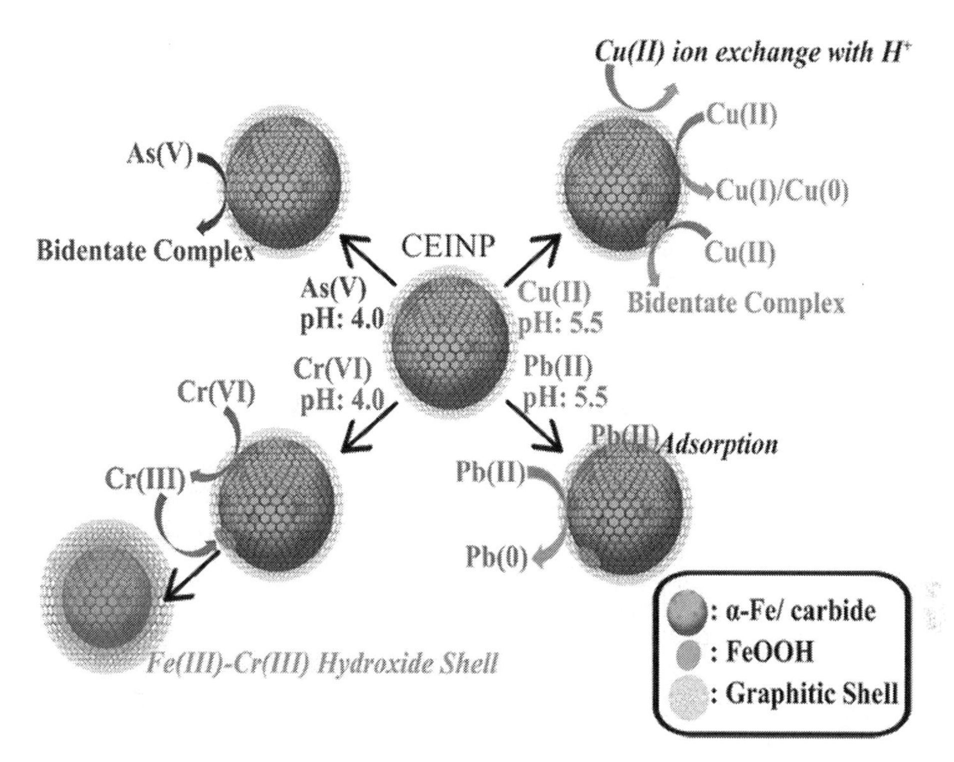


**FIGURE 9.3** Adsorption mechanism of Cu(II), Cr(VI), As(V), and Pb(II) by using carbon-encapsulated iron nanoparticles. (Data reprinted with permission of Elsevier 2020 [26].)

carboxylated have been widely utilized as nanoadsorbents in the extraction of toxic auramine O (AO), an organic cationic dye [27]. CNC flakes have also shown their efficient adsorptive behaviour in the better removal of methylene blue (MB) than the native CNC powder [28]. In another article, CNCs prepared from waste printed papers were used as green adsorbents for the removal of azo dyes [hydroxynaphtol blue (HB) and Congo red (CR)] from water [29]. Zhao et al. reported nanocomposite beads composed of carboxylated GO/chitosan/cellulose (GCCSC) which showed high mechanical strength for the adsorption of Cu(II) from wastewater and soil [30]. The synergistic effect of GO and cellulose helped in providing high mechanical strength and porous structure to the nanocomposite, which further helped in blocking direct soil and graphene contact, resulting in a high adsorption capacity of the nanocomposite.

### 9.3.2 Cellulose-based Nanomaterials as Nanomembranes

Nanocellulose plays a vital role in microfiltration, nanofiltration, and reverse osmosis for remediation of water and soil. Many researchers have reviewed its role in nanomembranes, where these CNCs and CNFs could be embedded in a nanocomposite matrix having an interconnected frame [31]. Polyvinylidene fluoride (PVDF) has been used along with Meldrum's acid (2,2-dimethyl1,3-dioxane-4,6-dione) and CNF to form novel green Meldrum/CNF/PVDF membranes for $Fe_2O_3$ NP filtration and adsorption of cationic crystal violet (CV) dye from an aqueous solution [32]. This group compared the performance of PVDF electrospun membranes, CNF/PVDF unmodified membrane, and Meldrum/CNF/PVDF membranes. The Meldrum's acid worked as an esterification agent which enhanced the overall adsorption capacity of composite membrane towards CV dye. The results showed excellent filtration capacity along with adsorption ability of ecofriendly Meldrum/CNF/PVDF membranes against $Fe_3O_4$ NPs and CV dye. In another report, the surface modification

**TABLE 9.2**
**Comparative Study on the Applications of Water and Soil Remediation Using Cellulose-based Nanoadsorbents and Nanomembranes**

| Pollutant | Base Matrix | Maximum Adsorption Capacity (mg/g) | Key Findings | References |
|---|---|---|---|---|
| Cd (II)<br>Pb(II)<br>Cu(II) | PGMA | 53.4<br>52.0<br>69.6 | • Synthesis *via* γ-initiated grafting polymerization<br>• Natural-based material modified with covalently attached IDA groups | [23] |
| Au(III) | CMDEAEC | 15 | • Pseudo-second-order kinetics model<br>• Nanoscale particle size with good swell-ability<br>• Langmuir isotherm model fitted best ($R^2 = 0.995$) | [24] |
| La (III)<br>Co (II)<br>Cs(I) | Cellulose/ $HO_7Sb_3$ | — | • Synthesis via sol–gel mixing<br>• Highly selective adsorption of radionuclides La(III)<Co(II)<Cs(I) | [25] |
| Cr(VI)<br>Cu(II)<br>As(V)<br>Pb(II) | CEINPs | 18.5<br>42.2<br>38.5<br>17.3 | • Obtained results are from 10 wt.% iron-loaded CEINPs<br>• 40 wt.% CEINPs loaded with iron showed 688.6 mg/g max adsorption for As(V)<br>• Freundlich model fitted data best ($R^2 = 0.95–0.99$) | [26] |
| AO | Wood-based colloidal CNCs | 20 | • Sulphated CNC showed highest removal<br>• Optimum time 30 min<br>• Freundlich model fitted data best ($R^2 = 0.9217$) | [27] |
| MB | CNC flakes | 188.7 | • Optimum condition: adsorbent dosage- 0.7 g/L, at 25 °C and pH 6<br>• Langmuir model fitted data best ($R^2 = 0.999$) | [28] |
| HB<br>CR | CNCs | 0.1700 mmol/g<br>0.1564 mmol/g | • Endothermic nature of adsorption<br>• CNC yield from waste paper 30%<br>• Langmuir model fitted data best | [29] |
| Cu(II) | GCCSC beads | 22.4 | • Faster adsorption kinetics (3 h in equilibrium)<br>• Higher pH and lower temperature increases the adsorption | [30] |
| CV | Meldrum/CNF/PVDF<br>CNF/PVDF | 3.984<br>2.948 | • Meldrum's acid worked as an esterification agent<br>• High filtration ability against $Fe_3O_4$ NPs | [31] |
| TBC | Cellulose/GONS<br>Pristine cellulose membrane<br>PVDF | 93.6 ± 2.4%<br>27.4 ± 12.5%<br>1.368 | • Trans-membrane pressure increased after 24 h 55% with pristine cellulose; 6% with cellulose/GONS | [34] |

of nanocellulose membranes *via in situ* TEMPO functionalization of CNC was executed to achieve a porous network structure and excellent filtration performance in terms of high water permeability and functionality along with high mechanical stability [33]. The TEMPO functionalization introduced the carboxyl groups and hence hydrophilicity to the membrane which helped in effective adsorption of metal ions Cu(II) and Fe(II)/Fe(III).

Membrane fouling is the major challenge to the reverse osmosis (RO) process which results in higher energy costs of plants, and researchers always have this issue to face while aiming for effective distillation or water purification. To overcome the challenges in the RO process, Ibrahim and group reported a novel cellulose microfiltration membrane in combination with graphene oxide nanosheets (GONSs) [34]. These novel cellulose/GONS microfiltration membranes showed effective performance in terms of the removal of total bacterial count (TBC), high water flux capability, and enhanced biofouling resistance. The comparative study of prepared cellulose/GONS and pristine cellulose membranes proved the effectiveness of nanocomposite microfiltration membranes, which was proposed to be due to the synergistic combination of GO and the biodegradable cellulose substrate.

## 9.4 PROTEIN-BASED NANOCOMPOSITES

To decrease overall atmospheric pollution, there is always a growing demand for biodegradable, recyclable, environment- and eco-friendly green materials for water and soil remediation purposes. Therefore, waterborne natural materials-based nanocomposites are always in demand for environment remediation. The low cost, excellent biocompatibility, as well as biodegradability and easier processability are some of the inherent properties of these natural materials which give direction to researchers in finding new and innovative methods for their usage in different water and soil remediation techniques. The easily tuneable specific surface area of protein-based materials makes them flexible enough to be used as adsorbents, disinfectants, thin films, or nanofibrils, and even in the form of nanomembranes for water or soil filtration. The easier availability of these materials is another important aspect for using protein nanocomposites for remediation purposes. Protein wastes are available all over the world and they can be reused and recycled into efficient absorbers or membrane forms for remediation purposes, further suggesting their potential application in environment remediation. Many researchers have reviewed and explained different syntheses, aspects, and applications of these materials in water and soil treatments [35,36]. Table 9.3 provides a comparative study of the applications of water and soil remediation using protein- and hydrogel-based nanoadsorbents and nanomembranes.

### 9.4.1 Protein-based Nanoadsorbents

Protein-based adsorbents work remarkably well in the removal of metal ions and dyes from water and soil. Peydayesh et al. reported amyloid fibril aerogels as adsorbents for the effective extraction of model organic pollutants of water, such as bentazone, bisphenol A, and ibuprofen. β-Lactoglobulin, present in milk whey protein, was used to fabricate the described aerogel which worked remarkably well, with removal efficiencies of 92%, 78%, and 98% for the given pollutants without using extra energy or pressure [37]. Hence the outstanding pollutant removal capabilities of amyloid fibril aerogels point to its potential use in different pollutant removal applications. In another report, lysozyme protein along with long-chain polyethyleneimine (PEI) has been utilized to fabricate a novel hybrid highly selective adsorbent for the extraction of Pb(II) metal ions [38]. The abundance of primary and secondary amine functional groups present on PEI enables its strong metal complexation ability, which was further modified with a polydopamine (PDA) interface to achieve regenerative fibrous composites. Figure 9.4 schematically presents the PEI conjugation to the PDA-coated

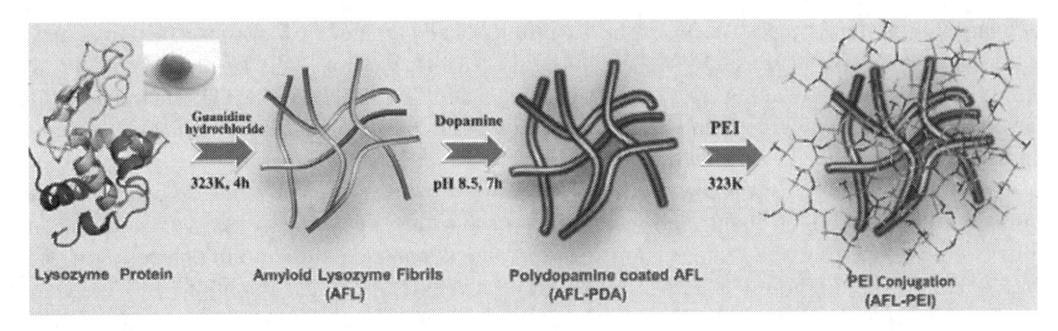


**FIGURE 9.4** Schematic representation of AFL-PEI nanofibril synthesis. (Data reprinted with permission of Elsevier [38].)

amyloid lysozyme fibrils (AFL-PDA) derived from lysozyme protein. The PEI/AFL-PDA composite exhibited superior adsorption capability and excellent selectivity for lead (II) removal, even in the presence of competitive Ca(II), Mg(II), and Na(II) ions. The high surface-to-volume ratio of the composite also enabled the faster kinetics of the matrix.

Proteins and CNFs together in composite form also work remarkably well in the adsorption of toxic heavy metal ions from aqueous solutions. In a recent report, lysozyme nanofibrils (LNFs) were used along with CNFs to fabricate dual nanofibrillar-based free-standing biosorbent films for the chemisorptive adsorption of Hg(II) from spring waters [39]. The nanocomposite film exhibited good mechanical strength, thermal properties, and low-concentrated Hg(II) ion removal efficiency in aqueous solution. Recently, a group of researchers reported 3D hierarchical BSA-$NiCo_2O_4$@$MnO_2$/C-based bioinspired multifunctional micromotors having dual oxidase/peroxidase kind of activity for the spectrophotometric determination of ongoing activity of enzymes and then its usage in Cu(II) ions removal from an aqueous solution [40]. The synergistic combination of C microtubes and $NiCo_2O_4$@$MnO_2$ nanosheets helped in providing sufficient accessible reactive sites to the adsorbent, resulting in an excellent detection process through this unique system. The fatty acid derivatives have also been used in the soil remediation process. An article reported the removal of PAHs described as aged polycyclic aromatic hydrocarbons from polluted soil by adding common fatty acids (sodium palmitate and sodium linoleate) present in plant root exudates to the PAHs [41]. The results indicated that both of these proteins led to a transformation in the soil bacterial community. The specific enrichment of the available microbes from the g-proteobacteria was ascribed to the high removal efficiency of nanocomposite after the sodium linoleate amendment PAHs than in the sodium palmitate amendment PAH. Among these, the high-molecular-weight PAHs exhibited better efficiency than the lower-molecular-weight PAHs.

### 9.4.2 Protein-based Nanomembranes

Protein-based nanomembranes work remarkably well in the broad field of environment remediation. The high performance of these hybrid nanomembranes in water or soil treatment is due to the high surface area of proteins, which further provides high metal binding capacity and mechanical strength to the membrane. Whey protein fibrils (WPFs) along with activated carbon (AC) materials were used to prepare hybrid membranes for Hg(II) and Cr(VI) removal from water [42]. In this combination, the low-cost feedstock whey reduced the overall cost of the WPF-AC hybrid membrane, whereas AC improved the adsorption capacity and mechanical strength. Amyloid–carbon-based hybrid membranes have also exhibited their potential in metal ion removal from water or soil because of the availability of multiple binding sites on the amyloid fibrils. One article reported on β-lactoglobulin amyloid fibrils with $ZrO_2$ NPs (<10 nm) to prepare hybrid membranes for efficient and

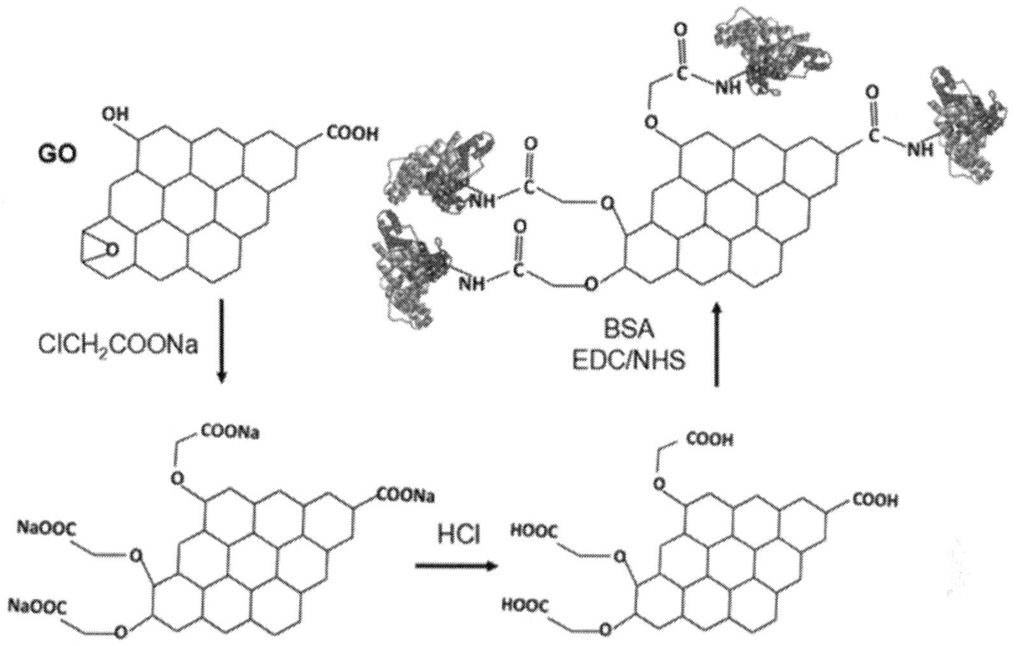

FIGURE 9.5 Schematic representation of the functionalization and GO-BSA nanohybrid membrane formation via covalent attachment. (Data reprinted with permission of MDPI [45].)

highly selective removal of fluoride from polluted water [43]. High selectivity and high efficiency with 99.5% removal of fluoride proved the importance of amyloid fibrils in the preparation of hybrid membranes. Peydayesh et al. have also reported an amyloid-activated carbon hybrid nanomembrane by utilizing β-lactoglobulin amyloid native monomers and amyloid fibrils for the adsorptive removal of Cr and Ni from electroplating industrial wastewater [44]. The results proved that the adsorption of metal ions on the β-lactoglobulin surface was based on the strong exothermic process. High adsorption capacity, long lifetime working ability, and excellent reusability for up to 10 cycles are some of the important features of this amyloid-AC hybrid nanomembrane, which helped it in scoring an outstanding performance in the water treatment process. Bovine serum albumin (BSA) is also one of the most important proteins used in various hybrid membranes for water and soil remediation. GO and covalently attached BSA-based hybrid membrane was reported for selective metal ions detection from wastewater [45]. The metal-binding ability of BSA enabled the selective recognition of Au(III) and Co(II) metal ions through the hybrid membrane. Figure 9.5 schematically presents the functionalization of GO sheets and the formation of GO-BSA nanohybrid membrane via covalent immobilization.

## 9.5 HYDROGELS

Hydrogels are widely used in the water and soil remediation process because of their easier regeneration, separation, and handling capabilities. The polymeric hydrogels possess different important characteristics such as excellent gel volume, good swelling-ability, hydrophilicity, as well as hydrophobic surface properties, enabling their applications in environmental remediation. The excellent swelling capacity, owing to the cross-linking within the polymer strands, along with the porous structure of engineered hydrogels, further provides the diffusing possibility to solute into the hydrogel's structure, resulting in their good adsorption capabilities for contaminant removal.

These advanced hydrogels are reusable up to multiple cycles. They can be generated on large-scale productions but their applications remain limited to the laboratory scale and many innovative futuristic research approaches are required for their real-life applications to take place. Many researchers have reviewed their fabrication process, applications, and undergoing mechanisms in the removal of various pollutants from the environment [46,47].

Moharramia and Motamedi have reported a biobased hydrogel nanocomposite with magnetite-functionalized CNCs (MCNCs) as nanoadsorbent for the effective removal of the cationic dyes crystal violet (CV) and methylene blue (MB) [48]. In this report the CNCs were produced *via* acid hydrolysis of enzymatic-mediated sugar-beet pulp and then the $Fe_3O_4$ NPs were anchored over CNCs to form MCNC-based hydrogel. The as-prepared hydrogel nanocomposite showed excellent, selective, and reusable adsorption of cationic dyes. Multilayered inorganic–organic hybrid nanomembranes having negatively charged layers and multiple interaction sites were prepared for adsorbing heavy metal ions and dyes in a one-step process from wastewater [49]. These hybrid nanomembranes were prepared by utilizing dichalcogenide nanosheets [metallic molybdenum disulphide ($MoS_2$)] and silk nanofibrils ($MoS_2$-SNF hybrid nanomembrane). During the filtration process, it has been observed that the precious metal ions were easily reduced to their NP form without requiring any additional thermal or chemical process. The degraded soil has an intrinsic property of water retention, and so a group of researchers utilized this important property by converting domestic animal wastes into hydrogels for water remediation [50]. In another report, Cr(VI)-contaminated soil was innovatively processed by utilizing nanosized FeS-coated humic acid (HA)-based nanocomposite material (CMC-FeS@HA) and Cr-resistant microflora [51]. The group improved the soil quality along with its chemical remediation and microbial remediation resulting in an innovative approach to the multiple-technology involvement in the environmental purpose. A chelator-mimetic multifunctionalized low-cost hydrogel, synthesized by a simple conjugation process, also exhibited high affinity and reusability against Cd(II), Pb(II), and As(II) pollutants removal from wastewater [52]. The as-prepared innovative hydrogel based upon dihydroxybenzoic

**TABLE 9.3**
**Comparative Study on the Applications of Water and Soil Remediation Using Protein- and Hydrogels-based Nanoadsorbents and Nanomembranes**

| Pollutant | Base Matrix | Maximum Adsorption Capacity (mg/g) | Key Findings | References |
|---|---|---|---|---|
| Bentazone<br>Bisphenol A<br>Ibuprofen | Amyloid fibril aerogels | 54.2<br>50.6<br>69.6 | • 92%, 78%, and 98% removal of bentazone, bisphenol A, and ibuprofen<br>• Reusable up to 3 cycles | [37] |
| Pb(II) | PEI/AFL-PDA | 493 | • Faster kinetics in 2 min<br>• Regenerated system | [38] |
| Hg(II) | CNFs/LNFs biobased films | 99% | • Preparation *via* simple vacuum filtration method of water suspensions<br>• pH dependent, optimum time 24 h<br>• Sorption kinetics based on pseudo-2nd-order and Elovich models | [39] |
| Cu(II) | BSA-$NiCo_2O_4$@ $MnO_2$/C | — | • Multifunctional micromotors with dual oxidase/peroxidase kind of activity<br>• Spectro-photometric determination of ongoing enzyme activity | [40] |

**TABLE 9.3 (Continued)**
**Comparative Study on the Applications of Water and Soil Remediation Using Protein- and Hydrogels-based Nanoadsorbents and Nanomembranes**

| Pollutant | Base Matrix | Maximum Adsorption Capacity (mg/g) | Key Findings | References |
|---|---|---|---|---|
| PAH | Common fatty acids | — | • Soil remediation by fatty acids present in plant root exudates<br>• Sodium linoleate amendment showed higher efficiency than sodium palmitate | [41] |
| Hg(II)<br>Cr(VI) | WPF-AC hybrid membrane | 25<br>18 | • Optimum conditions 74°C, 7 h and 3.8% whey protein<br>• 81% and 57% removal of Hg(II) and Cr()<br>• Active in more than 10 filtration cycles | [42] |
| Fluoride | β-Lactoglobulin amyloid fibrils-$ZrO_2$ NPs | — | • Highly selective hybrid membrane<br>• High efficiency with 99.5% removal | [43] |
| Cr<br>Ni<br>Ag<br>Pt | Amyloid-AC hybrid membrane | >99% | • 99% removal of Cr and Ni with $2.92 \times 10^{-16}$ $m^2$ permeability<br>• 10 wt.% amyloid fibrils showed excellent performance | [44] |
| Au(III)<br>Co(II) | GO-BSA membranes | — | • Selective recognition of Au(III) and Co(II) from Au(III), Co(II), Fe(II), and Cu(II) concentrated aqueous solution | [45] |
| CV<br>MB | MCNCs/starch-g-(AMPS-co-AA) hydrogel | 2500.0<br>1428.6 | • Acid hydrolysis of enzymatic-mediated sugar-beet pulp utilized for CNC production<br>• $Fe_3O_4$ NPs were anchored over CNCs | [48] |
| Heavy metal ions<br>Dyes | $MoS_2$–SNF hybrid membrane | ~99% | • Heavy metal ions removal, i.e. Hg(II), Pb(II), Cu(II)<br>• Nobel metal ions removal, i.e. Pd(II), Au(III), Ag(I) | [49] |
| Cd(II)<br>Pb(II)<br>As(I) | PAAm/TGA/DHBA hydrogel | — | • Rapid removal (~50%) in 5 min<br>• Chelator-mimetic multifunctionalized hydrogel<br>• Freundlich model fits data at best | [52] |
| Cr(VI) | $Fe_3O_4$NPs/CS/glyoxal hydrogel film | 80–90% | • Langmuir model fits data at best along with Pseudo 2nd-order kinetics ($R^2 > 0.99$)<br>• Optimum conditions: pH 4, contact time = 110 min | [53] |

acid-thioglycolic acid modified polyallylamine (PAAm/TGA/DHBA) cross-linked polymer showed rapid (~50%) removal of contaminants within 5 min of the process with good reusability for up to five cycles. Glyoxal cross-linked magnetic chitosan hydrogel film ($Fe_3O_4$NPs/CS/glyoxal) has also shown its effectiveness in the removal of toxic metal Cr(VI) ions (80–90% removal) from wastewater [53].

## 9.6 CONCLUSION

Water and soil remediation has become an essential requirement to safeguard the environment. Here, the various methods/strategies and materials used for water purification as well as soil remediation are summarized and discussed. It has been observed that nanomaterials have played an important role in the effective removal of pollutants, whether in the form of nanoadsorbents or nanomembranes. Although the reported nanomaterials, such as cellulose, protein, hydrogels, and carbon-based materials, have successfully demonstrated laboratory-scale applications, the up-scaling to low-cost, green, reusable, recyclable, and commercialization to industrial-scale modules is still required and on-going. This has become a big challenge to the scientific community and hence combined and collaborative efforts are required from research institutions and industries. The research community needs to work on the next-generation regenerative nanomaterials to overcome the current issues in water and soil remediation. The selectivity along with cost-effectiveness of the nanoadsorbents/nanomembranes should also be addressed to improve the overall quality of water. The molecular-level design of nanocomposites also plays an important role in preparing defect-free, thin, and selective layers on porous microstructure support. The fouling resistance of nanomembranes is another big challenge which needs to be overcome. The fabrication of highly selective multifunctional nanomaterials and nanomembranes for organic/inorganic pollutants removal from water and soil with high innovation and potential is really a great challenge which can be met by working upon innovative research methods to realize them in real-life applications. This can be achieved only by a joint venture of the research community and commercial industrial engineers.

## REFERENCES


1. Niu, B., Loaiciga, H. A., Wang, Z., Zhan, F. B., & Hong, S. (2014). Twenty years of global groundwater research: A Science Citation Index Expanded-based bibliometric survey (1993–2012). *Journal of Hydrology*, 519, 966–975. https://doi.org/10.1016/j.jhydrol.2014.07.064
2. Rajesh, Gao, Z., Charlie Johnson, A. T., Puri, N., Mulchandani, A., & Aswal, D. K. (2021). Scalable chemical vapor deposited graphene field-effect transistors for bio/chemical assay. *Applied Physics Reviews*, 8(1), 011311. https://doi.org/10.1063/5.0024508
3. Puri, N., Niazi, A., Biradar, A. M., Mulchandani, A., & Rajesh. (2014). Conducting polymer functionalized single-walled carbon nanotube based chemiresistive biosensor for the detection of human cardiac myoglobin. *Applied Physics Letters*, 105(15), 153701. https://doi.org/10.1063/1.4897972
4. Uddin, M. K., & Baig, U. (2019). Synthesis of $Co_3O_4$ nanoparticles and their performance towards methyl orange dye removal: Characterisation, adsorption and response surface methodology. *Journal of Cleaner Production*, 211, 1141–1153. https://doi.org/10.1016/j.jclepro.2018.11.232
5. Khan, S. U., Zaidi, R., Hassan, S. Z., Farooqi, I. H., & Azam, A. (2016). Application of Fe-Cu binary oxide nanoparticles for the removal of hexavalent chromium from aqueous solution. *Water Science and Technology*, 74(1), 165–175. https://doi.org/10.2166/wst.2016.172
6. Kumari, U., Siddiqi, H., Bal, M., & Meikap, B. C. (2020). Calcium and zirconium modified acid activated alumina for adsorptive removal of fluoride: Performance evaluation, kinetics, isotherm, characterization and industrial wastewater treatment. *Advanced Powder Technology*, 31(5), 2045–2060. https://doi.org/10.1016/j.apt.2020.02.035
7. Dehghani, M. H., Kamalian, S., Shayeghi, M., Yousefi, M., Heidarinejad, Z., Agarwal, S., & Gupta, V. K. (2019). High-performance removal of diazinon pesticide from water using multi-walled carbon nanotubes. *Microchemical Journal*, 145, 486–491. https://doi.org/10.1016/j.microc.2018.10.053
8. Puri, N., Mishra, S. K., Niazi, A., & Srivastava, A. K. (2015). Physicochemical characteristics of reduced graphene oxide based Pt-nanoparticles-conducting polymer nanocomposite film for immunosensor applications. *Journal of Chemical Technology & Biotechnology*, 90(9), 1699–1706. https://doi.org/10.1002/jctb.4480
9. Hammad, M., Fortugno, P., Hardt, S., Kim, C., Salamon, S., Schmidt, T. C., ... & Wiggers, H. (2021). Large-scale synthesis of iron oxide/graphene hybrid materials as highly efficient photo-Fenton catalyst

for water remediation. *Environmental Technology & Innovation*, 21, 101239. https://doi.org/10.1016/j.eti.2020.101239

10. Nhlane, D., Richards, H., & Etale, A. (2021). Facile and green synthesis of reduced graphene oxide for remediation of Hg (II)-contaminated water. *Materials Today: Proceedings*, 38, 737–742. https://doi.org/10.1016/j.matpr.2020.04.163
11. Yakout, A. A., & Khan, Z. A. (2021). High performance Zr-$MnO_2$@ reduced graphene oxide nanocomposite for efficient and simultaneous remediation of arsenates As (V) from environmental water samples. *Journal of Molecular Liquids*, 334, 116427. https://doi.org/10.1016/j.molliq.2021.116427
12. Tara, N., Siddiqui, S. I., Bach, Q. V., & Chaudhry, S. A. (2020). Reduce graphene oxide-manganese oxide-black cumin based hybrid composite (rGO-MnO2/BC): A novel material for water remediation. *Materials Today Communications*, 25, 101560. https://doi.org/10.1016/j.mtcomm.2020.101560
13. Huang, Y., Wang, M., Li, Z., Gong, Y., & Zeng, E. Y. (2019). In situ remediation of mercury-contaminated soil using thiol-functionalized graphene oxide/Fe-Mn composite. *Journal of Hazardous Materials*, 373, 783–790. https://doi.org/10.1016/j.jhazmat.2019.03.132
14. Onyancha, R. B., Aigbe, U. O., Ukhurebor, K. E., & Muchiri, P. W. (2021). Facile synthesis and applications of carbon nanotubes in heavy-metal remediation and biomedical fields: A comprehensive review. *Journal of Molecular Structure*, 1238, 130462. https://doi.org/10.1016/j.molstruc.2021.130462
15. Liu, B., Song, W., Zhang, W., Zhang, X., Pan, S., Wu, H., ... & Xu, Y. (2021). $Fe_3O_4$@ CNT as a high-effective and steady chainmail catalyst for tetracycline degradation with peroxydisulfate activation: Performance and mechanism. *Separation and Purification Technology*, 273, 118705. https://doi.org/10.1016/j.seppur.2021.118705
16. Wang, Y., Li, W., Li, H., Ye, M., Zhang, X., Gong, C., ... & Yu, C. (2021). Fe/$Fe_3C$@ CNTs anchored on carbonized wood as both self-standing anode and cathode for synergistic electro-Fenton oxidation and sequestration of As (III). *Chemical Engineering Journal*, 414, 128925. https://doi.org/10.1016/j.cej.2021.128925
17. Alsohaimi, I. H., Khan, M. A., Alothman, Z. A., Khan, M. R., & Kumar, M. (2015). Synthesis, characterization, and application of Fe-CNTs nanocomposite for $BrO_3$ – remediation from water samples. *Journal of Industrial and Engineering Chemistry*, 26, 218–225. https://doi.org/10.1016/j.jiec.2014.11.033
18. Rahman, M. M. (2020). In-situ preparation of cadmium sulphide nanostructure decorated CNT composite materials for the development of selective benzaldehyde chemical sensor probe to remove the water contaminant by electrochemical method for environmental remediation. *Materials Chemistry and Physics*, 245, 122788. https://doi.org/10.1016/j.matchemphys.2020.122788
19. Zhao, J., Huang, Y., Wang, G., Qiao, Y., Chen, Z., Zhang, A., & Park, C. B. (2021). Fabrication of outstanding thermal-insulating, mechanical robust and superhydrophobic PP/CNT/sorbitol derivative nanocomposite foams for efficient oil/water separation. *Journal of Hazardous Materials*, 418, 126295. https://doi.org/10.1016/j.jhazmat.2021.126295
20. Liu, C., Xia, J., Gu, J., Wang, W., Liu, Q., Yan, L., & Chen, T. (2021). Multifunctional CNTs-PAA/MIL101 (Fe)@ Pt composite membrane for high-throughput oily wastewater remediation. *Journal of Hazardous Materials*, 403, 123547. https://doi.org/10.1016/j.jhazmat.2020.123547
21. Seddiqi, H., Oliaei, E., Honarkar, H., Jin, J., Geonzon, L. C., Bacabac, R. G., & Klein-Nulend, J. (2021). Cellulose and its derivatives: Towards biomedical applications. *Cellulose*, 1–39. https://doi.org/10.1007/s10570-020-03674-w
22. Sharma, A., Thakur, M., Bhattacharya, M., Mandal, T., & Goswami, S. (2019). Commercial application of cellulose nano-composites–A review. *Biotechnology Reports*, 21, e00316. https://doi.org/10.1016/j.btre.2019.e00316
23. Barsbay, M., Kavaklı, P. A., Tilki, S., Kavaklı, C., & Güven, O. (2018). Porous cellulosic adsorbent for the removal of Cd (II), Pb (II) and Cu (II) ions from aqueous media. *Radiation Physics and Chemistry*, 142, 70–76. https://doi.org/10.1016/j.radphyschem.2017.03.037
24. Guo, W., Yang, F., Zhao, Z., Liao, Q., Cai, C., Zhang, Y., & Bai, R. (2018). Cellulose-based ionic liquids as an adsorbent for high selective recovery of gold. *Minerals Engineering*, 125, 271–278. https://doi.org/10.1016/j.mineng.2018.06.012
25. Abdel-Galil, E. A., Moloukhia, H., Abdel-Khalik, M., & Mahrous, S. S. (2018). Synthesis and physico-chemical characterization of cellulose/$HO_7Sb_3$ nanocomposite as adsorbent for the removal of some

radionuclides from aqueous solutions. *Applied Radiation and Isotopes*, 140, 363–373. https://doi.org/10.1016/j.apradiso.2018.07.022

26. Neeli, S. T., Ramsurn, H., Ng, C. Y., Wang, Y., & Lu, J. (2020). Removal of Cr (VI), As (V), Cu (II), and Pb (II) using cellulose biochar supported iron nanoparticles: A kinetic and mechanistic study. *Journal of Environmental Chemical Engineering*, 8(5), 103886. https://doi.org/10.1016/j.jece.2020.103886
27. Pinto, A. H., Taylor, J. K., Chandradat, R., Lam, E., Liu, Y., Leung, A. C., ... & Sunasee, R. (2020). Wood-based cellulose nanocrystals as adsorbent of cationic toxic dye, Auramine O, for water treatment. *Journal of Environmental Chemical Engineering*, 8(5), 104187. https://doi.org/10.1016/j.jece.2020.104187
28. Tan, K. B., Reza, A. K., Abdullah, A. Z., Horri, B. A., & Salamatinia, B. (2018). Development of self-assembled nanocrystalline cellulose as a promising practical adsorbent for methylene blue removal. *Carbohydrate Polymers*, 199, 92–101. https://doi.org/10.1016/j.carbpol.2018.07.006
29. Putro, J. N., Santoso, S. P., Soetaredjo, F. E., Ismadji, S., & Ju, Y. H. (2019). Nanocrystalline cellulose from waste paper: adsorbent for azo dyes removal. *Environmental Nanotechnology, Monitoring & Management*, 12, 100260. https://doi.org/10.1016/j.enmm.2019.100260
30. Zhao, L., Yang, S., Yilihamu, A., Ma, Q., Shi, M., Ouyang, B., ... & Yang, S. T. (2019). Adsorptive decontamination of $Cu^{2+}$-contaminated water and soil by carboxylated graphene oxide/chitosan/cellulose composite beads. *Environmental Research*, 179, 108779. https://doi.org/10.1016/j.envres.2019.108779
31. Sharma, P. R., Sharma, S. K., Lindström, T., & Hsiao, B. S. (2020). Nanocellulose-enabled membranes for water purification: perspectives. *Advanced Sustainable Systems*, 4(5), 1900114. https://doi.org/10.1002/adsu.201900114
32. Gopakumar, D. A., Pasquini, D., Henrique, M. A., de Morais, L. C., Grohens, Y., & Thomas, S. (2017). Meldrum's acid modified cellulose nanofiber-based polyvinylidene fluoride microfiltration membrane for dye water treatment and nanoparticle removal. *ACS Sustainable Chemistry & Engineering*, 5(2), 2026–2033. https://doi.org/10.1021/acssuschemeng.6b02952
33. Karim, Z., Hakalahti, M., Tammelin, T., & Mathew, A. P. (2017). In situ TEMPO surface functionalization of nanocellulose membranes for enhanced adsorption of metal ions from aqueous medium. *RSC Advances*, 7(9), 5232–5241. https://doi.org/10.1039/C6RA25707K
34. Ibrahim, Y., Banat, F., Yousef, A. F., Bahamon, D., Vega, L. F., & Hasan, S. W. (2020). Surface modification of anti-fouling novel cellulose/graphene oxide (GO) nanosheets (NS) microfiltration membranes for seawater desalination applications. *Journal of Chemical Technology & Biotechnology*, 95(7), 1915–1925. https://doi.org/10.1002/jctb.6341
35. Gautam, S., Sharma, B., & Jain, P. (2021). Green Natural Protein Isolate based composites and nanocomposites: A review. *Polymer Testing*, 99, 106626. https://doi.org/10.1016/j.polymertesting.2020.106626
36. Peydayesh, M., & Mezzenga, R. (2021). Protein nanofibrils for next generation sustainable water purification. *Nature Communications*, 12(1), 1–17. https://doi.org/10.1038/s41467-021-23388-2
37. Peydayesh, M., Suter, M. K., Bolisetty, S., Boulos, S., Handschin, S., Nyström, L., & Mezzenga, R. (2020). Amyloid fibrils aerogel for sustainable removal of organic contaminants from water. *Advanced Materials*, 32(12), 1907932. https://doi.org/10.1002/adma.201907932
38. Liu, M., Jia, L., Zhao, Z., Han, Y., Li, Y., Peng, Q., & Zhang, Q. (2020). Fast and robust lead (II) removal from water by bioinspired amyloid lysozyme fibrils conjugated with polyethyleneimine (PEI). *Chemical Engineering Journal*, 390, 124667. https://doi.org/10.1016/j.cej.2020.124667
39. Silva, N. H., Figueira, P., Fabre, E., Pinto, R. J., Pereira, M. E., Silvestre, A. J., ... & Freire, C. S. (2020). Dual nanofibrillar-based bio-sorbent films composed of nanocellulose and lysozyme nanofibrils for mercury removal from spring waters. *Carbohydrate Polymers*, 238, 116210. https://doi.org/10.1016/j.carbpol.2020.116210\
40. Yang, W., Li, J., Lyu, Y., Yan, X., Yang, P., & Zuo, M. (2021). Bioinspired 3D hierarchical BSA-$NiCo_2O_4$@ $MnO_2$/C multifunctional micromotors for simultaneous spectrophotometric determination of enzyme activity and pollutant removal. *Journal of Cleaner Production*, 309, 127294. https://doi.org/10.1016/j.jclepro.2021.127294
41. Wang, Q., Hou, J., Yuan, J., Wu, Y., Liu, W., Luo, Y., & Christie, P. (2020). Evaluation of fatty acid derivatives in the remediation of aged PAH-contaminated soil and microbial community and degradation gene response. *Chemosphere*, 248, 125983. https://doi.org/10.1016/j.chemosphere.2020.125983

42. Ramírez-Rodríguez, L. C., Díaz Barrera, L. E., Quintanilla-Carvajal, M. X., Mendoza-Castillo, D. I., Bonilla-Petriciolet, A., & Jiménez-Junca, C. (2020). Preparation of a hybrid membrane from whey protein fibrils and activated carbon to remove mercury and chromium from water. *Membranes*, 10(12), 386. https://doi.org/10.3390/membranes10120386
43. Zhang, Q., Bolisetty, S., Cao, Y., Handschin, S., Adamcik, J., Peng, Q., & Mezzenga, R. (2019). Selective and efficient removal of fluoride from water: in situ engineered amyloid fibril/$ZrO_2$ hybrid membranes. *Angewandte Chemie*, 131(18), 6073–6077. https://doi.org/10.1002/ange.201901596
44. Peydayesh, M., Bolisetty, S., Mohammadi, T., & Mezzenga, R. (2019). Assessing the binding performance of amyloid–carbon membranes toward heavy metal ions. *Langmuir*, 35(11), 4161–4170. https://doi.org/10.1021/acs.langmuir.8b04234
45. Yu, X., Sun, S., Zhou, L., Miao, Z., Zhang, X., Su, Z., & Wei, G. (2019). Removing metal ions from water with graphene–bovine serum albumin hybrid membrane. *Nanomaterials*, 9(2), 276. https://doi.org/10.3390/nano9020276
46. Weerasundara, L., Gabriele, B., Figoli, A., Ok, Y. S., & Bundschuh, J. (2020). Hydrogels: Novel materials for contaminant removal in water—A review. *Critical Reviews in Environmental Science and Technology*, 51, 1–45. https://doi.org/10.1080/10643389.2020.1776055
47. Van Tran, V., Park, D., & Lee, Y. C. (2018). Hydrogel applications for adsorption of contaminants in water and wastewater treatment. *Environmental Science and Pollution Research*, 25(25), 24569–24599. https://doi.org/10.1007/s11356-018-2605-y
48. Moharrami, P., & Motamedi, E. (2020). Application of cellulose nanocrystals prepared from agricultural wastes for synthesis of starch-based hydrogel nanocomposites: Efficient and selective nanoadsorbent for removal of cationic dyes from water. *Bioresource Technology*, 313, 123661. https://doi.org/10.1016/j.biortech.2020.123661
49. Zhao, F., Peydayesh, M., Ying, Y., Mezzenga, R., & Ping, J. (2020). Transition metal dichalcogenide–silk nanofibril membrane for one-step water purification and precious metal recovery. *ACS Applied Materials & Interfaces*, 12(21), 24521–24530. https://doi.org/10.1021/acsami.0c07846
50. Mwangi, I., Kiriro, G., Swaleh, S., Wanjau, R., Mbugua, P., & Ngila, J. C. (2019). Remediation of degraded soils with hydrogels from domestic animal wastes. *International Journal of Recycling of Organic Waste in Agriculture*, 8(2), 159–170. https://doi.org/10.1007/s40093-019-0242-1
51. Tan, H., Wang, C., Li, H., Peng, D., Zeng, C., & Xu, H. (2020). Remediation of hexavalent chromium contaminated soil by nano-FeS coated humic acid complex in combination with Cr-resistant microflora. *Chemosphere*, 242, 125251. https://doi.org/10.1016/j.chemosphere.2019.125251
52. Mohammadi, Z., Shangbin, S., Berkland, C., & Liang, J. T. (2017). Chelator-mimetic multi-functionalized hydrogel: Highly efficient and reusable sorbent for Cd, Pb, and As removal from waste water. *Chemical Engineering Journal*, 307, 496–502. https://doi.org/10.1016/j.cej.2016.08.121
53. Mirabedini, M., Kassaee, M. Z., & Poorsadeghi, S. (2017). Novel magnetic chitosan hydrogel film, cross-linked with glyoxal as an efficient adsorbent for removal of toxic Cr (VI) from water. *Arabian Journal for Science and Engineering*, 42(1), 115–124. https://doi.org/10.1007/s13369-016-2062-1

# 10 Environmental Adequacy of Green Polymers and Biomaterials

*Shilpa Borehalli Mayegowda, Bhoomika S., Kavita Nagshetty, and Manjula N.G.*

## CONTENTS



## 10.1 INTRODUCTION

The human population has exponentially increased in the last few decades and with progressive developments this has also led to various types of pollution that have adverse effects on the atmosphere and living systems. One example is the accumulation of massive amounts of non-degradable wastes that have changed the biosphere affecting the potential survival capabilities of life. Non-degradable synthetic polymers have been used for many decades and have been accumulated in the environment. One of these is plastics, derived from the Greek word "plastikos," that can be molded into different forms [1]. It has been observed that, worldwide, an estimated 335 million tons of synthetic polymers have been manufactured worldwide in 2016 with the majority of this production being polyethylene (PE), polypropylene (PP), polyurethane (PUR), polystyrene (PS), polyethylene terephthalate (PET), and polyvinyl chloride (PVC). These synthetic polymers have been extensively used in various industrial, medical, as well as domestic applications [2,3], as shown in Figure 10.1. Considering the adverse effects of these synthetic polymers in the environment, their degradation has been attempted by different processes that include chemical, frictional, and thermo-oxidative techniques [5,6]. However, these processes take a relatively long time to decompose them from the environment, which can vary from 50 years to more than 100 years, increasing the pollution load. These issues have compelled researchers to devise and promote programs that can design novel strategies to be directed and aimed at facilitating the replacement of synthetic polymers with degradable polymers.

DOI: 10.1201/9781003240884-10

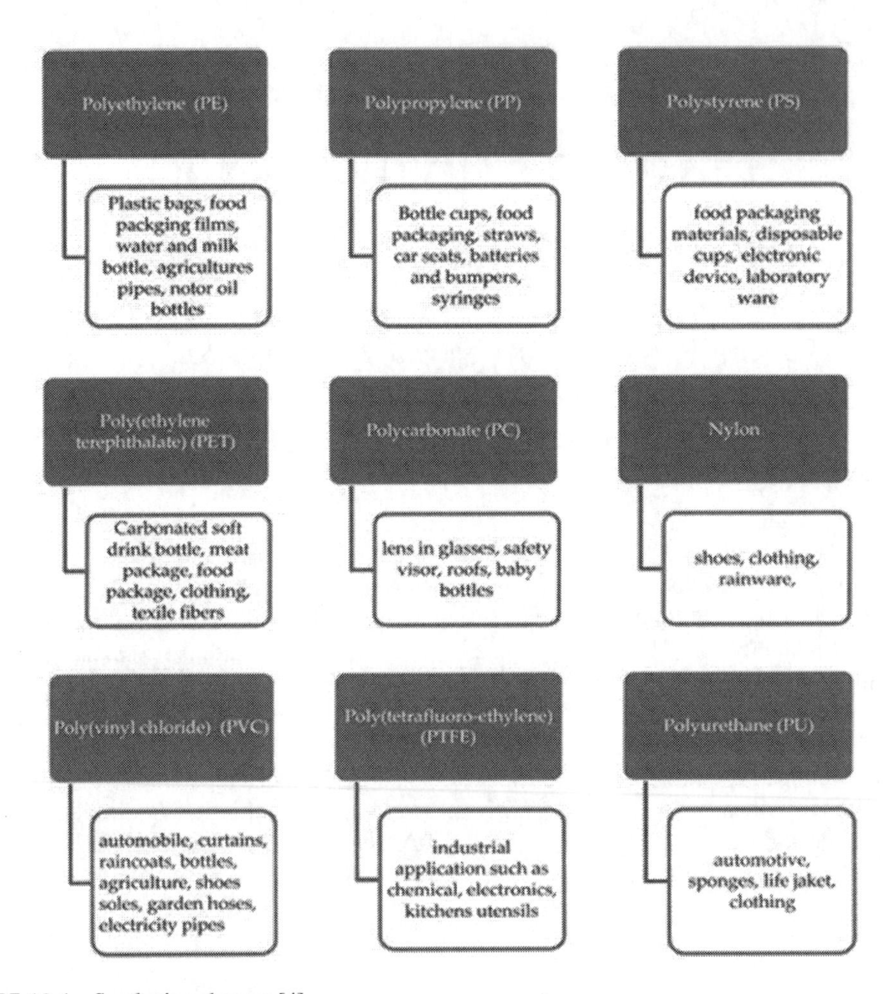


**FIGURE 10.1** Synthetic polymers [4].

Living matter constitutes the substances that act as the building blocks for biobased materials. The biopolymers that help to strengthen natural fibers are identified as "green composites," as shown in Figure 10.2. Ecologically, biopolymers are biodegradable by the action of ecological parameters that include oxygen, temperature, light, or microorganisms. Green polymers are produced by a wide range of microbes and require specific nutrients and environmental parameters [8]. Basically, these lipid polymers are storage materials that are accumulated in microbes in various forms of liquid granules or amorphous compounds that are mobile, allowing the cells to survive under stress [9,10]. These granules, with a molecular weight ranging between 50–1000 kDa, were observed by conventional techniques like gel-permeation chromatography, sedimentation analysis, light scattering, and intrinsic viscosity measurements [11].

Biopolymers are environmentally friendly as they are produced from renewable resources or substrates from other industrial wastes. Most of these biopolymers are made up of monomers that have various β-oxidation intermediates [(R)-3-hydroxyacyl–CoAs] and are polymerized by enzymes through condensation of the carboxy function of CoA thioester with the 3-hydroxy group (or the thiol group) of the next one [10,12,13]. Meanwhile, some are made up of unusual monomers like glutamic acid through different pathways that are used for synthesis, suggesting that these biopolymers undergo strain-specific processes. Though the process of biodegradable plastics has been developed already, there remains a need for certain modifications to be done and also to reduce

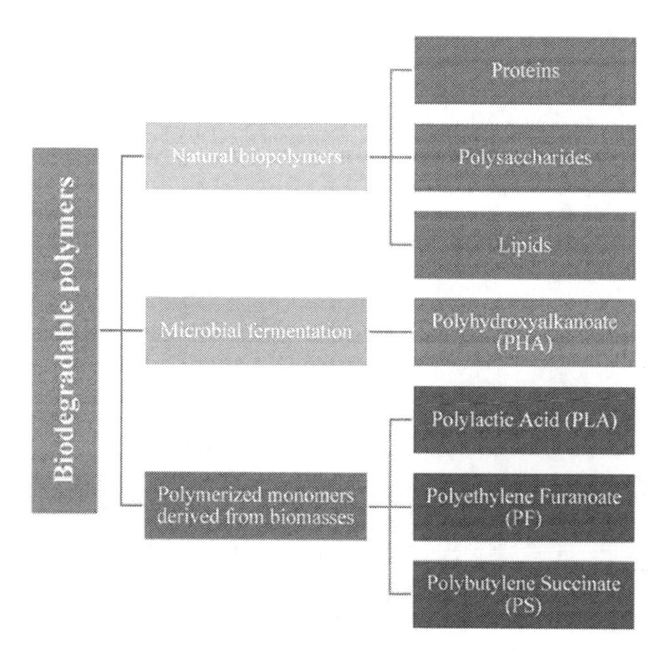


**FIGURE 10.2** Biodegradable polymers from different sources [7].

the recovery costs to cover the complete commercial production in order to replace the synthetic forms originating from petroleum sources [14].

Green polymers were first described by Lemoigne in 1926 [15] in *Bacillus megaterium* used as bioplastics that were classified as poly-3-hydroxybutyrate (PHB). In the mid-1970s, the world observed a petroleum crisis that led to scientific attempts at discovering alternative resources, such as biobased plastics, that are more sustainable due to their lower associated carbon emissions than petroleum extraction and refinement [16]. Over 300 microbial species have been found to be efficient producers of biopolymers that can grow aerobically and anaerobically. These microbes have been basically derived from the group of archaebacteria, photosynthetic bacteria, and lower eukaryotes, and are able to amass and metabolize these polyesters [13,17–19]. Polyhydroxyalkanoate (PHA) and polyhydroxybutyrate (PHB) are environmentally sustainable biopolymers that are efficiently biodegradable and so can help in replacing synthetic polymers with their applications in various fields [20]. Currently, there has been a lot of focus on biopolymer synthesis by genetic manipulation and metabolic engineering for several recombinant microbes that favors high productive yields with a reduction in overall costs.

## 10.2 SYNTHETIC POLYMERS

One of the most advanced inventions is the synthetic polymers/plastics, with approximately 20 different kinds being used worldwide [2]. With respect to the dependence on synthetic polymers it has been drastically increased with their production and usage over the last few decades due to their durability, cost effectiveness, flexibility, and versatility. These synthetic biopolymers have been applied in various fields that include medical, food, and agricultural areas due to their soft, transparent, flexible properties. They have been invariably used in biomedical areas such as drug-delivery systems, sutures, clot-removal devices for aneurysms, vascular stents, and orthodontic therapy [21,22]. Its use has been extensive in food packaging, pharmaceuticals, cosmetics, detergents, and chemicals. It has been estimated that the worldwide usage of synthetic polymers is over 300 million

**TABLE 10.1**
**Major Synthetic Thermopolymers and Their Applications**

| Synthetic Plastic | Uses | Life Span in Years |
|---|---|---|
| **Polyethylene** | Plastics used to make milk cartons, water bottles, toys, and bags. Used in food packaging, irrigation, and drainage tubes | 600 |
| **Polystyrene** | Disposable cups, materials used in packaging, laboratory apparatus, electronic uses | 80 |
| **Polyurethane** | Refrigerator insulation wires, tires, gaskets, jackets used for life saving | 50 |
| **Polyvinyl chloride** | Seat covers for automobiles, used for making curtains for showers, raincoats, water hoses, electricity pipes, etc. | 150 |
| **Polypropylene** | Disposable syringes, straws for drinking, automobile seats, batteries, bottles used in medicine, etc. | 600 |
| **PET—Polyethylene terephthalate** | Used as packaging for processed meat, in canning industry, bottle jars, textile fibers, soft drink bottles, especially carbonated ones | 450 |
| **Nylon and its derivatives** | Helmets, inks, cloth used for parachutes, cellophane, speedometer gears, windshield wipers in vehicles | 30–40 |
| **Polycarbonate** | Used in street lights, glass lenses, rear lights for vehicles, nozzles in paper industries, safety visors, baby bottles, etc. | 25–30 |
| **PTFE—Polytetrafluoroethylene** | Nonstick utensils as coating, used in electronics and bearings, specific chemical industries | 12–15 |

*Source:* (25).

metric tons worldwide annually (23) and about 50% are disposed of within a year of their purchase (24). However, the most concerning problem is associated with the proper disposal and most importantly a degradation process without harmful effects (Table 10.1).

### 10.2.1 Effects of Synthetic Polymers on Human Health

Synthetic polymers have an adverse effect on the health of humans. It has been observed that polymeric plastic and even the building blocks (such as the monomeric units of bisphenol A) along with their additives, e.g., plasticizers that are formed by the amalgamation with two molecules such as antimicrobial polycarbonate, have an adverse effect on both human health and the environment [26]. One of the major monomeric compounds of numerous synthetic plastics (polycarbonate) is bisphenol A, a compound first described in 1891 and used as a main additive in the synthetic polymer production of PVC, etc. [27]. The estimated annual production of BPA was around 2.2 million metric tons in 2003, with it being extravagantly used in the food and beverage packaging industries. During polymerization it leaves some monomeric units unbound and these are released into food and beverage containers that come into contact with humans, and this can have an effect on the hormonal cycle [28]. It has been observed that BPA can mimic the reproductive "estrogen" hormone, causing ovarian chromosomal abnormalities, reduced sperm production, rapid puberty onset, and an altered immune system with the onset of diseases such as type-2 diabetes, cardiovascular disorders, and obesity. It has been well noted that higher concentrations lead to a risk of various cancers such as breast and prostate cancers, pain, metabolic disorders, etc. However, its effects are more prominent in pregnant women who are at a higher risk, with it causing disorders in developing fetuses and young children [29]. Another synthetic component that includes di-(2-ethylhexyl) phthalate has a turnover of 2 million tons/year with its maximum use in medical applications for various devices. However, it is alarming to note that these phthalates can make their way into human body systems during medical procedures that include kidney dialysis, extracorporeal membrane oxygenation, and blood transfusions, or by ingesting food that is contaminated and inhaling of house dust, etc. [30–32].

### 10.2.2 Effects of Synthetic Polymers on the Environment

The non-degradable polymers have been globally distributed and are consistently persistent due to various factors such as urbanization, industrialization, and trade routes, and they have been found to be present in soil, ocean currents, etc. Humans act as the main vector for the accumulation of contaminants, especially chemicals like heavy metals and organic contaminants, which are the major pollutants in the environment [5]. These chemicals seep through the ecosystem via water, land, and air that include compounds like benzene found in polystyrene, vinyl chloride (in PVC), dioxins (in PVC), plasticizers such as phthalates (in PVC and others), and bisphenol-A or BPA and formaldehyde (in polycarbonate). Notably, these components present in synthetic plastics are carcinogenic, neurotoxic, and disruptive to the hormone cycle. These compounds come under the category of persistent organic pollutants, producing a high level of toxicity and instigating irreparable loss to both terrestrial and aquatic life. Further, they produce toxic gases like dioxin, carbon monoxide, and hydrogen cyanide that are devastating to both humans and animals causing lung and respiratory diseases, disorders of the nervous system, along with reduced function of the immune system. Chlorinated plastics release toxins and chemicals that enter the groundwater level and destroy reserved water resources used for drinking. Researchers have revealed the presence of polymers in oceans and seas that have triggered global warming causing shaded surfaces on top of the water bodies and aquatic vegetation to perish. Noteworthy examples include the huge damage caused in the water bodies in the megacity of Dhaka, polluted by plastic bottles, canes, bags, etc. [33].

## 10.3 BIOAVAILABILITY

Modern technology has been a boon, but at the same time the increase in industrialization and urbanization has also exacerbated several pollution factors. Biopolymers are derived from sources that include animals, plants, and microorganisms that are abundantly present in nature and can be used as a renewable resource in innumerable applications. Biopolymers have been commercially produced on a large scale in industries. Biopolymers possess an array of various properties that are almost identical to those of synthetic polymers currently in the market. There has been an increase in non-degradable waste, with adverse impacts, which has resulted in the development of degradable bioplastics by the microbes and also biopolymers that have efficient degradability. Microorganisms synthesize different classes of biopolymers that are made up of polysaccharides (basically made up of monosaccharides, disaccharides, and/or sugar acids that are linked by glycosidic linkages/bonds), polyamides (amino acids linked by polypeptide bonds), polyesters (made up of lipid components, hydroxy fatty acids cross-linked with ester bonds), and polyphosphates (polyPs or inorganic phosphates bonded by anhydride linkages). The focus on understanding the biosynthetic pathways has led to various roles of these microbial biopolymers in agriculture and medicine, including their pathogenicity in biofilms and their persistence. With the production of biopolymers from microbes, PHAs, for example, have an advantage as they are synthesized as carbon storage intracellularly by various bacteria under stress (30–80% of dry cell weight). The most versatile benefit of these biopolymers is that they have similar properties to conventional synthetic plastics [34,35]. Different biopolymers are produced by microbial species including polylactides, polycaprolactone, PHA, polysaccharides, aliphatic polyesters, and copolymers, or a combination of these polymers. The most important blended polymers include poly(3-hydroxybutyrate-co-3-hydroxyvalerate) and PHBs that have been synthesized commercially [36]. They have also been genetically engineered to produce biopolymers that have the potential to replace the presently used synthetic polymers [37]. Microbial biopolymers have been applied in various fields for their sustainable properties in the environment. The advantages of using microorganisms for the production of biopolymers are illustrated in Figure 10.3.

Although there has been a lot of investigation into biopolymers along with their applications, with microbial producers that exceed more than 250 species, in reality only a very few are being used for commercial production. Microbial biopolymers have been variously studied in *Hyphomicrobacterium* spp. and *Spirillum* spp. that are able to synthesize in limiting conditions of certain nutrients such as carbon. Other microbes used include *Caulobacter crescentus* and *Rhodobacter rubrum* for phosphates, and *Ralstonia eutropha*, *Pseudomonas oleovoran*, and *Alcaligenes latus* for nitrogen [39]. Polymers such as PHA and PHBs are synthesized by various prokaryotes and most frequently studied in *Aeromonas* spp., *Bacillus* spp., *Cupriavidus* spp., and *Pseudomonas* spp., as they are able to utilize inexpensive raw materials that contain carbon as a sole source for microbial growth that is cost effective. The most extensively studied microbe, *Cupriavidus necator* (formerly known as *Waustersia eutrophus*, *Rasstonia eutropha*, and *Alcaligenes eutropha*), is an autotroph that accumulates PHA using $CO_2$ as the sole carbon source which helps in energy creation to carry out the cellular activities. The commercial production of this PHA is being marketed as BiopolTM [40,41]. It has been reported that *Bacillus megaterium*, *Ralstonia eutropha*, *Azotobacter vinelandii*, and *Alcaligenes latus* along with several strains of Methylotrophs had higher PHB production. An estimated 84% production was observed in the case of *Bacillus megaterium* that created interest in these microbes for researchers [42]. Recently, marine bacteria have gained consideration as potentially beneficial candidates for the production of PHAs with the main benefit of biodegradation by anaerobes that form water, methane, and carbon dioxide in the atmosphere like sea, soil, lake water, and sewage, that can be done with eco-friendly disposal in the environment [43].

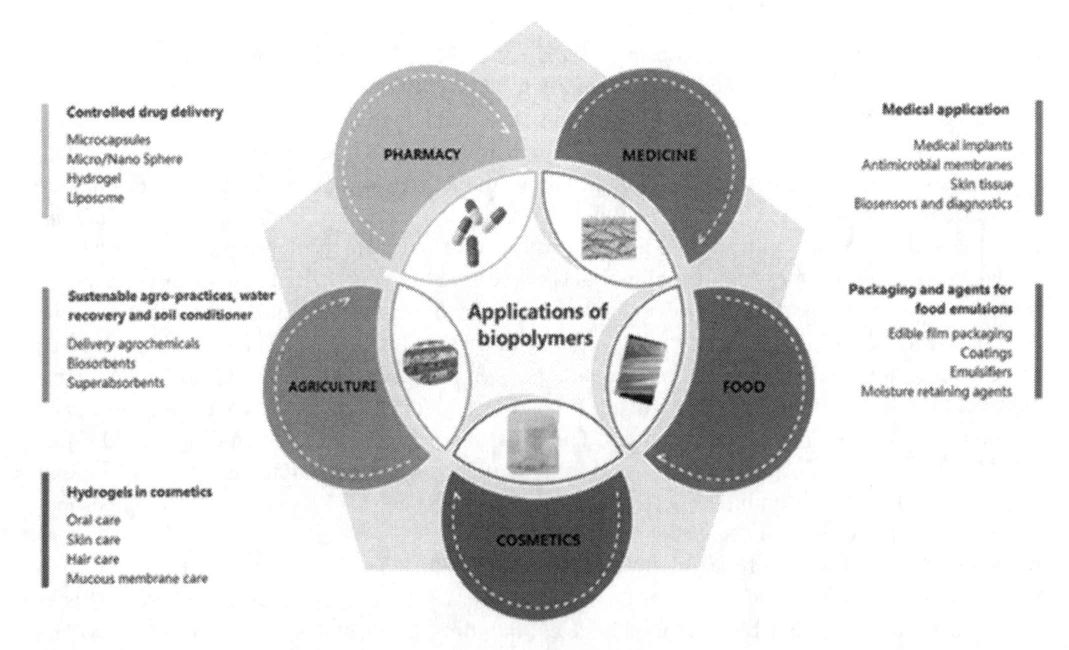


**FIGURE 10.3** Applications of biopolymers in various fields [38].

It is noteworthy that investigation so far for PHA production by a few marine bacteria has efficiently produced biopolymers that have been carried out under marine conditions. However, the mechanism for the production of PHAs needs to be analyzed in marine microbes that have been shown to produce significant biopolymers with efficiency that include *Haloferax*, *Halobacterium*, *Halorubrum*, *Halococcus*, *Haloarcula*, *Natrialba*, *Haloquadratum*, *Haloterrigena*, *Natronococcus*, and *Natronobacterium*. It has been interesting to note that these microbes are found to be efficient producers of PHB [44]. The isolated marine bacteria from sediments off the marine coast coming from the genera *Vibrio* and *Beneckea* were the first to be reported as potential producers of PHA [45]. Other important microbes that synthesize and accumulate PHBs as energy reserves when nitrogen supplies are reduced in the nutrients include *Bacillus megaterium*, *Rhodospirillum rubrum*, *Azotobacter chroococcum*, *Pseudomonas oleovorans*, *Zoogloea ramigera*, and *Alcaligenes eutrophus* [46,47]. It has been well noted that genetically engineered *Escherichia coli* has been found to be suitable for the commercial production of PHA on a large scale [48,49]. The most important aspect relies on the microbial pathway for varying the synthesis of biopolymers. There has been extensive research into these pathways to help the synthesis of the components that are required for green polymers. Furthermore, *Pseudomonas* species isolated from an Antarctic origin helped researchers to understand the mechanism that led to the biosynthesis of PHB along with the role of enzymes and their properties [50,51]. *Pseudomonas mediterranea* has been transformed and characterized for biopolymers mediated by glycerol, which was found to have created a unique capacity with a very filmable polymer as an alternative to the unusual sticky MCL-PHB. It was also remarkable to notice its application in polymeric mixes for medical devices and packaging [52]. PHB is used in the production of several food containers and shampoo bottles; it also provides effective use of dishwashers due to its strong material impact and temperature resistance; and it is also used in disposable utensils, composite glial growth factors, PHB-coated papers, disposable razors, fishing nets, drug carriers, golf tees, and many others [53], as mentioned in Table 10.2.

## 10.4 SUSTAINABILITY

Pollution from non-biodegradable polymers can affected the behavior along with the morphological and physiological characteristics of an individual, with adverse effects on the distribution and abundance of populations, and the association of populations that has further changed the dynamics of ecosystems in both the aquatic and terrestrial milieus [54]. There has been a drastic increase in the use of these synthetic polymers and their material derivatives, which has a very negative influence on the environment due to their corrosive nature, weathering, and deterioration [55–57]. Therefore, the idea of sustainable development has gained influence to guide proper planning and transition in policy making for sustainable development both at the national and international levels on such issues as that of synthetic polymers [58]. Providentially, sustainable polymers are attractively increasing due to the role of the science of industrial ecology, through quantitative life-cycle analysis (LCA) that has enabled us to gain a better understanding of the environmental impacts of synthetic materials. Basically, LCA is defined as the scope of a specific material flows and their impacts along with advanced improvised strategies for a process [59].

Biopolymers are synthesized by plants and microbes by their enzymes that help to build by using blocks like amino acids, sugars, or hydroxy fatty acids to yield high-molecular-weight molecules. Moreover, biopolymers are characterized as thermoplastics with specific characteristics that are analogous to those of the petroleum-based polymers/synthetic polymers [60]. A major factor of biopolymers over synthetic ones is their complete biodegradation that is achieved by bacteria, fungi, and algae yielding carbon dioxide, water, and compost. Biopolymers such as PHA, synthesized by microbes, are biodegradable and can have features similar to synthetic ones with respect to stiffness, brittleness, or being flexible [61]. Subsequently, *Bacillus megaterium* has been used to enable PHA production that has resulted in an upsurge in the study of PHA and other polymers, and

**TABLE 10.2**
**Microbes Involved in Biopolymers and Their Uses**

| Bioplastics | Uses | Microbes |
|---|---|---|
| **Polyglycolic acid (PGA)** | Implantable composites; parts used for bone fixation; controlled drug-release system | *Bacillus pumilus, Laceyella sacchari, Bacillus smithii, Pseudomonas aeruginosa, Sphingobacterium* spp, *Chryseobacterium* spp. |
| **Polylactic acid (PLA)** | Sustainable release vectors for pesticides and fertilizers; bags used for compost and waste bags; packaging in food and other industries; paper coatings | *Amycolatopsis* spp, *Bacillus brevis, Rhizopus delemar, Penicillium roquefort, Pseudomonas geniculate* |
| **Polycaprolactone (PCL)** | Mulch and agricultural films; herbicides with fibers used to control weeds; seedling containers; used as carriers for controlled-release drug systems | *Clostridium botulinum, Firmicutes, Pseudomonas stutzeri, Aspergillus flavus, Aspergillus niger, Rhizopus delemar, Rhizopus arrizus, Fusarium* |
| **Polyhydroxybutyrate (PHB)** | Used as carriers in controlled drug-release systems; disposable nappies; bottles; bags; as tissue engineering scaffold; wrapping films, etc. | *Pseudomonas lemoignei, Penicillium funiculosum, Firmicutes, Schlegelella thermodepolymerans, Streptomyces, Alcaligenes faecalis,* |
| **Polyhydroxyvalerate (PHBV)** | Used in biomedical devices; carriers in sustained-release systems for medical drugs; medicines; insecticides; for drug delivery in cattle | *Clostridium botulinum, Clostridium acetobutylicum, Streptomyces* spp, *Streptoverticillium kashmirense*<br>*Colwellia* spp, *Desulfobacterium autotrophicum, Desulfosarcina variabilis, Desulfosarcina variabilis* |
| **Polyvinyl alcohol (PVOH)** | Packaging materials like laundry detergents; pesticides; hospital washables; applications that dissolve in water to release products | *Pseudomonads, Sphingomonads, Geotrichum fermentans, Penicillium, Sphingopyxis* spp, *Novosphingobium spp, Xanthobacter flavus.* |
| **Polyvinyl acetate (PVAc)** | Adhesives; packaging; paper bags; paper lamination; tube winding | *Geotrichum fermentans, Alcaligenes faecalis, Brevibacillus laterosporus, Geobacillus tepidamans, Bacillus megaterium, Achromobacter cholinophagum, Streptomyces venezuelae.* |

also more than a few hundred other known microorganisms have been shown to generate potential biopolymers [62]. Numerous bacteria have been studied for their polymers, including *Azotobacter vinelandii*, *Alcaligenes latus*, *Pseudomonas oleovorans*, *Herbaspirillum seropedicae*, and *Wautersia eutropha*. These microorganisms were efficient enough for commercial production, due to their fast and easy imitation and growing time that helped to amass a large amount of granules consisting of PHA molecules inside microbial cells, that can account for up to a total biomass of 90% [46, 48].

Biopolymers have been efficaciously produced on a commercial scale in industries, however there are various parameters that need to be controlled, such as the cost of the substrate precursor, yield of the product, with high-volume production, and most importantly, the cost involved in downstream processing for product purification with biological retention with the pure product. There are various techniques involved in the development of sustainable plastics that have been followed successfully on a large scale. Of these, the photodegradable synthetic polymers contain both balanced antioxidants and photodegradation enzymes. These molecules enable them to be produced in controlled conditions along with increased performance efficiency after the photo-initiated degradation stage has occurred [63]. It benefits by compensating the performance in comparison to conventional synthetic polymers due to its cost-effective structures. The drawbacks of oxo-degradable constituents (at present) are the continuous use of non-renewable resources or fossil fuels, that are not able to completely metabolize to form carbon dioxide and water in soil. Polymers that are degradable have been developed that do not contain pro-oxidants or antioxidants, which allows them to undergo slow degradation over a period of time. However, comparable existing advantages and disadvantages have been proposed to photodegradable polymers with reference to the cost-effectiveness, method efficiency, parameters, performance properties, non-renewable resources usage, and degradation products that are formed other than $CO_2$ and $H_2O$, which includes products like esters, alcohols, aldehydes, ketones, and branched alkenes [64]. Ever since, with these drawbacks and no possibility to overcome these issues, increasing interest in both biodegradability (i.e., compostability) and renewable contents to achieve the desirable goals in developing sustainable biopolymers has been encouraged. Biobased and biodegradable polyesters like PHAs are produced aerobically by bacteria from various carbon sources derived from waste substrates that have tunable elastomeric or thermoplastic properties and belong to monomer composition. However, after a great deal of effort, biopolymers need to be studied and researched for their high potential. Research and development require more energy as it is cost-effective fermentative systems which need to be understood properly for production needs.. This hindrance can be minimized in future for the exploitation of these biopolymers as commodity materials. Various approaches are being studied to solve these issues, such as one of the main limitations that can be resolved by cultivation methods using mixed microbial cultures (MMCs) instead of single pure cultures, using cheap biodegradable raw materials as a carbon source instead of chemically selected and pure substrates, and by improvising downstream processing steps that are involved in the extraction and purification processes. This green and sustainable extraction process is an aid for the replacement of toxic and non-environmentally friendly solvents and additives that can have delirious effects on health and the environment. Further, MMC can be genetically manipulated for the production of PHAs and other biopolymers in an even cheaper and more sustainable way [65]. As of now, not much data are available on characterization of the biopolymers extracted from MMCs and on the efficiency of polymer recovery in the extraction steps compared to that from single-strain cultures [66,67]. However, a lot of investigation needs to be carried out on green polymers that is being progressed for a new opening for biopolymer applications.

Bioengineering along with microbiology aims to help improvise the upstream process involving microbes for the production of biopolymers (low-cost substrates and fermentation processes) and the optimization of downstream processes for recovery of the desirable product with minimum cost. Despite various techniques available there remains a lot of research to be carried out on biopolymer production by tracking new generating strains, optimizing culture techniques, and engineering these biopolymers using recombinant microbial strains with a variety of tactics to cut production costs and enhance productivity [68,69]. Biopolymers are essential alternatives to unsustainable synthetic goods because of their biodegradability, environmentally acceptable manufacturing techniques, and extensive application ranges. It becomes of utmost importance to increase efforts in obtaining sustainable resources from microbes like biopolymers and displacing traditional plastics. In turn this will help to reinstate our environment that has been severely damaged by the undiscerning practice of such synthetic polymers and to help prevent further ill effects. Further, we need to work on enhancing the long-term sustainability of materials with production processes; substrate materials used should not compete with our normal food resources consumed by animals and humans that might also cause a reduction in long-term dependency on non-renewable resources.

## 10.5 BIODEGRADATION

The most important aspects of these microbes rely on the biodegrading role of various xenobiotic compounds in the environment. Microbial degradation involving bacteria and fungi on the natural and synthetic plastics has been variously studied [70]. However, microbial degradability depends on the different properties exhibited by them provided by the optimal growth parameters of the environment [71]. Some of the characteristics that are accounted for include both the chemical and physical properties of polymers, specific and consortia microbes involved, and polymer pretreatment [72,73]. These polymers are potential substrates for the heterotrophic microbes that are used as carbon sources to obtain energy. During the process, large polymers are depolymerized to smaller monomers by physical or biological factors to be absorbed, degraded, and mineralized by microbial cells [3,74]. The efficiency of biodegradation can be enhanced by various physical factors such as heating or cooling, freezing or thawing, and wetting or drying processes that illustrate the mechanical damage of these polymers [75]. It has been found that fungal penetration and growth on polymers helps to cause small-scale swelling and bursting [76]. Further, these synthetic polymers are depolymerized with the help of microbial enzymes into monomers that are engrossed into the cells and mineralized [77].

The importance of microbial biodegradation with the role of enzymes has been well documented for synthetic polymers, like polystyrene, polypropylene, polyethylene PE, polyvinyl chloride, PET, and PU. Research shows that the increased molecular weight of the synthetic polymers reduces the effective degradability due to less solubility, making them difficult to assimilate by microbes. Meanwhile, it can be noted that the polymers are made up of monomers, dimers, and oligomers that are in repeating units that are degraded and mineralized by enzymes produced by microbes. These microbial enzymes are unique biocatalysts that play an important role in degrading high-molecular-weight polymers to make them soluble so that they can easily assimilate as substrate through the cell membrane by the microbes and be reduced further by their cellular enzymes. The extracellular and intracellular depolymerases produced by the microbial cells help actively in the biodegradation of synthetic polymers [78], as shown in Figure 10.4.

In microbes, exoenzymes act on the complex polymers, breaking them into smaller molecules like oligomers, dimers, and monomers for recycling. Through semipermeable outer membranes, smaller subunits enter into the microbial cells and are utilized as a sole source of carbon and energy, a progression referred to as depolymerization. The end-products after the process include $CO_2$, $H_2O$, and $CH_4$ with the associated production of other higher-value bioproducts. This biodegradable process is referred to as mineralization [79,80]. Microbial enzymes that play a significant role in the degradation of synthetic polymers involve manganese peroxidase, lipase, laccase, lignin peroxidase, esterase, and amylase and are potential catalysts [81,82]. The three main important microbial enzyme systems that help in degradation are lignin peroxidase, manganese peroxidase, and laccases [83,84]. Lignolytic enzymes further include the phenol oxidase or laccase, and heme peroxidase consists of lignin peroxidase, manganese peroxidase, and versatile peroxidase [83,85].

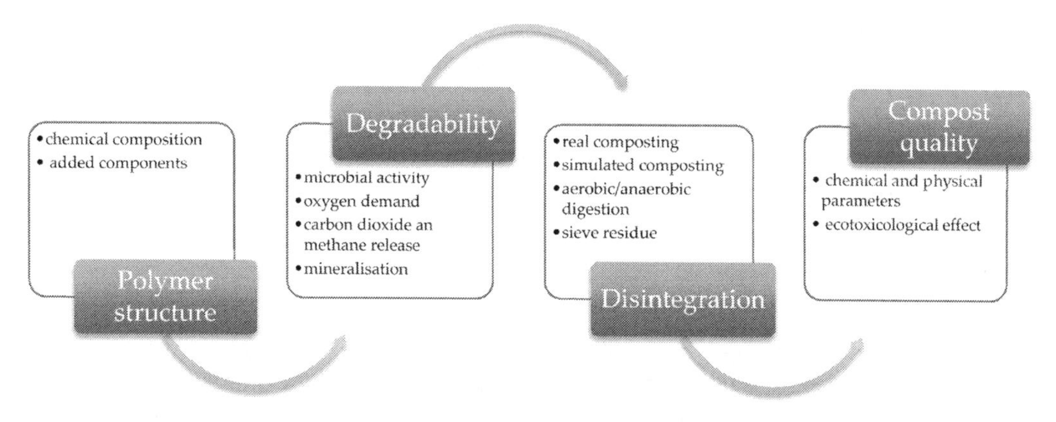


**FIGURE 10.4** Degradation pathways of biodegradable polymers [4].

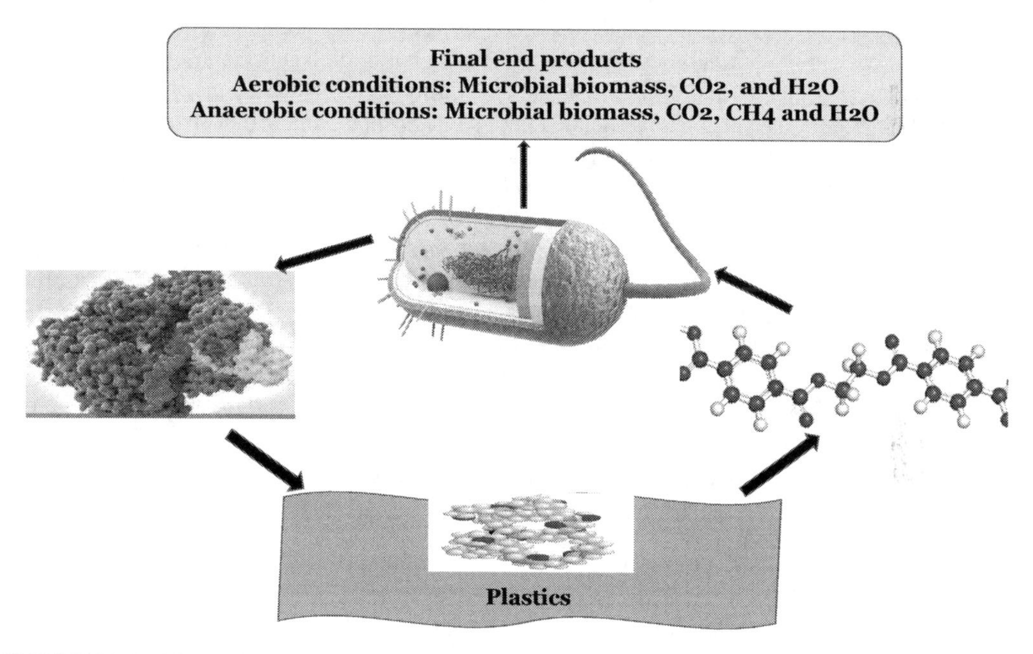


**FIGURE 10.5** Mechanism of enzymatic biodegradation of plastics under aerobic conditions [89].

Biodegradation of polymers involves a very complex method that depends on various parameters that include the surface characteristics of the polymers, substrate availability, morphological characteristics, and the molecular weight of the polymers [86–88]. Further, the process can be measured by a number of other parameters/variables such as weight loss in the substrate, changes to mechanical properties and/or polymer chemical structure, and the carbon dioxide percentage that has been emitted. Previously, microbial degradation endeavored to show that the activity resulted in the modification of physical parameters, such as stretchable strength, water uptake, and crystallinity. However, these microbial degradative pathways are frequently determined by environmental parameters. Under aerobic conditions as shown in Figure 10.5 [89], they utilize oxygen and degrade complex polymers with the formation of water, carbon dioxide, and microbial biomass as the final end-products. However, under anaerobic or anoxic conditions it can be observed that the consortia of anaerobes plays a significant role in microbial degradation with primary products that include the microbial biomass, $CO_2$, $CH_4$, and $H_2O$. These anaerobes are mostly noted to play an important role in landfills or compost with an anaerobic environment mainly consisting of methanogenic bacteria that includes *Methanosarcina barkeri*, *Methanobacterium thermoautotrophicum*, and *Methanobacterium wolfei* [90]. Further, genetic engineering and metabolic pathway studies are a boon that will help to characterize these polymer-degrading microbes with their enzymes, and will provide an opportunity to improve recycling with modification and, thereby, reduce the pollution caused by such synthetic polymers in the environment. Microbes that help in biodegradation are listed in Table 10.3.

## 10.6 WASTE MANAGEMENT

The synthetic polymers are highly resistant to degradation by chemical and physical methods with the process causing discarded polymers to accumulate in the environment. These synthetic polymers in the environment have been a cause of soil, water, and air pollution. Disposal of these polymers has been a major issue as they accumulate in the environment as solid wastes. When discarding such municipal solid wastes, landfilling is the most suitable technique. However, these methods are not sustainable if proper management and treatment methods are not adopted, along with increased recycling and educating communities. Further, increasing pollution has become of utmost importance in the present scenario, and proper disposal and waste management can be achieved with the use of green polymers. The problem associated with solid waste treatment can also be overcome in the environment by replacing synthetic polymers with biodegradable polymers for various products. Green polymers are easily decomposable and waste managing strategies can be applied to control waste through the process of composting. This, in turn, provides a biodegradable material, humus, also referred to as compost. Microbes, like bacterial mesophiles and thermophiles, produce enzymes that catalyze reactions though an oxidation process producing water, methane, carbon dioxide, water, minerals, and compost/humus that is made up of stabilized organic matter. The most remarkable advantage of this process is the removal of pathogens that are killed by heat. For efficacious plastic waste management, a concoction of sludge and solid waste can be operated. During the process of composting, these polymers must disintegrate or degrade in such a way that the remaining polymer is not distinguishable from the other organic matter in the finished compost/manure. The disintegration of synthetic polymer materials or products is considered as acceptable, if the processed compost is found to have not more than 10% of its original weight after decomposition and can easily filter through a 2.0 mm diameter space. The process is measured as efficient when 90% of the test sample is completely able to filter through the pore size of a 2.0 mm sieve when it undergoes filtration [115,116].

Degradation of plastics by microbes was premeditated in the 1980s, making it the most sustainable method for an active setting without any ill effects on humans or the environment. It helped to pave a way for innovative methods for managing waste through novel strategies by the process of biodegrading the xenobiotic compounds like synthetic polymers by utilizing environmental parameters in communities with industrial biological waste treatment facilities [117]. Solid waste treatment in precise conditions can be controlled in anaerobic digesters or composting facilities and has valued techniques for the treatment and recycling of organic waste materials [118]. Currently, there has been an increase in the number of landfilling sites that can contaminate soil and water, hence affecting human health, that could be reduced with the use of biopolymers and biodegradation processes. Composting treatment can help to recycle compostable material through biological treatment; however, its efficiency is reduced due to non-compatible materials that further decrease the quality of the compost and impair its marketable value. Composting is also denoted by various environmental parameters as it requires a very high temperature of about 58 °C under aerobic conditions with moisture (about 50%). The polymer compost is used as a nutrient source as it has various growth factors for the thermophilic microbes that grow and helps in further degradation to smaller molecules. The efficiency of this technique can be determined by the net $CO_2$ produced

**TABLE 10.3**
**List of Microbes Degrading Different Types of Synthetic Polymers**

**A Wide Range of Microbes Found to Be Degrading Various Types of Synthetic Polymers**

| Synthetic Polymers | Microbes Involved in Degradation | References |
|---|---|---|
| **Polyethylene** | *Brevibacillus borstelensis* | 91 |
| | *Rhodococcus ruber* | 92, 93 |
| | *Penicillium simplicissimum YK* | 94 |
| **Polyurethane** | *Comamonas acidovorans* TB-35 | 95 |
| | *Curvularia senegalensis* | 96 |
| | *Fusarium solani*, *Aurebasidium pullulans*, *Cladosporium* spp., *Pseudomonas chlororaphis* | 97 |
| **Poly vinyl chloride** | *Pseudomonas putida* AJ | 98 |
| | *Ochrobactrum* TD, *Pseudomonas fluorescence* B-22, *Asperigillus niger* van Tieghem F-1119 | 99 |
| **Plasticized polyninyl chloride** | *Aureobasidium pullulans* | 100 |
| **BTA—Copolyster** | *Thermomonspora fusca* | 101 |
| **Natural plastics** | | |
| **Poly (3-hydroxybutyrate-CO-3-mercaptopropionate)** | *Schlegelella thermodepolymerans* *Pseudomonas indica* K2 | 102 |
| **Poly (3-hydroxybutyrate)** | *Pseudomonas lemoignei* *Streptomyces* sp. SNG9 | 103, 104 |
| **Poly (3-hydroxybutyrate-CO-3-hydroxy)** | *Streptomyces* sp. SNG9 | 104 |
| **Poly(3-hydroxybutyrate-co-3-hydroxypropionate)** | *Ralstonia pikettii* T1 *Acidovorax* sp. TP4 | 104, 105 |
| **Poly(3-hydroxybutyrate) poly(3-hydroxypropionate) poly(4-hydroxybutyrate) poly (ethylene succinate) poly (ethylene adipate)** | *Alcaligenes faecalis*, *Pseudomonas stutzeri*, *Comamonas acidovorans* | 106 |
| **Poly (3-hydroxybutyrate)** | *Alcaligenes faecalis*, *Schlegelella thermodepolymerans*, *Caenibacterium thermophilum* | 107, 108 |
| **Poly(3-hydroxybutyrate-CO-3-hydroxyvalerate)** | *Clostridium botulinum*, *Clostridium acetobutylicum* | 109 |
| **Polycaprolactone** | *Clostridium botulinum*, *Clostridium acetobutylicum*, *Fusarium solani* | 109, 110 |
| **Polylactic acid** | *Fusarium moniliforme*, *Penicillium roqueforti*, *Amycolatopsis* sp., *Bacillus brevis*, *Rhizopus delemer* | 112–114 |

without toxins in comparison to the compost of polymer concoction minus $CO_2$ evolved from the unamended compost (blank) tested in various reactors [119].

## 10.7 RECYCLING

Geyer et al. [120] analyzed plastics and reported on the cumulative wastes that had been primarily produced and recycled between 1950–2015 globally, that had reached 6300 million tons and an estimated 79% of this was landfilled and ended up in the natural habitat. However, synthetic polymers that have been traditionally produced using older methods with fossil oil are rarely closed-loop recycled [20]. These synthetic polymers are not sustainable and not biodegradable within a short period of time and further, they cause delirious effects to humans and ecosystems [121]. However, this also led to increasing ill effects on human health and the environment, therefore the challenging difficulties faced by synthetic polymers for proper disposal and recycling needed to be addressed.

Polymer production uses about 4% of the total worldwide oil as a feedstock and over a third of this is used by packaging industries, with the products being rapidly discarded. With declining fossil fuels and limited landfills for waste disposal this makes synthetic polymer applications an important priority as they are simply not sustainable [122]. With the increasing population, use of easily produced, more available, affordable synthetic polymers has been directed to the massive accumulation of tons of waste in the form of non-biodegradable polymers in the environment. It takes many decades to degrade or decompose these conventional polymers and, even worse, the toxin production has an immense effect on the health of humans and animals, and also on the ecosystem. These encountered problems can be counteracted by using biopolymeric substances that can be willingly eliminated and degraded from the environment in an eco-friendly way [123]. Synthetic degradation is very slow process in nature that depends on the various parameters and the type and role of wild microbes (124–126). However, these plastics can be a renewable resource with the help of microbes, as depicted in Figure 10.6 [127].

Most synthetic polymers, such as aliphatic polyesters, poly(vinyl alcohol), poly(lactic acid), polycaprolactone, and polyamides have been recycled with the help of microorganisms and their enzymes. There are several oligomeric components that are biodegradable including styrene, isoprene, butadiene, oligomeric ethylene, acrylonitrile, and acrylate [128]. Synthetic polymers and waste management can be controlled by material reduction, design for end-of-life recyclability, increased recycling capacity, and the development of biobased feedstocks, strategies to reduce

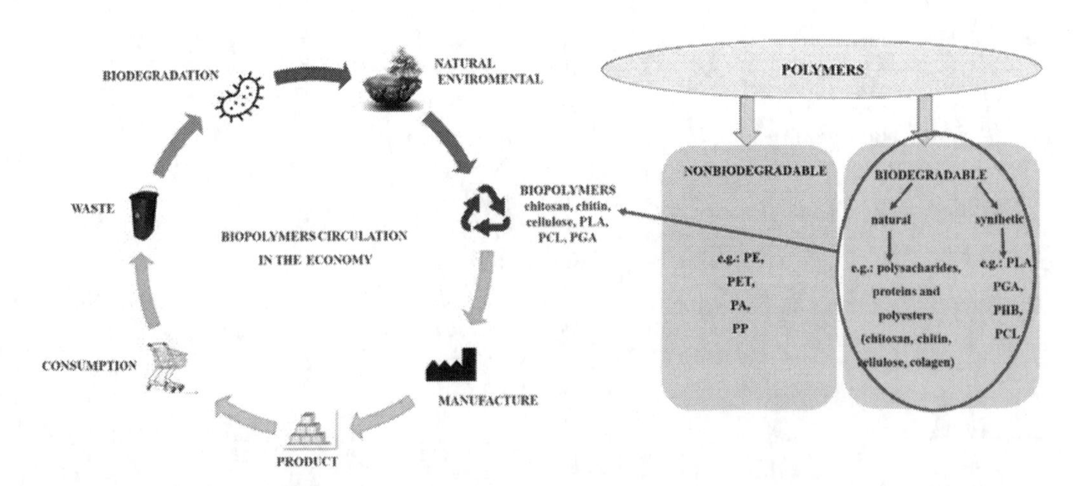


**FIGURE 10.6** Life cycle of biodegradable polymers within the context of a circular economy [127].

littering, and applications in the form of green chemistry, life-cycle analysis, and revised risk assessment and management approaches [122].

The role of microbial degradation of synthetic plastics or even bioplastics has been a boon to the environment that helps to recycle these products. Since these biopolymers are made of biomass involving natural products like corn and sugarcane they are eco-friendly, save fossil fuels, reduce $CO_2$ emissions, and degrade compounds from plastic wastes. In recent years, biopolymers have increased in importance and have rapidly increased their use in food industries. Of these polymers, PHAs have gained greatest acceptance, as the microbial production is based on renewable resources that are agricultural substrates providing fatty acids and sugars as a source of carbon and energy for microbial growth [129]. PHAs and other biopolymers are synthesized, the biodegradation of such polymers is completely compatible with the carbon cycle, and they are based on renewable compounds [130]. There are various other factors that play a very key role in the polymers biodegradation. These factors include the chemical constituents, morphological characteristics, molecular weight, mechanical attributes of these biocomposites, and the techniques that are used for processing, which all play important roles that are dependent on the biopolymer parts of the composites [131]. These biopolymers are specifically employed in fields that depend on the cost, availability, moisture absorption, biocompatibility, and degradation stability [132].

## 10.8 FUTURE CHALLENGES

The production of biopolymers by microorganisms is a challenging process due to the high viscosity that is present in the culture media where the diffusion rate is reduced due to ATP and dissolved oxygen production. Hence, there is a need to strategize to develop anaerobic conditions tolerable by these microorganisms or to enhance the systems that generate energy and help to improve the productivity rate. Biopolymers can be produced by non-pathogens or Generally Recognized as Safe (GRAS) microbes that are generally considered as safe products for an extensive range of applications or products. Despite there having been a greater advancement in the production of biopolymers as well as that of tailored biopolymers, some challenges remain to be worked on. There has been an overabundance of interacting compounds with numerous feedback loops in microbial cells that are complex, however the use of genetically engineered novel "GRAS microbes" that are certified by the Food and Drug Administration (FDA) to be considered as cell factories for biopolymers remains a challenge.

The complexity can be reduced by the use of system biology to help improve the metabolic genome models, networks in metabolic modeling, and large data sets for computational simulations that help with further innovative work into synthetic biology approaches. In the future, the computational biology approach will pave a way to provide us with better information and technology for efficient genetic engineering strategies for an accurate cell factory for the developmental processes.

## 10.9 CONCLUSION

Currently, we are living in a milieu that is affecting health and the environment due to very high amounts of pollutants from various sources, with one of the highest pollutants in the present scenario being synthetic polymers. Their antagonistic effects have led to the development of suitable, eco-friendly, economical, recyclable substitutes that have directed the synthesis and use of biopolymers, a suitable remedial approach towards pollution caused by the petroleum origin of synthetic polymers. The technology involving recombinant DNA and genetic engineering has exponentially helped us to use microbes and their enzyme systems for their novel degrading properties. Recombinant microbes producing novel green polymers are a need, wherein these technologies do not impose limitations in the biosynthetic pathways as in the original strains. Novel microorganisms with special characteristics, metabolic pathways, and enzyme systems play an important role in the milieu for the production of different biopolymers, with vast applications and also carry out their degradation. Nevertheless, their significant role in the biodegradation and recycling of synthetic polymers, with an extremely promising future, should not to be forgotten. Furthermore, investigations need to be carried out to reduce the production costs and make them available at an affordable price that can compete with synthetic polymers. An utmost prerequisite for all this will be renewable resources or utilizing $CO_2$ directly as the main carbon source for biopolymer synthesis using microorganisms. The discovery of novel biopolymers with microbial sources will surely help to exhibit features not shown by synthetic polymers, which will be of prime importance and extra value in the coming years.

## REFERENCES

1. Fried, J. R. (2014). *Polymer science and technology*. Pearson Education.
2. Plastics Europe. (2017). *Plastics—The Facts 2017: An Analysis of European Plastics Production, Demand and Waste Data 2017*. Plastics Europe.
3. Shah, A. A., Hasan, F., Hameed, A., & Ahmed, S. (2008). Biological degradation of plastics: a comprehensive review. *Biotechnology advances*, *26*(3), 246–265.
4. Siracusa, V. (2019). Microbial degradation of synthetic biopolymers waste. *Polymers*, *11*(6), 1066.
5. Barnes, D. K., Galgani, F., Thompson, R. C., & Barlaz, M. (2009). Accumulation and fragmentation of plastic debris in global environments. *Philosophical transactions of the royal society B: biological sciences*, *364*(1526), 1985–1998.
6. Brown, M. E., & Gallagher, P. K. (2011). Handbook of thermal analysis and calorimetry: recent advances, techniques and applications.
7. Khodaei, D., Álvarez, C., & Mullen, A. M. (2021). Biodegradable packaging materials from animal processing co-products and wastes: an overview. *Polymers*, *13*(15), 2561.
8. Madison, L. L., & Huisman, G. W. (1999). Metabolic engineering of poly (3-hydroxyalkanoates): from DNA to plastic. *Microbiology and molecular biology reviews*, *63*(1), 21–53.
9. Barnard, G. N., & Sanders, J. K. (1989). The poly-β-hydroxybutyrate granule in vivo: a new insight based on NMR spectroscopy of whole cells. *Journal of biological chemistry*, *264*(6), 3286–3291.
10. Sudesh, K., Abe, H., & Doi, Y. (2000). Synthesis, structure and properties of polyhydroxyalkanoates: biological polyesters. *Progress in polymer science*, *25*(10), 1503–1555.
11. Lee, S. Y., & Choi, J. (1999). Polyhydroxyalkanoates: biodegradable polymer. *Manual of industrial microbiology and biotechnology*, *2*, 616–627.
12. Steinbüchel, A., & Füchtenbusch, B. (1998). Bacterial and other biological systems for polyester production. *Trends in biotechnology*, *16*(10), 419–427.
13. Zinn, M., Witholt, B., & Egli, T. (2001). Occurrence, synthesis and medical application of bacterial polyhydroxyalkanoate. *Advanced drug delivery reviews*, *53*(1), 5–21.
14. Witholt, B., & Kessler, B. (1999). Perspectives of medium chain length poly (hydroxyalkanoates), a versatile set of bacterial bioplastics. *Current opinion in biotechnology*, *10*(3), 279–285.
15. Lemoigne, M. (1926). Products of dehydration and of polymerization of β-hydroxybutyric acid. *Bull Soc Chem Biol*, *8*, 770–782.
16. Masnadi, M. S., El-Houjeiri, H. M., Schunack, D., Li, Y., Englander, J. G., Badahdah, A., ... & Brandt, A. R. (2018). Global carbon intensity of crude oil production. *Science*, *361*(6405), 851–853.
17. Chanprateep, S. (2010). Current trends in biodegradable polyhydroxyalkanoates. *Journal of bioscience and bioengineering*, *110*(6), 621–632.
18. Keshavarz, T., & Roy, I. (2010). Polyhydroxyalkanoates: bioplastics with a green agenda. *Current opinion in microbiology*, *13*(3), 321–326.
19. Suriyamongkol, P., Weselake, R., Narine, S., Moloney, M., & Shah, S. (2007). Biotechnological approaches for the production of polyhydroxyalkanoates in microorganisms and plants—a review. *Biotechnology advances*, *25*(2), 148–175.
20. MacArthur, E., & Waughray, D. (2016). *Intelligent Assets: Unlocking the Circular Economy Potential.* Cowes, UK: Ellen MacArthur Foundation, 1–35.
21. Lendlein, A., Behl, M., Hiebl, B., & Wischke, C. (2010). Shape-memory polymers as a technology platform for biomedical applications. *Expert review of medical devices*, *7*(3), 357–379.

22. Serrano, M. C., & Ameer, G. A. (2012). Recent insights into the biomedical applications of shape-memory polymers. *Macromolecular bioscience*, *12*(9), 1156–1171.
23. Halden, R. U. (2010). Plastics and health risks. *Annual review of public health*, *31*, 179–194.
24. Souhrada, L. (1988). Reusables revisited as medical waste adds up. *Hospitals*, *62*(20), 82.
25. Ojeda, T. (2013). Polymers and the Environment. *Polymer Science*, *23*, 1–34.
26. Rahman, M., & Brazel, C. S. (2004). The plasticizer market: an assessment of traditional plasticizers and research trends to meet new challenges. *Progress in polymer science*, *29*(12), 1223–1248.
27. Dodds, E. C., & Lawson, W. (1936). Synthetic strogenic agents without the phenanthrene nucleus. *Nature*, *137*(3476), 996.
28. Wilson, N. K., Chuang, J. C., Morgan, M. K., Lordo, R. A., & Sheldon, L. S. (2007). An observational study of the potential exposures of preschool children to pentachlorophenol, bisphenol-A, and nonylphenol at home and daycare. *Environmental research*, *103*(1), 9–20.
29. Vandenberg, L. N., Hauser, R., Marcus, M., Olea, N., & Welshons, W. V. (2007). Human exposure to bisphenol A (BPA). *Reproductive toxicology*, *24*(2), 139–177.
30. Kamrin, M. A. (2009). Phthalate risks, phthalate regulation, and public health: a review. *Journal of Toxicology and Environmental Health, Part B*, *12*(2), 157–174.
31. Meeker, J. D., Sathyanarayana, S., & Swan, S. H. (2009). Phthalates and other additives in plastics: human exposure and associated health outcomes. *Philosophical transactions of the royal society B: biological sciences*, *364*(1526), 2097–2113.
32. Sathyanarayana, S., Karr, C. J., Lozano, P., Brown, E., Calafat, A. M., Liu, F., & Swan, S. H. (2008). Baby care products: possible sources of infant phthalate exposure. *Pediatrics*, *121*(2), e260–e268.
33. Biello, D. (2011). Are Biodegradable Plastics Doing More Harm Than Good? *Scientific American*.
34. Hocking, P. J., & Marchessault, R. H. (1994). *Chemistry and Technology of Biodegradable Polymers*, G. J. L. Griffin (Ed.). Blackie Academic & Professional.
35. Steinbüchel, A., & Füchtenbusch, B. (1998). Bacterial and other biological systems for polyester production. *Trends in biotechnology*, *16*(10), 419–427.
36. Kessler, B., & Witholt, B. (2002). Poly (3-Hydroxyalkanoates). In Kessler B, Witholt B (eds) *Encyclopedia of Bioprocess Technology: Fermentation, Biocatalysis, and Bioseparation*. Wiley. Online ISBN: 9780471250586.
37. Agus, J., Kahar, P., Abe, H., Doi, Y., & Tsuge, T. (2006). Altered expression of polyhydroxyalkanoate synthase gene and its effect on poly [(R)-3-hydroxybutyrate] synthesis in recombinant *Escherichia coli*. *Polymer degradation and stability*, *91*(8), 1645–1650.
38. Gheorghita, R., Anchidin-Norocel, L., Filip, R., Dimian, M., & Covasa, M. (2021). Applications of biopolymers for drugs and probiotics delivery. *Polymers*, *13*(16), 2729.
39. Kim, M. N., Lee, B. Y., Lee, I. M., Lee, H. S., & Yoon, J. S. (2001). Toxicity and biodegradation of products from polyester hydrolysis. *Journal of Environmental Science and Health, Part A*, *36*(4), 447–463.
40. Schubert, P., Steinbüchel, A., & Schlegel, H. G. (1988). Cloning of the *Alcaligenes eutrophus* genes for synthesis of poly-beta-hydroxybutyric acid (PHB) and synthesis of PHB in *Escherichia coli*. *Journal of bacteriology*, *170*(12), 5837–5847.
41. Luzier, W. D. (1992). Materials derived from biomass/biodegradable materials. *Proceedings of the National Academy of Sciences*, *89*(3), 839–842.
42. Füchtenbusch, B., Wullbrandt, D., & Steinbüchel, A. (2000). Production of polyhydroxyalkanoic acids by *Ralstonia eutropha* and *Pseudomonas oleovorans* from an oil remaining from biotechnological rhamnose production. *Applied microbiology and biotechnology*, *53*(2), 167–172.
43. Prasanna, T., Babu, P. A., Lakshmi, P. D., Chakrapani, R., & Rao, C. S. V. R. (2011). Production of poly (3-hydroxybutyrates) by *Bacillus* species isolated form soil. *Journal of Pharmaceutical Review and Research*, *1*, 15–18.
44. Brandl, H., Gross, R. A., Lenz, R. W., & Fuller, R. C. (1988). Pseudomonas oleovorans as a source of poly (β-hydroxyalkanoates) for potential applications as biodegradable polyesters. *Applied and environmental microbiology*, *54*(8), 1977–1982.
45. Poli, A., Di Donato, P., Abbamondi, G. R., & Nicolaus, B. (2011). Synthesis, production, and biotechnological applications of exopolysaccharides and polyhydroxyalkanoates by archaea. *Archaea*, *2011*, 1–13.
46. Lee, S. Y. (1996a). Bacterial polyhydroxyalkanoates. *Biotechnology and bioengineering*, *49*(1), 1–14.

47. Kim, B. S., & Chang, H. N. (1998). Production of poly (3-hydroxybutyrate) from starch by *Azotobacter chroococcum*. *Biotechnology letters*, *20*(2), 109–112.
48. Lee, S. Y. (1996b). Plastic bacteria? Progress and prospects for polyhydroxyalkanoate production in bacteria. *Trends in biotechnology*, *14*(11), 431–438.
49. Lee, S. Y., & Chang, H. N. (1995). Production of poly (hydroxyalkanoic acid). In M. Hiroto, H. Nishimura, Y. Kodera, F. Kawai, P. Rusin, H. L. Ehrlich, S. Y. Lee, et al. (eds) *Microbial and Enzymatic Bioproducts*, 27–58. Springer.
50. Ayub, N. D., Pettinari, M. J., Ruiz, J. A., & López, N. I. (2004). A polyhydroxybutyrate-producing *Pseudomonas* sp. isolated from Antarctic environments with high stress resistance. *Current microbiology*, *49*(3), 170–174.
51. López, N. I., Pettinari, M. J., Stackebrandt, E., Tribelli, P. M., Põtter, M., Steinbüchel, A., & Méndez, B. S. (2009). *Pseudomonas extremaustralis* sp. nov., a poly (3-hydroxybutyrate) producer isolated from an Antarctic environment. *Current microbiology*, *59*(5), 514–519.
52. Pappalardo, F., Fragalà, M., Mineo, P. G., Damigella, A., Catara, A. F., Palmeri, R., & Rescifina, A. (2014). Production of filmable medium-chain-length polyhydroxyalkanoates produced from glycerol by *Pseudomonas mediterranea*. *International journal of biological macromolecules*, *65*, 89–96.
53. Kumaravel, S., Hema, R., & Lakshmi, R. (2010). Production of polyhydroxybutyrate (bioplastic) and its biodegradation by *Pseudomonas lemoignei* and *Aspergillus niger*. *E-journal of Chemistry*, *7*(S1), S536–S542.
54. Chaukura, N., Kefeni, K. K., Chikurunhe, I., Nyambiya, I., Gwenzi, W., Moyo, W., ... & Abulude, F. O. (2021). Microplastics in the aquatic environment—the occurrence, sources, ecological impacts, fate, and remediation challenges. *Pollutants*, *1*(2), 95–118.
55. Kakar, A., Jayamani, E., Bakri, M. K. B., & Rahman, M. R. (2018a). Biomedical and packaging application of silica and various clay dispersed nanocomposites. In Kakar, Akshay, Elammaran Jayamani, Muhammad Khusairy Bin Bakri, and Md. Rezaur Rahman (eds) *Silica and Clay Dispersed Polymer Nanocomposites* (pp. 109–136). Woodhead.
56. Kakar, A., Jayamani, E., Bakri, M. K. B., & Rahman, M. R. (2018b). Durability and sustainability of the silica and clay and its nanocomposites. In Kakar, Akshay, Elammaran Jayamani, Muhammad Khusairy Bin Bakri, and Md. Rezaur Rahman (eds) *Silica and Clay Dispersed Polymer Nanocomposites* (pp. 137–157). Woodhead.
57. Webb, H. K., Arnott, J., Crawford, R. J., & Ivanova, E. P. (2013). Plastic degradation and its environmental implications with special reference to poly (ethylene terephthalate). *Polymers*, *5*(1), 1–18.
58. Bakri, M. K., Omoregie, A., & Rahman, M. R. (2021). Environmental sustainability of biopolymers. *Academia letters*, Article 2924, 1–7.
59. Perlack, R. D., Wright, L. L., Turhollow, A. F., Graham, R. L., Stokes, B. J., & Erbach, D. C. (2005). *Biomass as Feedstock for a Bioenergy and Bioproducts Industry: The Technical Feasibility of a Billion-Ton Annual Supply*. US Department of Energy and US Department of Agriculture. Washington, DC: US Department of Energy.
60. Rahman, M. R., Taib, N. A. A. B., Bakri, M. K. B., & Taib, S. N. L. (2021). Importance of sustainable polymers for modern society and development. In Rahman, Md. Rezaur, Nur-Azzah Afifah Binti Taib, Muhammad Khusairy Bin Bakri, and Siti Noor Linda Taib (eds) *Advances in Sustainable Polymer Composites* (pp. 1–35). Woodhead.
61. Bin Bakri, M. K., Jayamani, E., & Kakar, A. (2018). Potential in the Development of Borneo Acacia Wood Reinforced Polyhydroxyalkanoates Bio-Composites. In Bakri, Muhammad Khusairy bin, Elammaran Jayamani, and Akshay Kakar (eds) *Key Engineering Materials* (Vol. 779, pp. 19–24). Trans Tech.
62. Mohapatra, S., Maity, S., Dash, H. R., Das, S., Pattnaik, S., Rath, C. C., & Samantaray, D. (2017). Bacillus and biopolymer: prospects and challenges. *Biochemistry and biophysics reports*, *12*, 206–213.
63. Graham, H. A. H., Decoteau, D. R., & Linvill, D. E. (1995). Development of a polyethylene mulch system that changes color in the field. *HortScience*, *30*(2), 265–269.
64. Khabbaz, F., Albertsson, A. C., & Karlsson, S. (1999). Chemical and morphological changes of environmentally degradable polyethylene films exposed to thermo-oxidation. *Polymer degradation and stability*, *63*(1), 127–138.
65. Salehizadeh, H., & Van Loosdrecht, M. C. M. (2004). Production of polyhydroxyalkanoates by mixed culture: recent trends and biotechnological importance. *Biotechnology advances*, *22*(3), 261–279.

66. Jacquel, N., Lo, C. W., Wei, Y. H., Wu, H. S., & Wang, S. S. (2008). Isolation and purification of bacterial poly (3-hydroxyalkanoates). *Biochemical engineering journal*, *39*(1), 15–27.
67. Serafim, L. S., Lemos, P. C., Torres, C., Reis, M. A., & Ramos, A. M. (2008). The influence of process parameters on the characteristics of polyhydroxyalkanoates produced by mixed cultures. *Macromolecular bioscience*, *8*(4), 355–366.
68. McAdam, B., Brennan Fournet, M., McDonald, P., & Mojicevic, M. (2020). Production of polyhydroxybutyrate (PHB) and factors impacting its chemical and mechanical characteristics. *Polymers*, *12*(12), 2908.
69. Surendran, A., Lakshmanan, M., Chee, J. Y., Sulaiman, A. M., Thuoc, D. V., & Sudesh, K. (2020). Can polyhydroxyalkanoates be produced efficiently from waste plant and animal oils?. *Frontiers in bioengineering and biotechnology*, *8*, 169.
70. Gu, J. D., Ford, T. E., Mitton, D. B., & Mitchell, R. (2000). Microbial corrosion of metals. In R. Winston Revie (ed.) *The Uhlig Corrosion Handbook* (2nd Edition, pp. 915–927). Wiley.
71. Pauli, N. C., Petermann, J. S., Lott, C., & Weber, M. (2017). Macrofouling communities and the degradation of plastic bags in the sea: an in situ experiment. *Royal Society open science*, *4*(10), 170549.
72. Artham, T., & Doble, M. (2008). Biodegradation of aliphatic and aromatic polycarbonates. *Macromolecular bioscience*, *8*(1), 14–24.
73. Gu, J. D. (2003). Microbiological deterioration and degradation of synthetic polymeric materials: recent research advances. *International biodeterioration & biodegradation*, *52*(2), 69–91.
74. Kolvenbach, B. A., Helbling, D. E., Kohler, H. P. E., & Corvini, P. F. (2014). Emerging chemicals and the evolution of biodegradation capacities and pathways in bacteria. *Current opinion in biotechnology*, *27*, 8–14.
75. Kamal, M., & Huang, B. (1992). Natural and artificial weathering of polymers. In *Handbook of Polymer Degradation*, Hamid, S. H., Amin, M. B., & Maadh, A. G. (Eds.) (pp. 127–168). Marcel Dekker.
76. Griffin, G. J. L. (1980). Synthetic polymers and the living environment. *Pure and applied chemistry*, *52*(2), 399–407.
77. Goldberg, D. (1995). A review of the biodegradability and utility of poly (caprolactone). *Journal of environmental polymer degradation*, *3*(2), 61–67.
78. Pathak, V. M. & Navneet (2017). Review on the current status of polymer degradation: a microbial approach. *Bioresources and Bioprocessing*, *4*, 15.
79. Grima, S., Bellon-Maurel, V., Feuilloley, P., & Silvestre, F. (2000). Aerobic biodegradation of polymers in solid-state conditions: a review of environmental and physicochemical parameter settings in laboratory simulations. *Journal of Polymers and the Environment*, *8*(4), 183–195.
80. Montazer, Z., Habibi Najafi, M. B., & Levin, D. B. (2020). Challenges with verifying microbial degradation of polyethylene. *Polymers*, *12*(1), 123.
81. Ganesh, P., Dineshraj, D., & Yoganathan, K. (2017). Production and screening of depolymerising enymes by potential bacteria and fungi isolated from plastic waste dump yard sites. *International journal applied research*, *3*, 693–695.
82. Matsumura, S. (2005). Mechanism of biodegradation S Matsumura, Keio University, Japan. *Biodegradable polymers for industrial applications*, *357*, 3–31.
83. Bhardwaj, H., Gupta, R., & Tiwari, A. (2012). Microbial population associated with plastic degradation. *Scientific reports*, *5*, 272–274.
84. Hofrichter, M., Lundell, T., & Hatakka, A. (2001). Conversion of milled pine wood by manganese peroxidase from *Phlebia radiata*. *Applied and Environmental Microbiology*, *67*(10), 4588–4593.
85. Dashtban, M., Schraft, H., Syed, T. A., & Qin, W. (2010). Fungal biodegradation and enzymatic modification of lignin. *International journal of biochemistry and molecular biology*, *1*(1), 36.
86. Ammala, A., Bateman, S., Dean, K., Petinakis, E., Sangwan, P., Wong, S., Yuan, Q., Yu, L., Patrick, C. & Leong, K.H. (2011). An overview of degradable and biodegradable polyolefins. *Progress in Polymer Science*, *36*, 1015–1049.
87. Albertsson, A. C., Andersson, S. O., & Karlsson, S. (1987). The mechanism of biodegradation of polyethylene. *Polymer degradation and stability*, *18*(1), 73–87.
88. Harrison, J. P., Boardman, C., O'Callaghan, K., Delort, A. M., & Song, J. (2018). Biodegradability standards for carrier bags and plastic films in aquatic environments: a critical review. *Royal Society open science*, *5*(5), 171792.

89. Teixeira, S., Eblagon, K. M., Miranda, F., R Pereira, M. F., & Figueiredo, J. L. (2021). Towards controlled degradation of poly (lactic) acid in technical applications. *C*, *7*(2), 42.
90. Barlaz, M. A., Schaefer, D. M., & Ham, R. (1989). Bacterial population development and chemical characteristics of refuse decomposition in a simulated sanitary landfill. *Applied and environmental microbiology*, *55*(1), 55–65.
91. Hadad, D., Geresh, S., & Sivan, A. (2005). Biodegradation of polyethylene by the thermophilic bacterium *Brevibacillus borstelensis*. *Journal of applied microbiology*, *98*(5), 1093–1100.
92. Hadar, Y., & Sivan, A. (2004). Colonization, biofilm formation and biodegradation of polyethylene by a strain of *Rhodococcus ruber*. *Applied microbiology and biotechnology*, *65*(1), 97.
93. Sivan, A., Szanto, M., & Pavlov, V. (2006). Biofilm development of the polyethylene-degrading bacterium *Rhodococcus ruber*. *Applied microbiology and biotechnology*, *72*(2), 346–352.
94. Yamada-Onodera, K., Mukumoto, H., Katsuyaya, Y., Saiganji, A., & Tani, Y. (2001). Degradation of polyethylene by a fungus, *Penicillium simplicissimum* YK. *Polymer degradation and stability*, *72*(2), 323–327.
95. Akutsu, Y., Nakajima-Kambe, T., Nomura, N., & Nakahara, T. (1998). Purification and properties of a polyester polyurethane-degrading enzyme from *Comamonas acidovorans* TB-35. *Applied and Environmental Microbiology*, *64*(1), 62–67.
96. Howard, G. T. (2002). Biodegradation of polyurethane: a review. *International Biodeterioration & Biodegradation*, *49*(4), 245–252.
97. Zheng, Y., Yanful, E. K., & Bassi, A. S. (2005). A review of plastic waste biodegradation. *Critical reviews in biotechnology*, *25*(4), 243–250.
98. Danko, A. S., Luo, M., Bagwell, C. E., Brigmon, R. L., & Freedman, D. L. (2004). Involvement of linear plasmids in aerobic biodegradation of vinyl chloride. *Applied and environmental microbiology*, *70*(10), 6092–6097.
99. Mogil'nitskii, G. M., Sagatelyan, R. T., Kutishcheva, T. N., Zhukova, S. V., Kerimov, S. I., & Parfenova, T. B. (1987). Disruption of the protective properties of the polyvinyl chloride coating under the effect of microorganisms. *Protection of Metals and Physical Chemistry of Surfaces (Engl. Transl.);(United States)*, *23*(1), 173–175.
100. Webb, J. S., Nixon, M., Eastwood, I. M., Greenhalgh, M., Robson, G. D., & Handley, P. S. (2000). Fungal colonization and biodeterioration of plasticized polyvinyl chloride. *Applied and environmental microbiology*, *66*(8), 3194–3200.
101. Kleeberg, I., Hetz, C., Kroppenstedt, R. M., Müller, R. J., & Deckwer, W. D. (1998). Biodegradation of aliphatic-aromatic copolyesters by *Thermomonospora fusca* and other thermophilic compost isolates. *Applied and Environmental Microbiology*, *64*(5), 1731–1735.
102. Elbanna, K., Lütke-Eversloh, T., Jendrossek, D., Luftmann, H., & Steinbüchel, A. (2004). Studies on the biodegradability of polythioester copolymers and homopolymers by polyhydroxyalkanoate (PHA)-degrading bacteria and PHA depolymerases. *Archives of microbiology*, *182*(2), 212–225.
103. Jendrossek, D., Frisse, A., Andermann, M., Kratzin, H. D., Stanislawski, T., Schlegel, H. G. (1995). Biochemical and molecular characterization of the *Pseudomonas lemoignei* polyhydroxyalkanoate depolymerase system. *Journal of Bacteriology*, *177*(3), 596–607.
104. Mabrouk, M. M., & Sabry, S. A. (2001). Degradation of poly (3-hydroxybutyrate) and its copolymer poly (3-hydroxybutyrate-co-3-hydroxyvalerate) by a marine *Streptomyces* sp. SNG9. *Microbiological research*, *156*(4), 323–335.
105. Wang, Y., Inagawa, Y., Saito, T., Kasuya, K. I., Doi, Y., & Inoue, Y. (2002). Enzymatic hydrolysis of bacterial poly (3-hydroxybutyrate-co-3-hydroxypropionate)s by poly (3-hydroxyalkanoate) depolymerase from *Acidovorax* sp. TP4. *Biomacromolecules*, *3*(4), 828–834.
106. Kasuya, T., Nakajima, H., & Kitamoto, K. (1999). Cloning and characterization of the bipA gene encoding ER chaperone BiP from *Aspergillus oryzae*. *Journal of bioscience and bioengineering*, *88*(5), 472–478.
107. Kita, K., Mashiba, S. I., Nagita, M., Ishimaru, K., Okamoto, K., Yanase, H., & Kato, N. (1997). Cloning of poly (3-hydroxybutyrate) depolymerase from a marine bacterium, *Alcaligenes faecalis* AE122, and characterization of its gene product. *Biochimica et Biophysica Acta (BBA)-Gene Structure and Expression*, *1352*(1), 113–122.

108. Romen, F., Reinhardt, S., & Jendrossek, D. (2004). Thermotolerant poly (3-hydroxybutyrate)-degrading bacteria from hot compost and characterization of the PHB depolymerase of *Schlegelella* sp. KB1a. *Archives of microbiology*, *182*(2), 157–164.
109. Abou-Zeid, D. M., Müller, R. J., & Deckwer, W. D. (2001). Degradation of natural and synthetic polyesters under anaerobic conditions. *Journal of biotechnology*, *86*(2), 113–126.
110. Benedict, C. V., Cameron, J. A., & Huang, S. J. (1983). Polycaprolactone degradation by mixed and pure cultures of bacteria and a yeast. *Journal of Applied Polymer Science*, *28*(1), 335–342.
111. Pranamuda, H., & Tokiwa, Y. (1999). Degradation of poly (L-lactide) by strains belonging to genus Amycolatopsis. *Biotechnology letters*, *21*(10), 901–905.
112. Pranamuda, H., Tokiwa, Y., & Tanaka, H. (1997). Polylactide degradation by an *Amycolatopsis* sp. *Applied and environmental microbiology*, *63*(4), 1637–1640.
113. Tomita, K., Kuroki, Y., & Nagai, K. (1999). Isolation of thermophiles degrading poly (L-lactic acid). *Journal of bioscience and bioengineering*, *87*(6), 752–755.
114. Torres, A., Li, S. M., Roussos, S., & Vert, M. (1996). Screening of microorganisms for biodegradation of poly (lactic-acid) and lactic acid-containing polymers. *Applied and Environmental Microbiology*, *62*(7), 2393–2397.
115. Degli-Innocenti, F., Tosin, M., & Bastioli, C. (1998). Evaluation of the biodegradation of starch and cellulose under controlled composting conditions. *Journal of environmental polymer degradation*, *6*(4), 197–202.
116. Körner, I., Redemann, K., & Stegmann, R. (2005). Behaviour of biodegradable plastics in composting facilities. *Waste management*, *25*(4), 409–415.
117. Augusta, J., Müller, R. J., & Widdecke, H. (1992). Biologisch abbaubare kunststoffe: Testverfahren und Beurteilungskriterien. *Chemie Ingenieur Technik*, *64*(5), 410–415.
118. Osazee, I. T. (2021). Landfill in a sustainable waste disposal. *European Journal of Environment and Earth Sciences*, *2*(4), 67–74.
119. Bellia, G., Tosin, M., Floridi, G., & Degli-Innocenti, F. (1999). Activated vermiculite, a solid bed for testing biodegradability under composting conditions. *Polymer degradation and stability*, *66*(1), 65–79.
120. Geyer, R., Jambeck, J. R., & Law, K. L. (2017). Production, use, and fate of all plastics ever made. *Science advances*, *3*(7), e1700782.
121. Chae, Y., & An, Y. J. (2018). Current research trends on plastic pollution and ecological impacts on the soil ecosystem: A review. *Environmental pollution*, *240*, 387–395.
122. Thompson, R. C., Moore, C. J., Vom Saal, F. S., & Swan, S. H. (2009). Plastics, the environment and human health: current consensus and future trends. *Philosophical transactions of the royal society B: biological sciences*, *364*(1526), 2153–2166.
123. Gross, R. A., & Kalra, B. (2002). Biodegradable polymers for the environment. *Science*, *297*(5582), 803–807.
124. Albertsson, A. C. (1980). The shape of the biodegradation curve for low and high density polyethenes in prolonged series of experiments. *European Polymer Journal*, *16*(7), 623–630.
125. Albertsson, A. C., Barenstedt, C., & Karlsson, S. (1994). Abiotic degradation products from enhanced environmentally degradable polyethylene. *Acta polymerica*, *45*(2), 97–103.
126. Cruz-Pinto, J. J. C., Carvalho, M. E. S., & Ferreira, J. F. A. (1994). The kinetics and mechanism of polyethylene photo-oxidation. *Die Angewandte Makromolekulare Chemie: Applied macromolecular chemistry and physics*, *216*(1), 113–133.
127. Dziuba, R., Kucharska, M., Madej-Kiełbik, L., Sulak, K., & Wiśniewska-Wrona, M. (2021). Biopolymers and biomaterials for special applications within the context of the circular economy. *Materials*, *14*(24), 7704.
128. Huang, S. J., Roby, M. S., Macri, C. A., & Cameron, J. A. (1992). The effects of structure and morphology on the degradation of polymers with multiple groups. *Special Publication-Royal Society of Chemistry*, *109*(1992), 149.
129. Kadouri, D., Jurkevitch, E., Okon, Y., & Castro-Sowinski, S. (2005). Ecological and agricultural significance of bacterial polyhydroxyalkanoates. *Critical reviews in microbiology*, *31*(2), 55–67.
130. Gavrilescu, M., & Chisti, Y. (2005). Biotechnology—a sustainable alternative for chemical industry. *Biotechnology advances*, *23*(7–8), 471–499.

131. Christian, S. J. (2020). Natural fibre-reinforced noncementitious composites (biocomposites). In Kent Harries, Bhavna Sharma (eds) *Nonconventional and Vernacular Construction Materials* (pp. 169–187). Woodhead.
132. George, A., Sanjay, M. R., Srisuk, R., Parameswaranpillai, J., & Siengchin, S. (2020). A comprehensive review on chemical properties and applications of biopolymers and their composites. *International journal of biological macromolecules*, *154*, 329–338.

# 11 Biorefinery Green Polymeric Approaches for Value-added Products

*Nabya Nehal, Priyanka Singh, and Pooja Agarwal*

## CONTENTS



DOI: 10.1201/9781003240884-11

## 11.1 INTRODUCTION

Over the past few years, the production of biobased products from living and non-living resources has facilitated biological conversion into biodiesel, bioplastics, bioethanol and other petroleum-based products that has shown noteworthy growth (Anugraha et al. 2016). In contrast to other catalysis processes, enzymatic action is most specific as it requires non-toxic and non-corrosive reactions and can be active in any condition (temperature, pH, pressure) (Dill et al. 1995). Enzymes are used in most of these techniques to catalyze the biochemical reactions of whole microorganisms or their cellular components which cause the reactions to occur without itself getting altered. After a series of reactions, the raw materials are chemically modified to obtain the desired end-product along with unwanted byproducts. The accessory enzymes, cellulases and hemicellulases are the three main classes of enzymes that can hydrolyze the cell wall material (Himmel et al. 2007). Continuous use of dyes, aromatic hydrocarbons, polymers of plastics and their derivatives have triggered a hazardous effect on environments (humans, animals, plants, soil and water bodies), however, substantial use of biorefineries and their advancement steps help to solve this problem (Pathak et al. 2014). However, the downstream and upstream processes are anticipated to derive purified forms of biodiesel, bioplastics, and other renewable energy resources.

The biorefineries approach has sparked great interest in overcoming socio-economic and environmental issues and reducing the capital costs of petroleum-based products. The biorefinery approach follows three main routes: generation of materials and energy, use of varied raw materials from living and non-living sources and the combination of mechanical and thermochemical processes (Hingsamer et al. 2019). As shown in Figure 11.1, different types of biorefineries have been characterized, such as whole crop, microalgal, two-platform, lignocellulosic and green biorefinery (de Jong et al. 2015, Diep et al. 2012).

However, the production of renewable energies is subsidized in many developed countries such as the USA, China, France, and Russia (Kraxner et al. 2013). Meanwhile, biomass in the biorefineries process has to be developed with the aim of improving the production of biofuels and bioenergy (Stichnothe et al. 2016). A major driver for the establishment of biorefineries is primary and secondary separation steps integrated with upstream and downstream processing of biomass. Sustainable and efficiently produced biomass can be composed from plants (fibers, sugars, oils and proteins), agro-industrial wastes, aquatic biomass, aquaculture, and organic and forest residues (de Jong et al. 2015), as shown in Figure 11.2. Forthwith, some biorefineries are frameworked on trading demands in paper and pulp industries, food and feed, and transportation fuels industries.

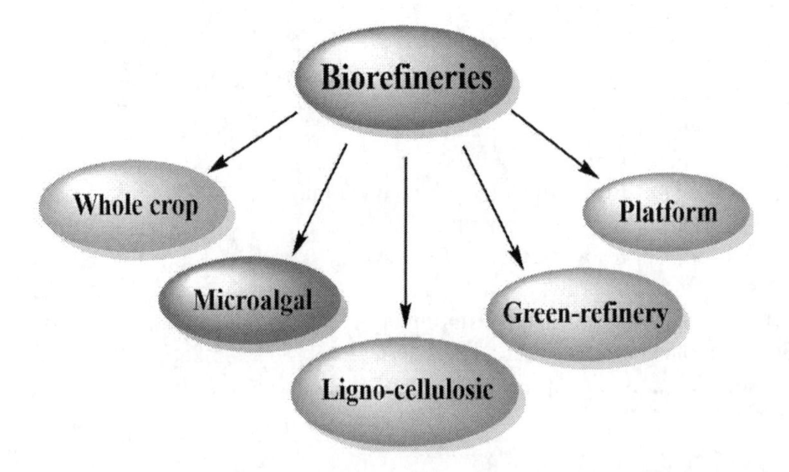


**FIGURE 11.1** Classification of biorefineries on the basis of the use of natural resources.

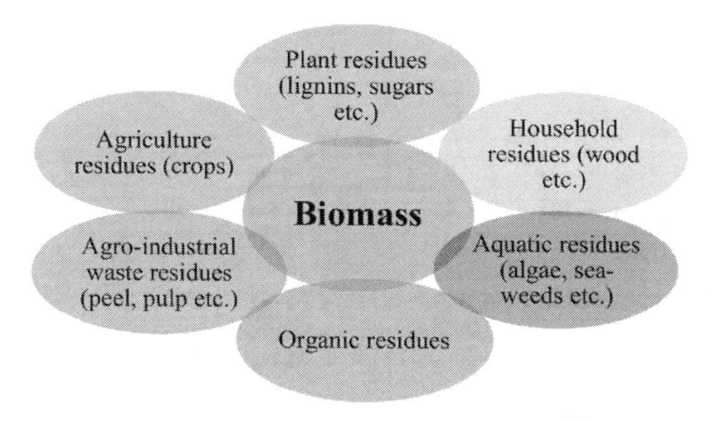


**FIGURE 11.2** Production of biomass from different types of residues through sustainable approaches.

These processes are in a nascent state and the essential objective of these technologies is to focus on pilot-scale production and the conversion cost of biomass (Lievonen 1999). The selection of pretreatment methods in the conversion of biorefineries is significantly important as they accomplish the production rate at the lowest possible expense (Anugraha et al. 2016). In this chapter, the production of various value-added products including biodiesel and bioplastics from renewable resources will be covered. An integrated microbial feedstock-based biorefinery approach improves overall production. For instance, different types of biorefinery resources (microalgal, starch, cellulose and edible and non-edible oils) and their pretreatment methods are described in detail.

## 11.2 HISTORICAL PERSPECTIVE

In the late 18th century, the term biorefinery was introduced and gained attention in the context of small-scale production (Mikkola et al. 2015) and, in 1987, Brundtland commission introduced the term sustainable development (Brundtland et al. 1987). The feedstock biomass is used to produce biofuel as an energy source. Initially, the nascent stage of biorefinery has its deep root in wood tar (oily sticky product erected from plants) (Matisons et al. 2012). In 2009, Khan introduced the concept of a microalgal biorefinery which was utilized as a renewable source of energy (Khan et al. 2009; Khan 2010). In 1853, Duffy and Patrick introduced the term trans-esterification. In early 1900, energy source inventor Dr. Rudolph Diesel designed the fuel engine derived from vegetable/edible oil. In the 1900s, he stated: "The diesel engine can be fed with vegetable oils and would help considerably in the development of agriculture of the countries which use it" (Ma et al. 1999; Zahan et al. 2018). Around 100 years ago, Dr. Diesel stated that, "The use of vegetable oils for engine fuels may seem insignificant today. But such oils may in course of time be as important as petroleum and the coal tar products of the present time" (Owolabi et al. 2012). In the 1860s, John Wesley Hyatt demonstrated cellulose from plant and tree resins and developed natural (man-made) polymers of plastics named "celluloid" that were used in plastics industries such as in men's collars and dental plates. In late 1862, Alexander Parkes introduced the first man-made polymers described as "Parkesine," which were derived from cellulose-based plant material (Pathak et al. 2014). In the early 19th century, modified man-made and synthetic plastics were introduced and used in various sectors because of their high temperature tolerance property. In 1953, German chemist, Karl Ziegler introduced polyethylene, while an Italian chemist, Giulio Natta, developed polypropylene and received the Nobel Prize in Chemistry in 1963 (Bellis 2011).

## 11.3 BIOREFINERY RESOURCES

### 11.3.1 Microalgal Biomass

On the basis of their morphology and size, microalgae are unicellular and microscopic, and also known as phytoplankton, whereas macroalgae are multicellular and ranging in size from 0.001 mm to 2 mm. Microalgae are portrayed as having excellent features such as easy to grow with a short maturation time, fast proliferation, and lower requirement of resources (land and water) (Goh et al. 2019). Algae use solar energy for food reservoirs, and capture sunlight and fix $CO_2$ for maturation and growth. These cells are endowed with some beneficial products such as fatty acids, proteins and lipids. Docosahexaenoic acids (DHA) and eicosatetraenoic acid (EPA) are highly accumulated products (Huang et al. 2010). Algae serve as a raw material for the production of value-added products (biofuel, biodiesel, bioethanol, butanol). On the basis of various factors such as pigmentation and morphology, microalgae are categorized into four groups, i.e., Chlorophyceae (green algae), Chrysophyceae (golden algae), Cyanophyceae (blue-green algae), and Bacillariophyceae (diatoms). Over the past few decades, microalgae have been cultivated to generate fossil fuels such as bioethanol, biodiesel, bio-oil and biohydrogen (Nan et al. 2015), as illustrated in Figure 11.3.

The estimated domestic consumption rate of energy produced by algal biomass is around 88% in the form of kerosene, biodiesel and gasoline (Guo et al. 2015). Mixotrophic algal culture uses glycerin, molasses and glucose in growth media and enhances productivity and reduces the cost of the media. Previously, in many reports, it has been stated that the use of molasses by *Arthrospira* algal organisms in mixotrophic cultivation enhances the growth rate and biomass as compared to autotrophic cultivation (Andrade et al. 2007).

### 11.3.2 Industrial Waste Residues

Agro-industrial wastes are the unwanted by-products derived from various agricultural sectors involving feedstocks and other beneficiary products with specific uses. It is well recognized that industrial waste is a rich source of bioactive compounds that can be used as food additives, nutraceuticals, value-added products and pharmaceuticals (Joana et al. 2013). They reduce

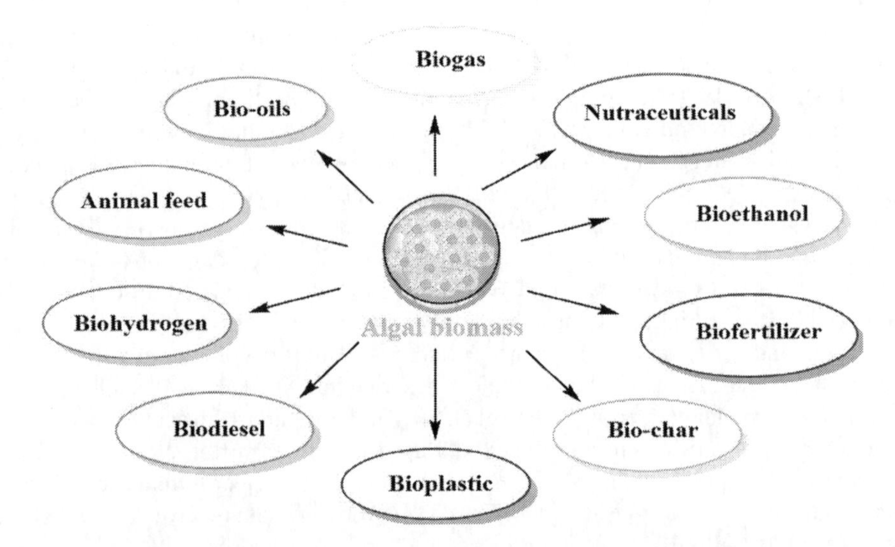


**FIGURE 11.3** Generation of different value-added products through utilization of microalgal biomass.

**TABLE 11.1**
**List of Industrial Waste Residues Used as Biocarriers in the Production of Value-added Products**

| Waste Used as Biocarrier | Microbial | Production | Renewable Energy | References |
|---|---|---|---|---|
| Eggshell | *Rhizopus oryzae* 1526 | Fumaric acid | — | Das et al. 2015 |
| Sugar-beet pulp | *Lactobacillus paracasei* NRRL-B 4564 | Lactic acid fermentation | Bioethanol and propylene glycol | Mladenovic et al. 2017 |
| Coconut shell activated carbon | *Actinobacillus succinogenes* | Succinic acid | — | Luthfi et al. 2017 |
| Oil palm frond bagasse | *Actinobacillus succinogenes* 130Z | Succinic acid | — | Luthfi et al. 2017 |
| Sweet sorghum bagasse | *Clostridium acetobutylicum* | Acetone-butanol-ethanol | Biofuels | Chang et al. 2014 |
| Corn silk | *Zymomonas mobilis* | Ethanol | Bioethanol | Todhanakasen et al. 2016 |
| Waste cassava tuber fiber | *Saccharomyces cerevisiae* | Ethanol | Bioethanol | Kunthiphun et al. 2017 |

health-related risks and are utilized in the treatment of diseases. For instance, vegetable and fruit wastes are among the most abundant sources of carbohydrates and secondary metabolites (phytochemicals) such as flavonoids, terpenoids, phenol and carotenoids. These phytochemicals are a rich source of vitamins and act as anti-oxidants, and reduce the risk of several diseases such as those used in the treatment of several types of cancer (Day et al. 2009). Generally, peel, pulp, seeds, skin, kernel and cherry husk of some perishable products such as tomato, mango, citrus fruits, grapes and coffee are not digestible by humans. Tomato peel contains higher levels of ascorbic acid, tocopherols and other metabolites in comparison to the seeds and pulp (George et al. 2004). Sharma et al. demonstrated a higher phenolic content in seeds of some seasonal fruits such as mango, jackfruit, avocado, citrus fruits and grapes as compared to the edible parts of these fruits. More than 50% of citrus fruit pulp mass contained flavonoids after their fruiting process (Sharma et al. 2017; Shrikhade 2000).

The list of different solid residues has been presented in Table 11.1. The immobilized green waste residues containing food and agricultural wastes are used as biocarriers in the bioconversion process. Sawdust, husk, molasses, wood chips and straw are carbon- and nitrogen-based biodegradable wastes that are predominantly used in immobilization techniques (Kadir et al. 2016; Schwarz et al. 2011; Tong et al. 2018). However, eggshells, corn silk, coconut shell activated carbon and sugar-beet pulps are agricultural waste materials that act as immobilization carriers. Eggshell is a rich source of calcium carbonate that is indirectly involved in the fermentation process for the synthesis of fumaric acid (Das et al. 2015). Therefore, the microbial or enzymatic catalyst involved in this process is eco-friendly, biocompatible, non-toxic and inexpensive.

## 11.4 PRETREATMENT METHODS

### 11.4.1 Upstream Processing

The upstream step of bioprocessing technology involves the large-scale production of biomolecules by carrying out the fermentation process using microbes, and animal and plant cell cultures in specific bioreactors (Groneymeyer et al. 2014). When the process reaches the desired density after carrying out the fermentation process in different modes such as batch, continuous and fed-batch, the biomolecules are extracted and passed on to the downstream segment of the bioprocess (Doran 1995). There are three main areas involving all factors and processes of fermentation.

- Related with the aspects of the producer microorganisms, the strategy of obtaining a suitable microorganism. Microbial strains help in increasing the yield and productivity and in maintaining the strain purity as well as for the preparation of a suitable inoculum of selected strains for increasing the economic efficiency.
- Related to fermentation media for selecting the most suitable cost-effective carbon and energy sources with other essential nutrients. This is to ensure maximization of yield and profit.
- Relates to fermentation conditions, to optimize the production of the targeted microbial products and their growth (Waites et al. 2009).

### 11.4.2 Downstream Processing

After production in the biological processes, extraction and purification remain the next most important pretreatment steps which are most difficult and expensive. In general, the recovery cost in biomass (microbial) processes contributes 60% of the total manufacturing cost. Therefore, downstream processing in the biological processes should be efficient and cost effective to make the process economically viable and reduce the combined cost of production by up to 20% (Mata et al. 2010).

Downstream processing of biochemical products requires recovery from a complex mixture of molecules, impurities and contaminants by making use of dedicated unit operations, responsible for bringing about a physical change that alters the product concentration along with the degree of purity. Downstream concentration and purification is an integral part of the production process and contributes significantly to the overall process costs as it depends on the nature of the product and its concentration in the reactor at the time of harvest (Clarke 2013). Some production parameters including temperature, pH, process handling and aseptic conditions also affect the downstreaming process. The production cost of extracellular and simple products is much lower than that of the intracellular products during recovery. The downstream processing involves harvesting and extraction of biomolecules (Waites et al. 2009). Extraction from microalgal biomass can be accomplished using various extraction techniques, including bead milling, autoclave, homogenization and ultrasound. Biological and chemical treatments (acids, enzymes, organic solvents and alkalis) accomplish the breakdown of cell walls. Physical methods (osmotic shock and deep freezing) have been used for the harvesting and extraction process. Figure 11.4 provides a schematic representation of downstream processing methods. Thick cell walls of microalgal are not expected to accomplish mechanical methods in the process of oil extraction (Lam et al. 2012b; Kim et al. 2013). Salim et al. (2012) used flocculating microalgae (*Ettlia texensis*, *Ankistrodesmus falcatus*, *Tetraselmis suecica* and *Neochloris oleoabundans*) to harvest non-flocculating microalgae (*Chlorella vulgaris* and *Scenedesmus obliquus*).

## 11.5 CULTIVATION AND HARVESTING

### 11.5.1 Separation Techniques

Typically, solid–liquid separation (SLS) is the first recovery operation for the separation of whole cells from culture broth, removal of cell debris, collection of protein precipitate, collection of inclusion bodies, etc. (Waites et al. 2009). These separations can be implemented in a variety of ways that are suitable for particular applications (Chisti 1998; Rousseau 1987). Centrifugation and filtration are the primary methods used to separate microbial cells, cell debris and other solid particles. In some cases, to increase the sedimentation rate, heat treatment or flocculation has been done.

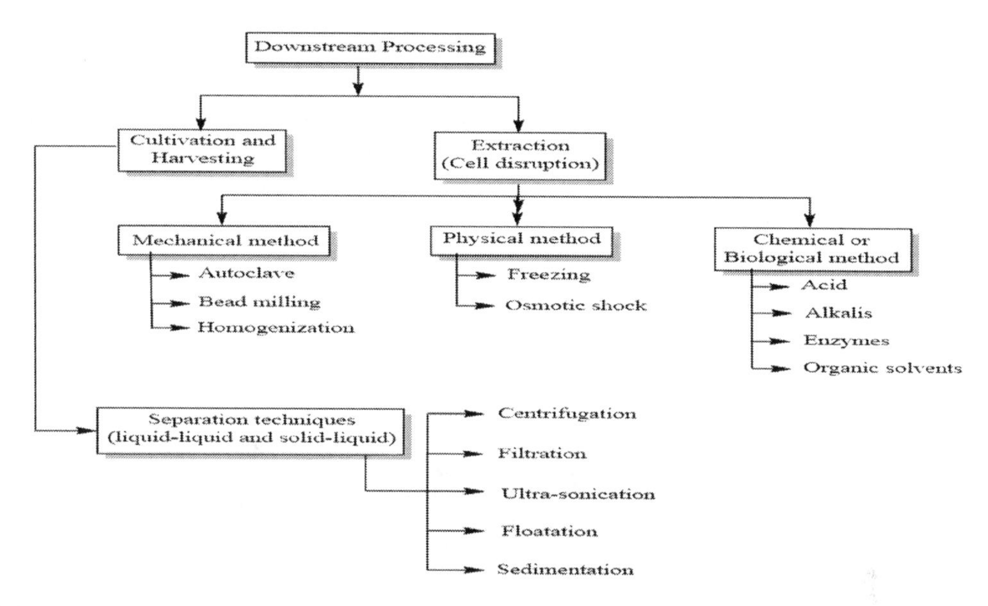


**FIGURE 11.4** Schematic representation of downstream processing method for purification of metabolites.

With the most recent approach, some modern techniques are also applied for the separation of solid materials from the fermented broth, namely, electrophoresis, ultrasonic treatment, precipitation, etc. (Weatherley 2013). Two aqueous phase extraction techniques are new aspects that have received much interest because of their easy scalability and cost effectiveness and so they could be used for these products (Bhattacharyya et al. 2008).

Some of the techniques used for solid–liquid separation are described below.

### 11.5.2 Filtration

One of the simplest and cheapest methods is to retain the particle size and residual nutrients while allowing the passage of the liquid through the filter and removing other contaminants (microbes) without the use of any chemicals (Ahmad et al. 2012). These techniques are generally useful for harvesting microalgal biomass and filamentous fungi, but they are less effective for collecting bacteria. The two main types of conventional filtration commonly used in industry are as follows:

1. Plate and frame filters or filter presses, which are industrial batch filtration systems. They are normally in the form of an alternating horizontal stack of porous plates and hollow frames.
2. Rotary vacuum filters are simple continuous filtration systems that are used in several industrial processes, particularly for harvesting microalgal biomass for the production of renewable energy (Waites et al. 2009; Kim et al. 2013).

Ultrafiltration is a form of filtration which is based on a membrane separation process driven by pressure gradient, in which the membrane fractionates the dissolved components of a liquid as a function of their solvated size and structure. Ultrafiltration is most commonly used to process a solution that has a mixture of components differing in molecular weight (Ghosh 2003). Ultrafiltration membranes allow passage of solutes depending on the nominal molecular weight cut-off (NMWCO) of the membrane being used (Singh et al. 2016).

### 11.5.3 Sedimentation

Sedimentation is extensively used for microbial biomass separation in the production of biodiesel, and in production of beverages and wastewater treatment. This low-cost technology is relatively slow and is suitable only for large flocs (greater than 100 μm diameter). The rate of particle sedimentation is a function of both size and density. Hence, the larger the size of the particle, the greater is its density and the faster is the rate of sedimentation (Waites et al. 2008).

### 11.5.4 Flotation

In the process of flotation, the microalgal cells are adsorbed on gas bubbles which get trapped in a foam layer and then the microalgal biomass is collected. This process can be done either by sparging the particulate feed with gas or by generating fine bubbles with dissolved gas or by electrolysis. Foam is stabilized by the addition of collector substances such as long-chain fatty acids or amines. Flotation efficiency is dependent on the size of the bubble: microbubbles (1–999 μm), nanobubbles (<1 μm) and fine bubbles (1–2 mm). As a result, small bubbles become attached to the cell walls and emphasize the longevity and carrying capacity (Zimmerman et al. 2011; Hanotu et al. 2012). The diameter of microalgal cells of between 10–500 μm is preferred for effective flotation (Henderson et al. 2010).

### 11.5.5 Ultrasonication

This is a process that can be used in harvesting and extraction methods in microalgal biomass. To break the cells at a small scale, ultrasonic waves are used. This involves cavitation, microscopic bubbles or cavities generated by pressure waves. It is performed by ultrasonic vibrators that produce a high-frequency sound with an acoustic wave frequency of approximately 2.1 MHz in cell disruption of *Monodus subterraneus* UTEX 151 microalgae (Bosma et al. 2003). The process relies on the cavitation generated due to the sonic waves and the shear force generated thereby. In terms of sonic waves, high electrical power is converted into mechanical energy, which propagates through a horn in the liquid generating cavitation. However, this technique also generates heat, which can denature thermo-labile proteins. Sonication is effective on a small scale but is not routinely used in large-scale operations due to problems with the power transmission and heat dissipation. There are several drawbacks to this system: high power consumption, heat generation, small operable volume, etc. In a further investigation, *Microcystis aeruginosa* was harvested by a combination of sonic waves and polyaluminum chloride (PAC) (Zhang et al. 2009).

### 11.5.6 Centrifugation

If, instead of simply using gravitational force to separate suspended particles, a centrifugal field is applied, the rate of SLS is significantly increased and much smaller particles can be separated. Centrifugation is used to separate particles as small as 0.1 μm diameter and is also suitable for some liquid–liquid separations (Waites et al. 2009). Different types of centrifuge machine are available with different dimensions and modes of application, such as capacities, speed, mode of loading and discharging.

- **Basket centrifuge** (perforated-bowl basket centrifuge) contains perforated bowls with a filter bag of nylon or cotton. It is used for the separation of mycelia mass, molds and crystalline compounds.
- **Tubular-bowl centrifuge** uses solid particle sediment on the wall of the rotor and the two liquid phases are separated into two distinct zones. This system is applied for the separation of particles having dimensions of 0.1 μm to 200 μm in diameter.

- **Disc-bowl centrifuge** consists of a central inlet pipe and a system of conical disc, made up of stainless steel arranged in stacks with a spacer. This system is highly efficient, has capacity for high-volume liquid handling, easy removal of solid and in situ cleaning facility (Bhattacharyya et al. 2008).
- **Multichamber centrifuge** system consists of multiple chambers mounted within the chamber of the rotor. The slurry is fed into the chambers through a system of spindles and travels through the system of chambers through a circuitous route.

## 11.6 EXTRACTION AND PURIFICATION

Solvent extraction is the most well-established liquid extraction operation, and relies on the preferential solubility of the product in an added organic phase, immiscible with the aqueous phase (Berk 2018). The efficiency of solvent extraction is measured in terms of a distribution coefficient which relates the concentrations of the solute in the different phases, depending largely on the suitability of the solvent. The two aqueous phases are immiscible due to small amounts of polymers added to one or both of the phases but they are still referred to as aqueous because they comprise mostly water (75 to 95%). Two common polymer additives are dextran and PEG, with each added to a separate phase. Other applications use a polymer in one phase with salt added in the other, such as PEG in one phase and potassium phosphate in the other. This form of liquid extraction is known as aqueous two-phase extraction and in all respects other than the nature of the phases, is governed by the same principles as solvent extraction (Yankov 2021). The distribution of the solvent is described by a partition coefficient. The partition coefficient is dependent on several variables, namely the nature of the solute, the nature of the phases, pH, ionic strength, polymer concentration and number of polymers (Rice 2014). For instance, the partition coefficient for the separation of hydrophobic and ionic proteins could be enhanced by using a polymer in one phase and a salt in the other so that the former would preferentially solubilize in the aqueous phase containing the polymer while the latter would preferentially solubilize in the aqueous phase containing the salt.

The biological product is either secreted to the extracellular environment or is retained intracellularly. However, the percentage of release to the outside environment is very low as compared to intracellular release (Chisti 1998). The recovery of intracellular products from the cell is obtained by the process of cell rupturing and separation of the extract from the resulting cell fragments. Chemical methods (acid, surfactants, alkalis and organic solvents) are used in the disruption of microalgal biomass. This is one of the currently popular and beneficial methods as it lowers the energy consumption as compared to other methods. Biological treatments (enzymes) help in the disruption of the cell envelope (Geciova et al. 2002). Braun et al. investigated *Chlorella* sp. cell walls that are 80% degraded after being incubated with a mixture of enzymes (cellulase, pectinase and hemicellulase) for up to 90 hours (Braun et al. 1975). In the extraction of arachidonic acid from *Mortierella alpine*, Young et al. (2011), applied a mixture of six enzymes (papain, snailase, pectinase, alcalase, neutrase, and cellulase). Neutrase enzyme increases the yield threefold. Among various possibilities, a mixture of enzymes (papain and pectinase) was selected because these enzymes degrade polysaccharides and proteins, respectively (Young et al. 2011).

### 11.6.1 Freeze Drying

In the process of drying, heat is transferred to the wet material and the moisture is removed as water vapor. This should be performed in such a way that it retains the biological activity of the product. The physical properties affecting the parameter of drying include the solid–liquid system, intrinsic properties of the solute, and conditions of the drying environment and heat transfer parameters (Ratti 2001). To recover the dried product, a thin film of solution is applied to the steam-heated surface of the drum which is scraped with a knife. The material to be dried is placed on heated shelves within a chamber to which a vacuum is applied placed in a vacuum tray. This lowers the temperatures to be used due to the lower boiling point of water at reduced pressures. It is suitable for small batches of expensive materials, such as some pharmaceuticals (Rey 2016). Freeze drying (lyophilization) is often used where the final products are live cells, as in starter culture preparations or for thermolabile products (Adams 2007). It is especially useful for some enzymes, vaccines and other pharmaceuticals where retention of biological activity is critical (Baertschi et al. 2016). In this method, frozen solutions of antibiotics, enzymes or microbial cell suspensions are prepared and the water is removed by sublimation under vacuum, directly from the solid to vapor state (Lopez-Quiroga et al. 2012). This method eliminates thermal and osmotic damage.

### 11.6.2 Osmotic Shock

The cell is disrupted by a sudden change in the salt concentration which changes the osmotic balance within the cells. Osmosis is a process in which movement of solvents occurs from a higher to a lower concentration. Two types of osmotic stresses can rupture cells: hyper- and hypo-osmotic stress (Prabakaran et al. 2011). This method is not very efficient for microbial cells with tough cell walls applied for mild release of the enzymes. The method has proved to be very efficient and unique for the release of mutant strains of *Chlamydomonas reinhardtii* (Yoo et al. 2012). Hyper-osmotic stress is the condition when the concentration of salt is higher outside the cell, and vice versa in hypo-osmotic stress (the concentration of salt is higher inside the cell). As a result, cells shrink and fluids are diffused outside the cells, which may lead to damage to the cell envelope causing cell death (Prabakaran et al. 2011).

### 11.6.3 Detergent Treatment

Detergents interact with lipoproteins so that they can damage the microbial cell membrane and release the intracellular content. The detergents used may be anionic, cationic or nonionic. The most widely used detergents include quaternary ammonium salts, sodium dodecyl sulfate (SDS), Triton X-100, etc. (Stanbury et al. 2017). During the application of these substances for the release of enzymes, protein denaturation can be caused and it is necessary to be removed from the cell-free extract before further purification (Bhattacharyya et al. 2008).

### 11.6.4 Alkalis

When the enzyme is highly alkali stable and can tolerate pH up to at least 11.5 this method is applied for cell disruption in very limited cases only. There are very few applications of this method which have been found for the release of the enzyme and one of the classical examples is for the release of L-asparaginase.

### 11.6.5 Organic Solvents

To release enzymes and other substances from microorganisms by creating channels through the cell membrane, organic solvents (butanol, chloroform and methanol) have been used.

### 11.6.6 Precipitation

Precipitation is mainly used for the concentration and separation of a protein mix from other products or for separation of different proteins, in which case it is called fractional precipitation. Fractional precipitation, the separation of individual proteins from a mix of proteins, is made possible by the fact that different proteins have different properties and, therefore, different solubility in a specific environment (Kim 2013).

### 11.6.7 Chromatography

The biological products of fermentation (proteins, pharmaceuticals, diagnostic compounds and research materials) are very effectively purified by chromatography. Chromatography is an analytical technique which separates the molecules on the basis of the differences in shape, size structure and/or composition. In general, it is a technique by which the components in a sample, carried by the liquid or gaseous phase, are resolved by sorption–desorption steps on the stationary phase. Chromatography usually consists of a stationary phase and a mobile phase. The stationary phase is the porous solid matrix packed in a column on which the mixture of compounds to be separated is loaded. The compounds are eluted by a mobile phase. A single mobile phase may be used continuously or it may be changed appropriately to facilitate the release of desired compounds (Bhattacharyya et al. 2008; Kim 2013). A number of factors are considered while choosing a chromatographic technique including molecular weight, isoelectric point, hydrophobicity and biological affinity, capacity, recovery and resolving power (selectivity) (Waites et al. 2009).

## 11.7 VALUE-ADDED PRODUCTS

### 11.7.1 Biodiesel

Diesel is non-polar (hydrophobic) in nature, with less solubility with polar molecules. The lower solubility and hydrophobicity factor of biodiesel helps to increase the rate of consumption (Trellu et al. 2016). Fossil fuels are used as an energy source throughout the world and therefore, and they had a great demand of around 1.8 million tons in 2007.We all know that these sources of energy are very limited in nature and are now on the verge of vanishing completely. This has led to an increase in demand for alternative energy sources. The fuel crisis increases daily, and researchers have found that biodiesel can fulfill the energy demand. Biodiesel is comprised of mono-alkyl esters of long-chain fatty acids derived by trans-esterification of animal fats and vegetable oils, and esters of fatty acids. The variant structural composition of free fatty acids can significantly affect the cost and production of biodiesel (Karmakar et al. 2010). The reaction is catalyzed by catalysts (homogeneous or heterogeneous) such as acid/base or enzyme (Xiao et al. 2009). Alcohol (methanol) and catalysts (such as sodium and potassium hydroxides) interact with animal fats and edible oils and help to generate biodiesel (Talha et al. 2016). One mole of triglycerides, 3 moles of alcohol and a catalyst are catalyzed together to produce 1 mole of glycerol and 3 moles of fatty esters (biodiesel) (Ma et al. 1999; Duran et al. 2014). Fatty acids, the primary product, are a hydrocarbon-rich fuel whereas glycerides, soaps and glycerol (secondary products) backbone need to be separated and purified for commercial use as glycerin in commercial value applications in cosmetics, biomedical, tobacco and food, etc. The produced biodiesel is further purified to remove fatty esters, and other residues (alcohol) are recovered or removed by an evaporation and distillation method. After adding all the raw materials in the reaction chamber, the byproducts of triglycerides are separated into two liquid phases. The sensor helps glycerol to settle down and the fatty ester content to float on top of the container; the counter chamber drags all the produced glycerol and converts it to glycerin. During purification of glycerin, some methanol byproducts are generated that can be used as fertilizers. For commercial production of an alternative fuel, a high rate of reaction (approximately 4000 times), easy availability and cost effectiveness are some irreplaceable factors to prefer alkali catalyst in comparison to acid (Atadashi et al. 2010). The chemical compositions of diesel and biodiesel differ. On the basis of alcohol moieties attached to free fatty acids chain, esters are distinguished in fatty acid methyl esters (FAME) and ethyl esters (FAEE). Fatty acids are categorized by the number of bonds and degree of saturation and length of chain. There is no double bond found in saturated fatty acids (e.g., stearic and palmitic), a single double bond makes MUFA (mono-unsaturated fatty acid e.g., oleic acid), a more than one double bond is needed to form polyunsaturated fatty acids (e.g., linoleic and linolenic acid) (Lanjekar et al. 2016). Trans-esterification is recommended for

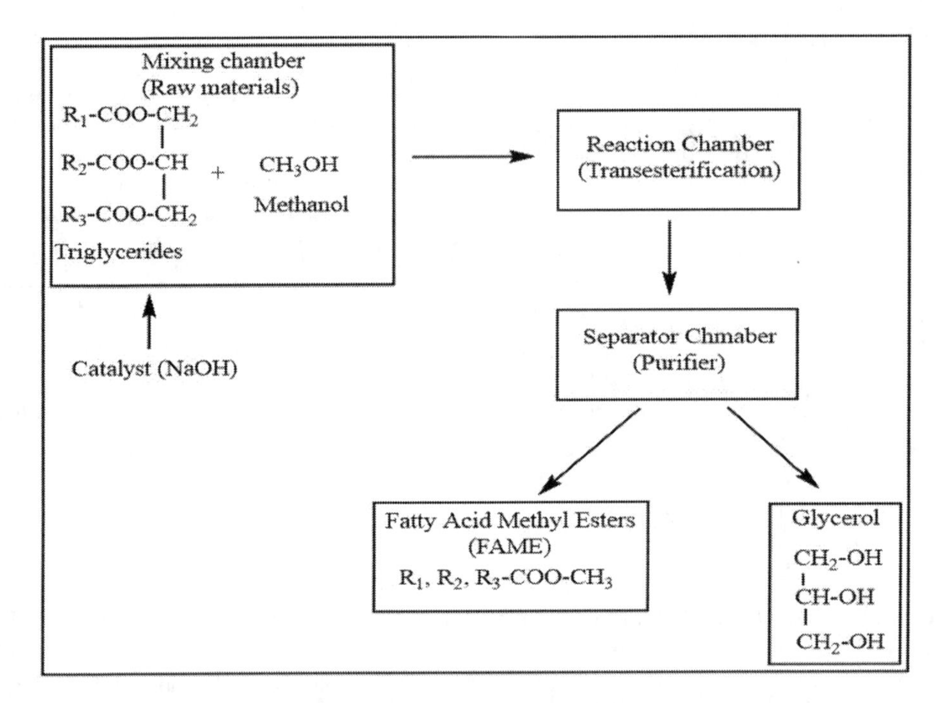


**FIGURE 11.5** Metabolic pathway for the production of fatty acid methyl esters via the trans-esterification process.

pilot-scale production. The production of fatty acid methyl esters via a trans-esterification process is shown in Figure 11.5.

Biodiesel has had great demand throughout the world over the last few decades. The consumption of renewable resources to fulfill the great demand for biodiesel may result in a huge crisis by 2030. The global demand for biodiesel may increase by up to 53% (International Energy Agency 2013). Biodiesel is mainly derived from renewable resources such as edible vegetable oils, animal fats, non-edible microalgae and waste cooking oil. It is non-toxic with a low pollution rate and is eco-friendly (Ahmad et al. 2011). There is a growing perceived economic and political need for the development of alternative fuel sources. This is due to general environmental, economic and geopolitical concerns around sustainability (Manienivan et al. 2009). Biodiesel has become one of the most promising alternatives to fossil fuels for diesel engines. It has become increasingly valuable due to the environmental consequences of petroleum-fueled diesel engines and the depletion of petroleum resources. Biodiesel can be produced by chemically combining any natural oil or fat with an alcohol such as methanol or ethanol. In the commercial production of biodiesel, methanol has been the most commonly used alcohol. A great deal of research on biodiesel has shown that fuel made using vegetable oil can be used successfully on diesel engines (Huang et al. 2012).

### 11.7.2 Physicochemical Properties

Density, cetane number, viscosity, flash point, pour point, cloud point, aniline point and diesel index are some standard properties used for biodiesel (Demirbas 2009). A variation in the chemical composition of fatty acids may alter the physicochemical properties of fuels produced using non-edible vegetable oil. Therefore, the standard parameter for fuel should be maintained before using biodiesel as an alternative fuel (Fattah et al. 2013). The ASTM (American Standard for Testing Materials) and EU (European Union) provide standard test methods (Zahan et al. 2018) that are generally used to check the purity of alternative fuels. The production and purification of good-quality biodiesel can be estimated using ASTM-D-6751-03 or EN-14214 (Atadashi et al. 2010).

- **Density:** EN ISO 3675 and ASTM D1298 are standard test methods used to measure the density of fuels. According to this method, 15°C temperature should be maintained for density measurement. The density of biodiesel is higher than that of other more common fuels (Masjuki 2010).
- **Cetane number (CN):** this is used to assess the quality and performance of fuels, and is also called the cetane rating, its name is derived from hydrocarbons i.e., n-hexadecane ($C_{16}H_{34}$) or cetane. The ignition rate of cetane is readily very high, thus, it is assigned as CN 100, meanwhile the CN of aromatic hydrocarbon (α-methylnaphthalene) is 0 because its auto-ignition rate is low (Yanowitz et al. 2017). Higher cetane numbers indicate better ignition of fuel within vehicles. The highest purity level of fuels is represented by the cetane rating, which means that fuel should be readily and easily ignited in a very short time period. It shows a relationship between fuel injection and fuel ignition due to compression (Christine et al. 2015). CN depends on the chemical bonds and molecular structure of hydrocarbons. Alkenes possess double bonds; hence, CN values are lower than those of alkanes. In addition, feedstocks such as tallow-, coconut- and palm-derived biodiesels ensure high CN values (Knothe 2014).
- **Viscosity:** this is used to assess the performance of the engine. A higher temperature may decrease the viscosity of fuels, i.e., they are inversely proportional.
- **Flash point:** This is the point that represents the hazardous flammable temperature present in the air. The flash point of non-edible biodiesel is higher in comparison to that of other fuels. This may detect ignition factors during storage, packing and exportation and importation of fuels.
- **Cloud and pour point:** These are the temperature at which oil or wax of fuel starts freezing and shows a cloudy appearance. Edible oil such as olive oil freezes at 4°C. In winter, olive oil-derived biodiesel starts thickening, creating waxy clumps and settling down in the container. In non-ionic surfactants, the lowest temperature at which the mixture starts separating in phases has been considered as 20–30°C. Accumulation of wax may form an oil–water emulsion type. The temperature at which fuel crystallizes and aggregates in the engine and ceases to flow is termed the pour point (Mejia et al. 2013). Adding solvents like ethanol may improve the pour point and viscosity of fuel but will not alter the cloud point properties (Ali et al. 2016). The physicochemical properties of diesel and biodiesel are elucidated in Table 11.2.

### 11.7.3 Biodiesel from Edible and Non-edible Oils

Around 95% production of biofuels around the world depends on edible oil. Many developed countries such as the United States, Philippines, Europe and Southeast-Asia and many others have used edible oil for pilot-scale production (Popp et al. 2016). In the production of biodiesel, vegetable oils (coconut oil, olive oil, sunflower oil) are used in a very high ratio (Yusuf et al. 2011). Such a huge amount of edible oil has some limitations, as a higher consumption rate of edible oil may affect the food chain and imbalance it globally. As the survey suggests, the enormous use of edible oil for the production of biodiesel may negatively impact the food chain globally. Therefore, the use of non-edible and waste cooking oil is necessary to reduce this potential crisis (Sirisomboonchai et al. 2015). Naturally, an enormous amount of energy crops are used as non-edible feedstocks. Non-edible feedstocks are significantly used because they are inexpensive and non-edible (containing toxic compounds) (Ahmad et al. 2011).

It is shown in Table 11.3 that castor, pongamia, karanja, mahua and jatropha are some of the non-edible feedstocks used for the production of biodiesel. Mahua is fermented non-edible oil used as a source of carbon for renewable energy production (Venkatesan et al. 2012).

**TABLE 11.2**
**Physicochemical Properties of Diesel and Biodiesel**

| Properties | Standard Method | Unit | Biodiesel | | Diesel | References |
|---|---|---|---|---|---|---|
| | | | Non-edible | Edible | | |
| **Density** | ASTM D1298 | $kg/m^3$ | Neem (912–965), jojoba (863–866) | Poppy (921), sesame (850) | 816–840 | Panwar et al. 2010; Ramadhas et al. 2004; Ferdous et al. 2012 |
| **Cetane number** | ASTM D613 | min | Rice bran seed oil (73.6), linseed oil (34.60) | Rapeseed (37.6), palm (62) | 47.0 | Murugesan et al. 2009; Ramadhas et al. 2004; Atabani et al. 2013; Karmakar et al. 2010 |
| **Viscosity** | ASTM D445 | Kg/m/s | Mahua oil (3.98), sea mango (29.57) | Poppy (56.1), walnut (3.88) | 4.3 | Murugesan et al.. 2009; Ramadhas et al. 2004; Moser 2012; Atabani et al. 2013; Kansedo et al. 2009 |
| **Flash point** | ASTM D93 | °C | Crambe oil (274), putranjiva oil (48) | Groundnut (176–202), soybean (141) | 71 | Murugesan et al. 2009; Atabani et al. 2013 |
| **Acid number** | ASTM D664 | mg KOH/g | — | — | — | Murugesan et al. 2009 |
| **Pour point** | ASTM D97–96a | °C | Rice bran seed oil (269), castor oil (–32) | Hazelnut (–13), rapeseed (NA) | –16 | Moser 2012; Ramadhas et al. 2004; Atabani et al. 2013; Gui et al. 2008 |
| **Cloud point** | ASTM D2500 | $^0$C | Jatropha (NA), jojoba (6–16) | — | –10 to –5 | Palash et al. 2014; Panwar et al. 2010; Ganapathy et al. 2011 |
| **Calorific value** | - | MJ/kg | Jojoba (42.76–47.38), jatropha (38.5–42) | Palm (33.5), sunflower (40.1–40.6) | 43.35 | Chauhan et al. 2010; Ganapathy et al. 2011; Ramadhas et al. 2004 |

**TABLE 11.3**
**Production of Biodiesel from Non-edible Feedstocks**

| Feedstocks | Originated from | Maturation Time | Oil Content (Weight %) | Fatty Acid Content | References |
|---|---|---|---|---|---|
| **Mahua** (*Madhuca indica*) | India | 10 yrs | Seed: 35–50<br>Kernel: 50 | Oleic acid (41.0–51.0)<br>Stearic acid (20.0–25.1) | Atabani et al. 2013 |
| **Cotton seed** | China, Europe, US | 6 weeks (after ball maturation) | Seed: 17–25 | Linoleic acid (55.2–55.5)<br>Oleic acid (19.2–23.26) | Ashraful et al. 2014 |
| **Polanga** (*Calophyllum inophyllum*) | Southeast Asia, India | NA | Seed: 65–75 | Oleic acid (34.09–37.57)<br>Linoleic acid (26.33–38.26) | Ashraful et al. 2014 |
| **Karanja** (*Pongamia pinnata*) | Southeast Asia | 4–7 yrs | Seed: 25–50<br>Kernel: 30–50 | Oleic acid (44.5–71.3)<br>Linoleic acid (10.8–18.3) | Atabani et al. 2013; Mahanta et al. 2006 |
| **Jojoba** (*Simmondsia chinensis*) | Mexico, California | 10–12 yrs | Seed: 40–50 | Oleic acid (43.5–66)<br>Linoleic acid (25.2–34.4) | Ashraful et al. 2014 |
| **Jatropha** (*Jatropha curcas*) | US, India, Mexico | 9–12 months | Seed: 20–60<br>Kernel: 40–60 | Oleic acid (34.3–44.7)<br>Linoleic acid (31.4–43.2) | Atabani et al. 2013; Banapurmath et al. 2008 |
| **Neem** (*Azadirachta indica*) | Asian countries | After 15 yrs | Seed: 20–30 | Oleic acid (25–54)<br>Palmitic acid (16–33) | Atabani et al. 2013 |
| **Tobacco oil** (*Nicotiana tabacum*) | India, Turkey, America | 4–5 months | Seed: 35–49<br>Kernel: 17 | Linoleic acid (69.49–75.58)<br>Oleic acid (11.24–14.54) | Ashraful et al. 2014 |
| **Linseed oil** (*Linum usitatissimum*) | Canada, India, Argentina | 30–40 days | Seed: 35–45 | Linolenic acid (46.10–51.12)<br>Oleic acid (20.17–24.05) | Atabani et al. 2013; Kasote et al. 2013; Acikgoz et al. 2007 |
| **Rubber seed oil** (*Hevea brasiliensis*) | Brazil | 8 yrs | Seed: 50–60<br>Kernel: 40–50 | Linoleic acid (39.6–40.5)<br>Oleic acid (17–24.6) | Atabani et al. 2013; Okieimen et al. 2002 |

### 11.7.4 Biodiesel from Animal Fats

Animal fats are the byproduct of slaughtered animals, e.g., beef tallow. They are distinct into edible and non-edible animal fat. The presence of FFA in edible fat may affect the cost and production of biodiesel. Animal fats are thick by nature, they make waxy clumps and aggregate together and can therefore block an engine (Canakci 2007).

### 11.7.5 Biodiesel from Microalgal Biomass

Microalgae as a raw material for biodiesel (third-generation biodiesel) has been reviewed extensively in recent years (Hannon et al. 2010). Cultivation of microalgae can produce 58.7 liters of oil per hectare, which is derived to generate 121.1 liters of biodiesel/hectare, making this a more promising method over conventional methods. Microalgae have long been recognized as potentially good sources for biofuel production because of their high oil content (more than 20%) and rapid biomass production. Algae biomass can play an important role in solving the problem between the production of food and that of biofuels. The cultivation of microalgae does not need much land as compared to terrestrial plants (Huang et al. 2015). Due to their high viscosity (about 10–20 times higher than diesel fuel) and low volatility, microalgae do not burn completely and form deposits in the fuel injector of diesel engines. The trans-esterification of microalgal oils will greatly reduce the original viscosity and increase the fluidity (Demirbas 2009) (Figure 11.6).

The commercial production of value-added products from microalgal biomass is a sequential process that follows upstream and downstream processing consisting of the extraction and purification of algal lipids. Typically, microalgal biorefinery technology has presented bottlenecks that are mainly associated with upstream processing (USP) and downstream processing (DSP).

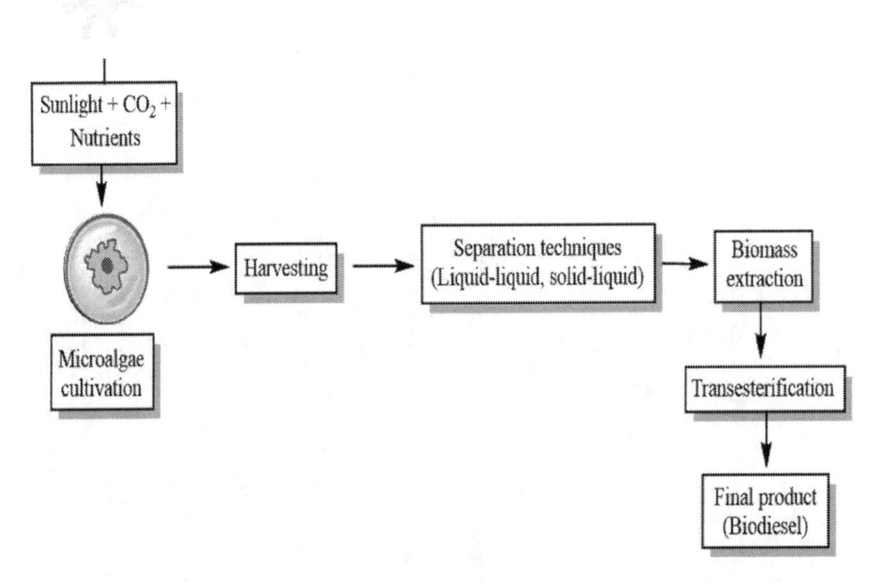


**FIGURE 11.6** Microalgal cultivation process for the production of biodiesel via the trans-esterification method.

## 11.8 BIOPLASTICS

At present, the synthesis of polyethylene is among the most problematic issues facing the environment as it releases harmful gases such as carbon monoxide, furan, dioxin and GHG (Das et al. 2016). The disposal of synthetic plastics (toxic, non-degradable) affects marine ecosystems, animals (consume plastics accidentally), soils (accumulation of toxic polymers in plants) and food systems (affecting the food chain) (Kassim et al. 2020). One of the aims of the use of bioplastics is furthermore to use all the synergies for efficient and sustainable production, to maximize or optimize the socio-economic and environmental benefits. Nowadays, bioplastics have become a daily necessity in several areas such as pharmaceuticals, food packaging, grocery bags, electronics and automotives, textiles, cosmetics and agriculture (Ashter 2016). The future expected growth of the naturally synthesized biobased plastic market and the development of new production processes for biodegradable plastics make it very fashionable.

Most biobased plastics are degradable [polylactic acid (PLA), polyhydroxyalkanoic acid (PHA), poly(butylene succinate) (PBS) and polycaprolactone (PCL) and PBAT], and other are non-degradable [biobased PTT, polyethylene terephthalate or polyester (PET), polypropylene (PP) and PVC] (Elias 1993). Green biorefinery (jute, starch, corn, vegetable oil, sugarcane and cassava), microalgal biorefinery, animal proteins (collagen, keratin, whey and gelatin) (Mekonnen et al. 2013) and microbial biorefinery (yeast) are the most sustainable substrates used in the production of biodegradable and biocompatible plastics.

On the basis of their origin, bioplastics are categorized into three groups:

- Chemical synthesis [PBS, PLA and polyglycolic acid (PGA)]
- Natural synthesis (chitin, starch and cellulose acetate)
- Microbial synthesis (curdlan, gellan gum and PHA) (Flieger et al. 2003).

There are three generic pathways to producing plastics from natural materials (Figure 11.7):

1. The extraction and modification of natural polymers from biomass
2. The polymerization of biobased monomers
3. The extraction of polymers produced in microorganisms.

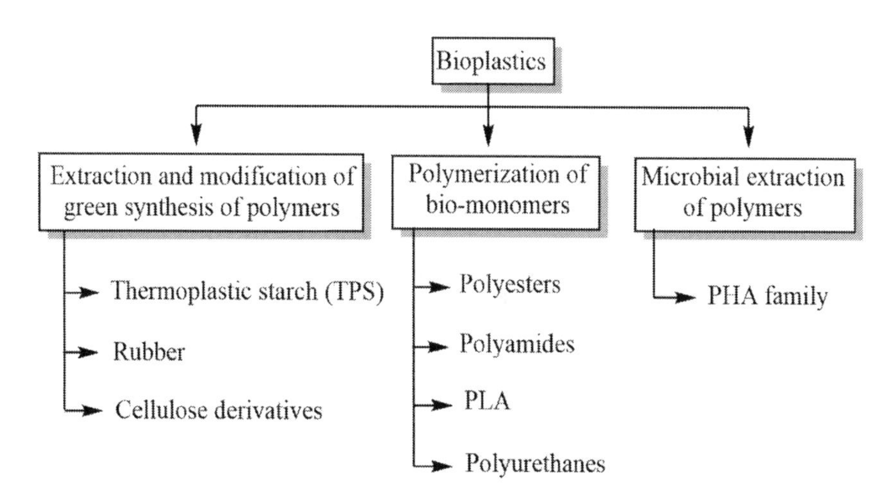


**FIGURE 11.7** Biological process for the synthesis of different types of bioplastics from natural materials.

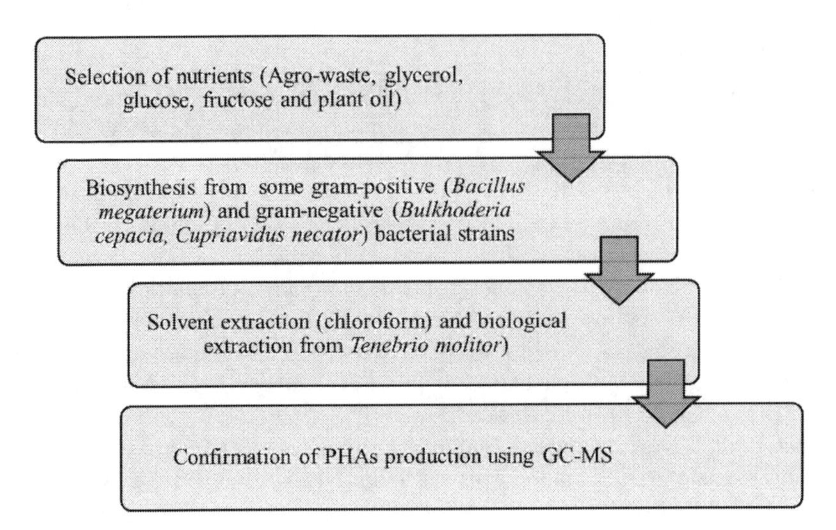


**FIGURE 11.8** Biological process for the production of polyhydroxyalkanoates (PHAs) from natural resources.

Generally, polyhydroxybutyrate (PHB) is widely used in biomedical industries due to its biodegradability, non-toxicity and biocompatibility. Bioplastics are eco-friendly, easy to degrade in a relatively short time, non-toxic and consume low energy. Polymers of polyesters are least degradable in comparison to polyacetals, and are used in drug-delivery areas. Figure 11.8 illustrates the generic pathway of polyhydroxyalkanoates (PHAs) production from microbial sources. Carbon- and nitrogen-enriched agro-industrial wastes are selected as a source of nutrition and further have been chemically extracted by chloroform and biologically by mealworm (*Tenebrio molitor*). Gas chromatography-mass spectrum studies have validated the production of PHAs (Kassim et al. 2020).

### 11.8.1 Bioplastics from Green Biorefinery

Banana peel, wheat, potatoes, rice and sugarcane are rich sources of starch (Shah et al. 2008). Starch is made up of two polysaccharide molecules: amylose and amylopectin, linked together with α-1,4 and α-1,6 glycosidic linkage (Figure 11.9). The ratio of these polysaccharides differs from species to species. Cellulose is one of the most abundant carbohydrates and can be widely extracted from the cell walls of certain bacteria.

### 11.8.2 Bioplastics from Microalgae

Microalgae are a highly enriched source of carbohydrate and proteins and can be used to produce biodegradable plastics (Wang et al. 2016). *Chlorella* and *Spirulina* are enriched sources of protein, and their biomass has been used to make biopolymers (Deprá et al. n.d.). Deprá et al. reported in their study that the leftover biomass of biofuel derivatives was chemically treated and used as feedstock to produce value-added products (biobased plastics). *Chlorella pyrenoidosa* can be used in a polymerization process to produce polyhydroxy butyrate (PHB) (Das et al. 2018). Bioplastics from algae are a by-product in algal refinery; this can in for generating other revenues along with biofuels. Microalgae-based plastic is in its infancy stage, and could play a vital role as an environmentally friendly future product. Zeller et al. (2013) proved that algae are a feasible alternative for bioplastic and thermoplastic production. Although they investigated *Chlorella* and *Spirulina*, the study in this chapter investigated the protein modification of catfish algae (planktonic algae)—considered a waste and nuisance for catfish farms—and *Nannochloropsis* (microalgae). The latter

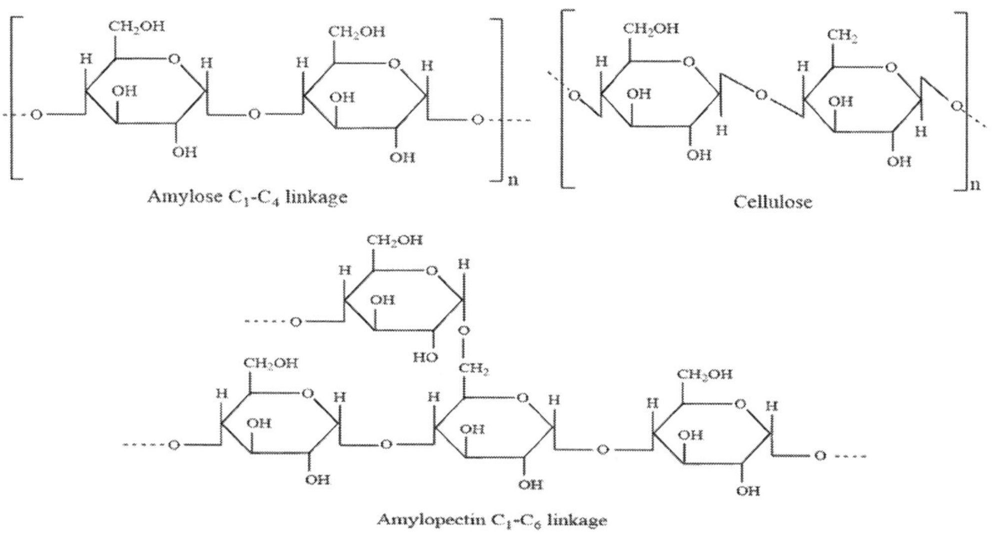

Amylose $C_1$-$C_4$ linkage

Cellulose

Amylopectin $C_1$-$C_6$ linkage

**FIGURE 11.9** Chemical structural representation of starch and cellulose linked glycosidic bond.

is a protein-rich coproduct that has been scaled commercially because of its ability to accumulate high levels of polyunsaturated fatty acids. Algae bioplastics and their thermoplastic blends were developed through thermomechanical processing and were evaluated for their thermal and dynamic mechanical properties. Additionally, one of the drawbacks to the protein modification of algae is odor, which either occurs naturally within the algae or is generated through thermoplastic processing because of heat and pressure. To commercialize these algae bioplastics, the odor-causing volatiles must be removed. A systematic design of the experiment approach has been undertaken to determine the influence of the factors, such as types of algae, scavenger materials (adsorbents), synthetic resin and compatibility on the odor of plastics (Wang et al. 2016; Zeller et al. 2013).

## 11.9 CONCLUSION

Biorefinery has many capabilities to be involved in the sustainable development of some value-added products such as bioethanol, biodiesel, bioplastics and biofuels. Biodiesel is non-toxic and does not affect human health, and also it can reduce emissions by about 65% of small particles of solid combustible products, carbon monoxide by approximately 50% and carbon dioxide by 78%. As discussed above, the use of green and microalgal biorefinery has a higher cetane number for biodiesel as compared to diesel. The combustion product should not be toxic when exhausted to the atmosphere. These requirements can be satisfied using liquid and gaseous fuels. The production of biodiesel from non-edible sources like jatropha, pongamia, mahua, neem, etc. and agro-industrial waste (molasses, bagasse) and freshwater microalgal meet the above engine performance requirement and therefore could offer a perfectly viable alternative to diesel oil in India. Among these, microalgal cells have been acclaimed as one of the best sources for the production of petroleum-based transportation fuels and biobased plastics. Through this study, some food and agricultural wastes such as eggshell, coconut shell, and the seeds, kernel and peel of fruits and vegetables have been used as biocarriers in immobilization techniques (feasible, cheapest and eco-friendly). In addition to cultivation, biomass is integrated with a pretreatment method that can develop purified by-products and effectively reduce costs. These processes and environmentally sustainable steps involved in harvesting, extraction and purification of algal biomass followed by the production of value-added products via trans-esterification are also discussed.

## ACKNOWLEDGMENT

The authors would like to thank the Department of Bioscience and Biotechnology of Banasthali Vidyapith, Rajasthan, India, for providing research support to carry out this research work.

## REFERENCES

Acikgoz, C., & Kockar, O. M. (2007). Flash pyrolysis of linseed (Linum usitatissimum L.) for production of liquid fuels. *Journal of Analytical and Applied Pyrolysis*, *78*(2), 406–412.

Adams, G. (2007). The principles of freeze-drying. *Cryopreservation and Freeze-Drying Protocols*, *368*, 15–38.

Ahmad, A. L., Yasin, N. M., Derek, C. J. C., & Lim, J. K. (2011). Microalgae as a sustainable energy source for biodiesel production: a review. *Renewable and Sustainable Energy Reviews*, *15*(1), 584–593.

Ahmad, A. L., Yasin, N. M., Derek, C. J. C., & Lim, J. K. (2012). Crossflow microfiltration of microalgae biomass for biofuel production. *Desalination*, *302*, 65–70.

Ali, O. M., Mamat, R., Abdullah, N. R., & Abdullah, A. A. (2016). Analysis of blended fuel properties and engine performance with palm biodiesel–diesel blended fuel. *Renewable Energy*, *86*, 59–67.

Andrade, M. R., & Costa, J. A. (2007). Mixotrophic cultivation of microalga Spirulina platensis using molasses as organic substrate. *Aquaculture*, *264*(1–4), 130–134.

Anugraha, T. S., Swaminathan, T., Swaminathan, D., Meyyappan, N., & Parthiban, R. (2016). Enzymes in platform chemical biorefinery. In *Platform Chemical Biorefinery* (pp. 451–469). Elsevier.

Ashraful, A. M., Masjuki, H. H., Kalam, M. A., Fattah, I. R., Imtenan, S., Shahir, S. A., & Mobarak, H. M. (2014). Production and comparison of fuel properties, engine performance, and emission characteristics of biodiesel from various non-edible vegetable oils: a review. *Energy Conversion and Management*, *80*, 202–228.

Ashter, S. A. (2016). *Introduction to Bioplastics Engineering*. William Andrew.

Atabani, A. E., Silitonga, A. S., Ong, H. C., Mahlia, T. M. I., Masjuki, H. H., Badruddin, I. A., & Fayaz, H. (2013). Non-edible vegetable oils: a critical evaluation of oil extraction, fatty acid compositions, biodiesel production, characteristics, engine performance and emissions production. *Renewable and Sustainable Energy Reviews*, *18*, 211–245.

Atadashi, I. M., Aroua, M. K., & Aziz, A. A. (2010). High quality biodiesel and its diesel engine application: a review. *Renewable and Sustainable Energy Reviews*, *14*(7), 1999–2008.

Baertschi, S. W., Alsante, K. M., & Reed, R. A. (Eds.) (2016). *Pharmaceutical Stress Testing: Predicting Drug Degradation*. CRC Press.

Banapurmath, N. R., Tewari, P. G., & Hosmath, R. S. (2008). Performance and emission characteristics of a DI compression ignition engine operated on Honge, Jatropha and sesame oil methyl esters. *Renewable Energy*, *33*(9), 1982–1988.

Bellis, M. (2011). *The History of Plastics*. About.com Inventors.

Berk, Z. (2018). *Food Process Engineering and Technology*. Academic press.

Bhattacharyya, M. S., Kamble, A., & Banerjee, U. C. (2008). *Down Stream Processing of Biologicals*. In Collections: Food and Industrial Microbiology http://nsdl.niscair.res.in/jspui/handle/123456789/676.

Bosma, R., Van Spronsen, W. A., Tramper, J., & Wijffels, R. H. (2003). Ultrasound, a new separation technique to harvest microalgae. *Journal of Applied Phycology*, *15*(2), 143–153.

Braun, E., & Aach, H. G. (1975). Enzymatic degradation of the cell wall of Chlorella. *Planta*, *126*(2), 181–185.

Brundtland, G., Khalid, M., Agnelli, S., Al-Athel, S., Chidzero, B., Fadika, L., ... & Okita, S. (1987). *Our Common Future: A Reader's Guide ('Brundtland report')*. IIED/Earthscan.

Canakci, M. (2007). The potential of restaurant waste lipids as biodiesel feedstocks. *Bioresource Technology*, *98*(1), 183–190.

Chang, Z., Cai, D., Wang, C., Li, L., Han, J., Qin, P., & Wang, Z. (2014). Sweet sorghum bagasse as an immobilized carrier for ABE fermentation by using Clostridium acetobutylicum ABE 1201. *RSC Advances*, *4*(42), 21819–21825.

Chauhan, B. S., Kumar, N., Du Jun, Y., & Lee, K. B. (2010). Performance and emission study of preheated Jatropha oil on medium capacity diesel engine. *Energy*, *35*(6), 2484–2492.

Chisti, Y. (1998). Strategies in downstream processing. *Bioseparation and Bioprocessing: A Handbook*, *2*, 3–30.

Christine & Scott Gable. (May 14, 2015). *Diesel Fuel Cetane Rating* [Online]. Available: http://alternativefuels.about.com/od/researchdevelopment/a/cetane.htm

Clarke, K. G. (2013). *Bioprocess Engineering: An Introductory Engineering and Life Science Approach.* Elsevier.

Das, R. K., Brar, S. K., & Verma, M. (2015). Valorization of egg shell biowaste and brewery wastewater for the enhanced production of fumaric acid. *Waste and Biomass Valorization*, *6*(4), 535–546.

Das, R. K., Brar, S. K., & Verma, M. (2016). Fumaric acid: production and application aspects. In *Platform Chemical Biorefinery* (pp. 133–157). Elsevier.

Das, S. K., Sathish, A., & Stanley, J. (2018). Production of biofuel and bioplastic from chlorella pyrenoidosa. *Materials Today: Proceedings*, *5*(8), 16774–16781.

Day, L., Seymour, R. B., Pitts, K. F., Konczak, I., & Lundin, L. (2009). Incorporation of functional ingredients into foods. *Trends in Food Science & Technology*, *20*(9), 388–395.

de Jong, E., & Jungmeier, G. (2015). Biorefinery concepts in comparison to petrochemical refineries. In *Industrial Biorefineries & White Biotechnology* (pp. 3–33). Elsevier.

Demirbas, A. (2009). Progress and recent trends in biodiesel fuels. *Energy Conversion and Management*, *50*(1), 14–34.

Deprá, M. C., dos Santos, A. M., Severo, I. A., Santos, A. B., Zepka, L. Q. & Jacob-Lopes, E. (n.d.). Microalgal biorefineries for bioenergy production: can we move from concept to industrial reality? *Bioenergy Research* 727–747.

Diep, N. Q., Sakanishi, K., Nakagoshi, N., Fujimoto, S., Minowa, T., & Tran, X. D. (2012). Biorefinery: concepts, current status, and development trends. *International Journal of Biomass and Renewables*, *1*(2), 1–8.

Dill, K. A., Bromberg, S., Yue, K., Chan, H. S., Ftebig, K. M., Yee, D. P., & Thomas, P. D. (1995). Principles of protein folding—a perspective from simple exact models. *Protein Science*, *4*(4), 561–602.

Doran, P. M. (1995). *Bioprocess Engineering Principles*. Elsevier.

Duran, E. A., Tinoco, R., Perez, A., Berrones, R., Eapen, D., & Sebastian, P. J. (2014). A comparative study of biodiesel purification with magnesium silicate and water. *Journal of New Materials for Electrochemical Systems*, *17*(2), 105–111.

Elias, H. G. (1993). *An Introduction to Plastics*. Weinheim: Wiley-VCH.

Fattah, I. R., Masjuki, H. H., Liaquat, A. M., Ramli, R., Kalam, M. A., & Riazuddin, V. N. (2013). Impact of various biodiesel fuels obtained from edible and non-edible oils on engine exhaust gas and noise emissions. *Renewable and Sustainable Energy Reviews*, *18*, 552–567.

Ferdous, K., Uddin, M. R., Khan, M. R., & Islam, M. A. (2012). Biodiesel from Sesame oil: base catalysed transesterification. *International Journal of Engineering & Technology*, *1*(4), 420–431.

Flieger, M., Kantorova, M., Prell, A., Řezanka, T., & Votruba, J. (2003). Biodegradable plastics from renewable sources. *Folia Microbiologica*, *48*(1), 27–44.

Ganapathy, T., Gakkhar, R. P., & Murugesan, K. (2011). Influence of injection timing on performance, combustion and emission characteristics of Jatropha biodiesel engine. *Applied Energy*, *88*(12), 4376–4386.

Geciova, J., Bury, D., & Jelen, P. (2002). Methods for disruption of microbial cells for potential use in the dairy industry—a review. *International Dairy Journal*, *12*(6), 541–553.

George, B., Kaur, C., Khurdiya, D. S., & Kapoor, H. C. (2004). Antioxidants in tomato (Lycopersium esculentum) as a function of genotype. *Food Chemistry*, *84*(1), 45–51.

Ghosh, R. (2003). *Protein Bioseparation Using Ultrafiltration: Theory, Applications and New Developments*. World Scientific.

Goh, B. H. H., Ong, H. C., Cheah, M. Y., Chen, W. H., Yu, K. L., & Mahlia, T. M. I. (2019). Sustainability of direct biodiesel synthesis from microalgae biomass: a critical review. *Renewable and Sustainable Energy Reviews*, *107*, 59–74.

Gronemeyer, P., Ditz, R., & Strube, J. (2014). Trends in upstream and downstream process development for antibody manufacturing. *Bioengineering*, *1*(4), 188–212.

Gui, M. M., Lee, K. T., & Bhatia, S. (2008). Feasibility of edible oil vs. non-edible oil vs. waste edible oil as biodiesel feedstock. *Energy*, *33*(11), 1646–1653.

Guo, Y., Yeh, T., Song, W., Xu, D., & Wang, S. (2015). A review of bio-oil production from hydrothermal liquefaction of algae. *Renewable and Sustainable Energy Reviews*, *48*, 776–790.

Hannon, M., Gimpel, J., Tran, M., Rasala, B., & Mayfield, S. (2010). Biofuels from algae: challenges and potential. *Biofuels*, *1*(5), 763–784.

Hanotu, J., Bandulasena, H. H., & Zimmerman, W. B. (2012). Microflotation performance for algal separation. *Biotechnology and Bioengineering*, *109*(7), 1663–1673.

Henderson, R. K., Parsons, S. A., & Jefferson, B. (2010). The impact of differing cell and algogenic organic matter (AOM) characteristics on the coagulation and flotation of algae. *Water Research*, *44*(12), 3617–3624.

Himmel, M. E., Ding, S. Y., Johnson, D. K., Adney, W. S., Nimlos, M. R., Brady, J. W., & Foust, T. D. (2007). Biomass recalcitrance: engineering plants and enzymes for biofuels production. *Science*, *315*(5813), 804–807.

Hingsamer, M., & Jungmeier, G. (2019). Biorefineries. In C. Lago, N. Caldes, Y. Lechon (eds) *The Role of Bioenergy in the Bioeconomy* (pp. 179–222). Academic Press.

Huang, D., Zhou, H., & Lin, L. (2012). Biodiesel: an alternative to conventional fuel. *Energy Procedia*, *16*, 1874–1885.

Huang, G., Chen, F., Wei, D., Zhang, X., & Chen, G. (2010). Biodiesel production by microalgal biotechnology. *Applied Energy*, *87*(1), 38–46.

Huang, J., Xia, J., Jiang, W., Li, Y., & Li, J. (2015). Biodiesel production from microalgae oil catalyzed by a recombinant lipase. *Bioresource Technology*, *180*, 47–53.

Joana Gil-Chávez, G., Villa, J. A., Fernando Ayala-Zavala, J., Basilio Heredia, J., Sepulveda, D., Yahia, E. M., & González-Aguilar, G. A. (2013). Technologies for extraction and production of bioactive compounds to be used as nutraceuticals and food ingredients: an overview. *Comprehensive Reviews in Food Science and Food Safety*, *12*(1), 5–23.

Kadir, A. A., Ismail, S. N. M., & Jamaludin, S. N. (2016, July). Food waste composting study from Makanan Ringan Mas. In *IOP Conference Series: Materials Science and Engineering* (Vol. 136, No. 1, p. 012057). IOP.

Kansedo, J., Lee, K. T., & Bhatia, S. (2009). Cerbera odollam (sea mango) oil as a promising non-edible feedstock for biodiesel production. *Fuel*, *88*(6), 1148–1150.

Karmakar, A., Karmakar, S., & Mukherjee, S. (2010). Properties of various plants and animals feedstocks for biodiesel production. *Bioresource Technology*, *101*(19), 7201–7210.

Kasote, D. M., Badhe, Y. S., & Hegde, M. V. (2013). Effect of mechanical press oil extraction processing on quality of linseed oil. *Industrial Crops and Products*, *42*, 10–13.

Kassim, M. A., Meng, T. K., Serri, N. A., Yusoff, S. B., Shahrin, N. A. M., Seng, K. Y., ... & Keong, L. C. (2020). Sustainable biorefinery concept for industrial bioprocessing. In Kuila A., Mukhopadhyay M. (eds) *Biorefinery Production Technologies for Chemicals and Energy* (pp. 15–53). Wiley & Sons.

Khan, S. A. (2010). Algal biorefinery: a road towards energy independence and sustainable future. *International Review of Chemical Engineering*, *2*(1), 63–68.

Khan, S. A., Hussain, M. Z., Prasad, S., & Banerjee, U. C. (2009). Prospects of biodiesel production from microalgae in India. *Renewable and Sustainable Energy Reviews*, *13*(9), 2361–2372.

Kim, J., Yoo, G., Lee, H., Lim, J., Kim, K., Kim, C. W., ... & Yang, J. W. (2013). Methods of downstream processing for the production of biodiesel from microalgae. *Biotechnology Advances*, *31*(6), 862–876.

Knothe, G. (2014). A comprehensive evaluation of the cetane numbers of fatty acid methyl esters. *Fuel*, *119*, 6–13.

Kraxner, F., Nordström, E. M., Havlík, P., Gusti, M., Mosnier, A., Frank, S., ... & Obersteiner, M. (2013). Global bioenergy scenarios–future forest development, land-use implications, and trade-offs. *Biomass and Bioenergy*, *57*, 86–96.

Kunthiphun, S., Phumikhet, P., Tolieng, V., Tanasupawat, S., & Akaracharanya, A. (2017). Waste cassava tuber fibers as an immobilization carrier of Saccharomyces cerevisiae for ethanol production. *BioResources*, *12*(1), 157–167.

Lam, M. K., & Lee, K. T. (2012). Microalgae biofuels: a critical review of issues, problems and the way forward. *Biotechnology Advances*, *30*(3), 673–690.

Lanjekar, R. D., & Deshmukh, D. (2016). A review of the effect of the composition of biodiesel on NOx emission, oxidative stability and cold flow properties. *Renewable and Sustainable Energy Reviews*, *54*, 1401–1411.

Lievonen, J. (1999). *Technological Opportunities in Biotechnology*. Valtion teknillinen tutkimuskeskus.

Lopez-Quiroga, E., Antelo, L. T., & Alonso, A. A. (2012). Time-scale modeling and optimal control of freeze–drying. *Journal of Food Engineering*, *111*(4), 655–666.

Luthfi, A. A. I., Jahim, J. M., Harun, S., Tan, J. P., & Mohammad, A. W. (2017). Potential use of coconut shell activated carbon as an immobilisation carrier for high conversion of succinic acid from oil palm frond hydrolysate. *RSC Advances*, *7*(78), 49480–49489.

Ma, F., & Hanna, M. A. (1999). Biodiesel production: a review. *Bioresource Technology*, *70*(1), 1–15.

Mahanta, P., Mishra, S. C., & Kushwah, Y. S. (2006). An experimental study of Pongamia pinnata L. oil as a diesel substitute. *Proceedings of the Institution of Mechanical Engineers, Part A: Journal of Power and Energy*, *220*(7), 803–808.

Manieniyan, V., Thambidurai, M., & Selvakumar, R. (2009). Study on energy crisis and the future of fossil fuels. *Proceedings of SHEE*, *10*, 2234–3689.

Masjuki, H. H., & Mofijur, M. (2010). *Biofuel Engine: A New Challenge*. Inaugural Lecture. Malaysia: University of Malaya.

Mata, T. M., Martins, A. A., & Caetano, N. S. (2010). Microalgae for biodiesel production and other applications: a review. *Renewable and Sustainable Energy Reviews*, *14*(1), 217–232.

Matisons, M., Joelsson, J. M., Tuuttila, T., Athanassiadis, D., & Räisänen, T. (2012). The Forest Refine project–development of efficient forest biomass supply chains for biorefineries. *NWBC*, *2012*, 51.

Mejía, J. D., Salgado, N., & Orrego, C. E. (2013). Effect of blends of Diesel and Palm-Castor biodiesels on viscosity, cloud point and flash point. *Industrial Crops and Products*, *43*, 791–797.

Mekonnen, T., Mussone, P., Khalil, H., & Bressler, D. (2013). Progress in bio-based plastics and plasticizing modifications. *Journal of Materials Chemistry A*, *1*(43), 13379–13398.

Mikkola, J. P., Sklavounos, E., King, A. W., & Virtanen, P. (2015). *The Biorefinery and Green Chemistry*.

Mladenović, D., Đukić-Vuković, A., Radosavljević, M., Pejin, J., Kocić-Tanackov, S., & Mojović, L. (2017). Sugar beet pulp as a carrier for Lactobacillus paracasei in lactic acid fermentation of agro-industrial waste. *Journal on Processing and Energy in Agriculture*, *21*(1), 41–45.

Moser, B. R. (2012). Preparation of fatty acid methyl esters from hazelnut, high-oleic peanut and walnut oils and evaluation as biodiesel. *Fuel*, *92*(1), 231–238.

Murugesan, A., Umarani, C., Chinnusamy, T. R., Krishnan, M., Subramanian, R., & Neduzchezhain, N. (2009). Production and analysis of bio-diesel from non-edible oils—a review. *Renewable and Sustainable Energy Reviews*, *13*(4), 825–834.

Nan, Y., Liu, J., Lin, R., & Tavlarides, L. L. (2015). Production of biodiesel from microalgae oil (Chlorella protothecoides) by non-catalytic transesterification in supercritical methanol and ethanol: process optimization. *The Journal of Supercritical Fluids*, *97*, 174–182.

Okieimen, F. E., Bakare, O. I., & Okieimen, C. O. (2002). Studies on the epoxidation of rubber seed oil. *Industrial Crops and Products*, *15*(2), 139–144.

Owolabi, R. U., Adejumo, A. L., & Aderibigbe, A. F. (2012). Biodiesel: fuel for the future (a brief review). *International Journal of Energy Engineering*, *2*(5), 223–231.

Palash, S. M., Kalam, M. A., Masjuki, H. H., Arbab, M. I., Masum, B. M., & Sanjid, A. (2014). Impacts of NOx reducing antioxidant additive on performance and emissions of a multi-cylinder diesel engine fueled with Jatropha biodiesel blends. *Energy Conversion and Management*, *77*, 577–585.

Panwar, N. L., Shrirame, H. Y., Rathore, N. S., Jindal, S., & Kurchania, A. K. (2010). Performance evaluation of a diesel engine fueled with methyl ester of castor seed oil. *Applied Thermal Engineering*, *30*(2–3), 245–249.

Pathak, S., Sneha, C. L. R., & Mathew, B. B. (2014). Bioplastics: its timeline based scenario & challenges. *Journal of Polymer and Biopolymer Physics Chemistry*, *2*(4), 84–90.

Popp, J., Harangi-Rákos, M., Gabnai, Z., Balogh, P., Antal, G., & Bai, A. (2016). Biofuels and their co-products as livestock feed: global economic and environmental implications. *Molecules*, *21*(3), 285.

Prabakaran, P., & Ravindran, A. D. (2011). A comparative study on effective cell disruption methods for lipid extraction from microalgae. *Letters in Applied Microbiology*, *53*(2), 150–154.

Ramadhas, A. S., Jayaraj, S., & Muraleedharan, C. J. R. E. (2004). Use of vegetable oils as IC engine fuels—a review. *Renewable Energy*, *29*(5), 727–742.

Ratti, C. (2001). Hot air and freeze-drying of high-value foods: a review. *Journal of Food Engineering*, *49*(4), 311–319.

Rey, L. (2016). Glimpses into the realm of freeze-drying: classical issues and new ventures. In Rey, L., May, J.C. (eds) *Freeze-Drying/Lyophilization of Pharmaceutical and Biological Products* (pp. 15–42). New York: CRC Press.

Rice, J. E. (2014). *Organic Chemistry Concepts and Applications for Medicinal Chemistry*. Academic Press.

Rousseau, R. W. (Ed.) (1987). *Handbook of Separation Process Technology*. John Wiley.

Salim, S., Vermuë, M. H., & Wijffels, R. H. (2012). Ratio between autoflocculating and target microalgae affects the energy-efficient harvesting by bio-flocculation. *Bioresource Technology*, *118*, 49–55.

Schwarz, M., & Bonhotal, J. (2011). *Composting at Home – The Green and Brown Alternative*. Cornell Waste Manage. Inst. Department Crop Soil Sci., 1–12.

Shah, A. A., Hasan, F., Hameed, A., & Ahmed, S. (2008). Biological degradation of plastics: a comprehensive review. *Biotechnology Advances*, *26*(3), 246–265.

Sharma, G., Gupta, A. K., Ganjewala, D., Gupta, C., & Prakash, D. (2017). Phytochemical composition, antioxidant and antibacterial potential of underutilized parts of some fruits. *International Food Research Journal*, *24*(3), 1167.

Shrikhande, A. J. (2000). Wine by-products with health benefits. *Food Research International*, *33*(6), 469–474.

Singh, R., & Hankins, N. P. (2016). Introduction to membrane processes for water treatment. *Emerging Membrane Technology for Sustainable Water Treatment*, 15–52.

Sirisomboonchai, S., Abuduwayiti, M., Guan, G., Samart, C., Abliz, S., Hao, X., ... & Abudula, A. (2015). Biodiesel production from waste cooking oil using calcined scallop shell as catalyst. *Energy Conversion and Management*, *95*, 242–247.

Stanbury, P. F., Whitaker, A., & Hall, S. J. (2017 *Principles of Fermentation Technology*. Elsevier.

Stevens, E. S. (2020). *Green Plastics*. Princeton University Press.

Stichnothe, H., Meier, D., & de Bari, I. (2016). Biorefineries: industry status and economics. In Lamers, P., E. Searcy, J.R. Hess, H. Stichnothe (eds) *Developing the Global Bioeconomy* (pp. 41–67). Academic Press.

Talha, N. S., & Sulaiman, S. (2016). Overview of catalysts in biodiesel production. *ARPN Journal of Engineering and Applied Sciences*, *11*(1), 439–442.

Todhanakasem, T., Tiwari, R., & Thanonkeo, P. (2016). Development of corn silk as a biocarrier for Zymomonas mobilis biofilms in ethanol production from rice straw. *The Journal of General and Applied Microbiology*, *62*(2), 68–74.

Tong, J., Sun, X., Li, S., Qu, B., & Wan, L. (2018). Reutilization of green waste as compost for soil improvement in the afforested land of the Beijing Plain. *Sustainability*, *10*(7), 2376.

Trellu, C., Mousset, E., Pechaud, Y., Huguenot, D., van Hullebusch, E. D., Esposito, G., & Oturan, M. A. (2016). Removal of hydrophobic organic pollutants from soil washing/flushing solutions: a critical review. *Journal of Hazardous Materials*, *306*, 149–174.

Venkatesan, M., Vikram, C. J., & Naveenchandran, P. (2012). Performance and emission analysis of pongamia oil methyl ester with diesel blend. *Middle East Journal of Scientific Research*, *12*(12), 1758–1765.

Waites, M. J., Morgan, N. L., Rockey, J. S., & Higton, G. (2009). *Industrial Microbiology: An Introduction*. John Wiley.

Wang, K., Mandal, A., Ayton, E., Hunt, R., Zeller, M. A., & Sharma, S. (2016). Protein byproducts In *Modification of Protein Rich Algal-Biomass to Form Bioplastics and Odor Removal.* (pp. 107–117). Academic Press.

Weatherley, L. R. (Ed.) (2013). *Engineering Processes for Bioseparations*. Elsevier.

Xiao, M. A. N., Mathew, S., & Obbard, J. P. (2009). Biodiesel fuel production via transesterification of oils using lipase biocatalyst. *Gcb Bioenergy*, *1*(2), 115–125.

Yankov, D. (2021). Aqueous two-phase systems as a tool for bioseparation–emphasis on organic acids. *Physical Sciences Reviews* 697–699.

Yanowitz, J., Ratcliff, M. A., McCormick, R. L., Taylor, J. D., & Murphy, M. J. (2017). *Compendium of Experimental Cetane Numbers* (No. NREL/TP-5400-67585). Golden, CO: National Renewable Energy Lab (NREL).

Yoo, G., Park, W. K., Kim, C. W., Choi, Y. E., & Yang, J. W. (2012). Direct lipid extraction from wet Chlamydomonas reinhardtii biomass using osmotic shock. *Bioresource Technology*, *123*, 717–722.

Young, G., Nippen, F., Titterbrandt, S., & Cooney, M. J. (2011). Direct transesterification of biomass using an ionic liquid co-solvent system. *Biofuels*, *2*(3), 261–266.

Yusuf, N. N. A. N., Kamarudin, S. K., & Yaakub, Z. (2011). Overview on the current trends in biodiesel production. *Energy Conversion and Management*, *52*(7), 2741–2751.

Zahan, K. A., & Kano, M. (2018). Biodiesel production from palm oil, its by-products, and mill effluent: a review. *Energies*, *11*(8), 2132.

Zeller, M. A., Hunt, R., Jones, A., & Sharma, S. (2013). Bioplastics and their thermoplastic blends from Spirulina and Chlorella microalgae. *Journal of Applied Polymer Science*, *130*(5), 3263–3275.

Zhang, G., Zhang, P., & Fan, M. (2009). Ultrasound-enhanced coagulation for Microcystis aeruginosa removal. *Ultrasonics Sonochemistry*, *16*(3), 334–338.

Zimmerman, W. B., Tesař, V., & Bandulasena, H. H. (2011). Towards energy efficient nanobubble generation with fluidic oscillation. *Current Opinion in Colloid & Interface Science*, *16*(4), 350–356.

# Index